Bacterial Evasion of Host Immune Responses

Our survival as multicellular organisms requires the constant surveillance of our internal and external (mucosal) environments by the multifarious elements of the innate and acquired systems of immunity. The objective of this surveillance, expensive as it is to the organisms, is to recognise and kill invading microorganisms. Over the past fifty years the cells and mediators involved in our immune defences have been painstakingly identified. However, it is only relatively recently that the ability of microorganisms to evade immunity has been recognised and investigated. *Bacterial Evasion of Host Immune Responses* introduces the reader to the mechanisms used by bacteria to evade both humoral and cellular immune responses, using systems ranging in complexity from the simple quorum sensing molecules – acyl homoserine lactones – to the supramolecular syringe-like devices of type III secretion systems. This book will be of interest to researchers and graduate students in microbiology, immunology, pharmacology, and molecular medicine.

BRIAN HENDERSON is professor of cell biology and runs the Cellular Microbiology Research Group at University College London. His research focuses on the role of molecular chaperones as microbial virulence factors and the mechanisms by which bacteria control host cytokine networks. He is the coauthor of the textbook *Bacterial Disease Mechanisms* (2002).

PETRA OYSTON is a principal scientist in the microbiology section of the Chemical and Biological Defence Sector, Porton Down, UK. She leads the Bacteriology Group there, where her work focuses on the regulation of genetic systems, vaccine development, genome sequencing, and pathogenicity of bacteria such as *Yersinia pestis*, Burkholderia spp. and *Francisella tularensis*.

Over the past decade, the rapid development of an array of techniques in the fields of cellular and molecular biology have transformed whole areas of research across the biological sciences. Microbiology has perhaps been influenced most of all. Our understanding of microbial diversity and evolutionary biology, and of how pathogenic bacteria and viruses interact with their animal and plant hosts at the molecular level, for example, has been revolutionized. Perhaps the most exciting recent advance in microbiology has been the development of the interface discipline of cellular microbiology, a fusion of classic microbiology, microbial molecular biology, and eukaryotic cellular and molecular biology. Cellular microbiology is revealing how pathogenic bacteria interact with host cells in what is turning out to be a complex evolutionary battle of competing gene products. Molecular and cellular biology are no longer discrete subject areas but vital tools and an integrated part of current microbiological research. As part of this revolution in molecular biology, the genomes of a growing number of pathogenic and model bacteria have been fully sequenced, with immense implications for our future understanding of microorganisms at the molecular level.

Advances in Molecular and Cellular Microbiology is a series edited by researchers active in these exciting and rapidly expanding fields. Each volume will focus on a particular aspect of cellular or molecular microbiology and will provide an overview of the area and examine current research. This series will enable graduate students and researchers to keep up with the rapidly diversifying literature in current microbiological research.

CELLULAR MICROBIOLOGY

ADVANCES IN MOLECULAR AND

Series Editors

Professor Brian Henderson
University College London

Professor Michael Wilson
University College London

Professor Sir Anthony Coates
St George's Hospital Medical School, London

Professor Michael Curtis
St Bartholemew's and Royal London Hospital, London

Advances in Molecular and Cellular Microbiology 2

Bacterial Evasion of Host Immune Responses

EDITED BY
Brian Henderson
University College London

Petra C. F. Oyston
DSTL Porton Down, UK

CAMBRIDGE
UNIVERSITY PRESS

PUBLISHED BY THE PRESS SYNDICATE OF THE UNIVERSITY OF CAMBRIDGE
The Pitt Building, Trumpington Street, Cambridge, United Kingdom

CAMBRIDGE UNIVERSITY PRESS
The Edinburgh Building, Cambridge CB2 2RU, UK
40 West 20th Street, New York, NY 10011-4211, USA
477 Williamstown Road, Port Melbourne, VIC 3207, Australia
Ruiz de Alarcón 13, 28014 Madrid, Spain
Dock House, The Waterfront, Cape Town 8001, South Africa

http://www.cambridge.org

First published 2003

Printed in the United Kingdom at the University Press, Cambridge

Typefaces FF Scala 9.5/13 pt. and Formata   *System* LaTeX 2_ε [TB]

A catalog record for this book is available from the British Library

Library of Congress Cataloging-in-Publication Data
Bacterial evasion of host immune responses / edited by Brian Henderson,
  Petra C. F. Oyston.
      p.   cm. – (Advances in molecular and cellular microbiology ; 2)
    Includes bibliographical references and index.
    ISBN 0-521-80173-7 (hardback)
    1. Virulence (Microbiology) – Molecular aspects.   2. Immunity.   3. Bacteria.
  I. Henderson, Brian, PhD.   II. Oyston, Petra C. F.   III. Series.
  QR175 .B333   2002
  616.07'9 – dc21   2002025621

ISBN 0 521 80173 7 hardback

Contents

CONTENTS

Contributors

Vickery Arcus
School of Biological Sciences
University of Auckland
Private Bag
92019 Auckland
New Zealand
v.arcus@auckland.ac.nz

Edward Baker
School of Biological Sciences
University of Auckland
Private Bag
92019 Auckland
New Zealand
e.baker@auckland.ac.nz

Richard Bellamy
Department of Medicine
Singleton Hospital
Sketty, Swansea SA20FB
United Kingdom
bellamyrj2000@yahoo.co.uk

Barrie Bycroft
Immune Modulation Research Group
School of Pharmaceutical Sciences and Institute of Infection and Immunity
University of Nottingham
Nottingham NG7 2RD
UK

Miguel Camara
Department of Molecular Microbiology
School of Pharmaceutical Sciences and Institute
    of Infection and Immunity
University of Nottingham
Nottingham NG7 2RD
UK

Benjamin M. Chain
Department of Immunology
Windeyer Institute of Medical Sciences
University College London
46 Cleveland Street
London W1P 6DB
UK
b.chain@ucl.ac.uk

Siri Ram Chhabra
Department of Medicinal Chemistry
School of Pharmaceutical Sciences and Institute
    of Infection and Immunity
University of Nottingham
Nottingham NG7 2RD
UK

Sek Chow
Immune Modulation Research Group
School of Pharmaceutical Sciences and Institute
    of Infection and Immunity
University of Nottingham
Nottingham NG7 2RD
UK

Stephen Diggle
Department of Molecular Microbiology
School of Pharmaceutical Sciences and Institute
    of Infection and Immunity
University of Nottingham
Nottingham NG7 2RD
UK

Michael S. Donnenberg
Division of Infectious Diseases
Department of Medicine
University of Maryland
Baltimore, Maryland 21201
USA
mdonnenb@umaryland.edu

Maria Fällman
Department of Molecular Biology
Umeå University
901 87 Umeå
Sweden

Åke Forsberg
Department of Medical Protection
Swedish Defence Research Agency
FOI
S-901 82 Umeå
Sweden
ake.fosberg@foi.se

John D. Fraser
School of Biological Sciences
Department of Molecular Medicine
University of Auckland
Private Bag
92019 Auckland
New Zealand
jd.fraser@auckland.ac.nz

Christopher Harty
Department of Medicinal Chemistry
School of Pharmaceutical Sciences and Institute
    of Infection and Immunity
University of Nottingham
Nottingham NG7 2RD
UK

Jürgen Heesemann
Max von Pettenkofer-Institut for Hygiene and Medical Microbiology
Pettenkofestr.9a
80336 Munich
Germany

Brian Henderson
Cellular Microbiology Research Group
Eastman Dental Institute
University College London
256 Gray's Inn Road
London WC1X 8LD
UK
b.henderson@eastman.ucl.ac.uk

Doreen Hooi
Immune Modulation Research Group
School of Pharmaceutical Sciences and Institute
    of Infection and Immunity
University of Nottingham
Nottingham NG7 2RD
UK

Michael A. Kerr
Department of Molecular and Cellular Pathology
University of Dundee
Ninewells Hospital Medical School
Dundee DD1 9SY
Scotland
m.a.kerr@dundee.ac.uk

Mogens Kilian
Department of Medical Microbiology and Immunology
University of Aarhus
The Bartholin Building
DK-800
Aarhus C
Denmark
Kilian@microbiology.au.dk

Jan-Michael Klapproth
Division of Infectious Diseases
Department of Medicine

University of Maryland
Baltimore, Maryland 21201
USA

Janusz Marcinkiewicz
Department of Immunology
Jagiellonian University
Krakow
Poland

Robert L. Modlin
Division of Dermatology and Department of Microbiology,
    Immunology and Molecular Genetics
School of Medicine
University of California, Los Angeles
Los Angeles, CA 90095
USA
Rmodlin@mednet.ucla.edu

Kayvan R. Niazi
Division of Dermatology and Department of Microbiology,
    Immunology and Molecular Genetics
School of Medicine
University of California, Los Angeles
Los Angeles, CA 90095
USA

Steven A. Porcelli
Department of Microbiology and Immunology
Albert Einstein College of Medicine
Bronx, NY 10461
USA

David Pritchard
Immune Modulation Research Group
School of Pharmaceutical Sciences and Institute
    of Infection and Immunity
University of Nottingham
Nottingham NG7 2RD
UK
David.Pritchard@nottingham.ac.uk

Thomas Proft
School of Biological Sciences
Department of Molecular Medicine
University of Auckland
Private Bag
92019 Auckland
New Zealand
t.proft@auckland.ac.nz

Roland Rosqvist
Department of Molecular Biology
Umeå University
901 87 Umeå
Sweden

Bruno Rouot
INSERM U-432
Montpellier
France

Klaus Ruckdeschel
INSERM U431
Universite Montpelier II
CC100, F34095
Montpelier
Cedex 05
France
rouot@crit.univ-montp3.fr

Nigel J Saunders
Molecular Infectious Diseases Group
Institute of Molecular Medicine
University of Oxford
Headington
Oxford OX3 9DS
UK
njsaunders@molbiol.ox.ac.uk

Gary Telford
Immune Modulation Research Group
School of Pharmaceutical Sciences and Institute of Infection and Immunity

University of Nottingham
Nottingham NG7 2RD
UK

Paul Williams
Department of Molecular Microbiology
School of Pharmaceutical Sciences and Institute of Infection and Immunity
University of Nottingham
Nottingham NG7 2RD
UK
paul.williams@nottingham.ac.uk

Preface

From birth we are protected from bacterial infections by the complex system of cells and cell products, which have functional and signalling properties, known collectively as immunity. The immune system has three major functions: (1) the ability to recognise infectious agents such as bacteria; (2) the capacity to kill these infecting organisms; and (3) the integration of (1) and (2) through specific cell–cell signalling. It is now recognised that the nature of our immune systems has been shaped in the crucible of evolution by interactions with infectious agents. It is also emerging that the various organisms that can infect us have evolved multiple mechanisms to evade both arms of our immune system – innate and adaptive immunity.

This book describes some of the emerging mechanisms employed by bacteria to evade both humoral and cellular immunity. The first section deals with novel aspects of the recognition of, and the response to, bacteria by a key cell population – dendritic cells (e.g., through Toll-like receptors), and by lymphocytes via the nonpolymorphic CD1 MHC molecules that recognise nonpeptidic antigens. The final chapter in this section describes natural resistance-associated macrophage protein (NRAMP), a metal ion transporter important in susceptibility to infection by mycobacteria. Mycobacteria also encode NRAMP-like proteins revealing another twist in the ongoing battle between bacteria and their hosts for essential metal ions such as iron and zinc.

In the second section attention switches to the ability bacteria have to evade humoral immunity. It has been known for many years that the bacterial capsule can protect against complement. However, over the past decade or so it has emerged that bacterial pathogens have evolved a plethora of selective mechanisms for evading the major mechanisms of complement-mediated killing. Another powerful mechanism for evading antibodies and antibody-mediated complement activation is the production of selective immunoglobulin-degrading proteases and immunoglobulin-binding proteins. A third way to avoid the deleterious actions of antibodies is to keep altering the cellular antigens by the processes of phase and antigenic variation.

The final section in this book deals with bacterial evasion of cellular immunity. The role of type III secretion systems in the inhibition of phagocytosis and in the inhibition of the key transcription factor, NF-κB, are detailed in two separate chapters. A small number of bacteria produce proteins, termed superantigens, which are able to stimulate a large proportion of the T cell repertoire but in the process remove or inactivate these cells. Superantigens thus have the potential to decrease overall T cell responsiveness. A fascinating finding is that the signals involved in bacterial quorum sensing – such as the acyl homoserine lactones – are also able to inhibit immune responses, including the induction of cytokine synthesis. The consequences of the immunoinhibitory actions of these molecules is discussed. The remaining chapters describe how bacteria interact with immune cells to control the synthesis of cytokines, proteins that act to integrate the functions of immune cells. The ability of bacteria to produce a vast range of molecules with cytokine-inducing (or in some cases, inhibiting) actions and the consequence of this for the physiological control of functional cytokine networks is reviewed. The role of enterotoxins as cytokine-modulating agents capable of acting as local adjuvants or local cytokine inhibitors is described and the consequences of this for the host are reviewed.

This volume brings together experts in bacteria-host interactions to explain how bacteria are recognised by the immune system and how this recognition and its consequences can be negated to enable the bacteria to survive.

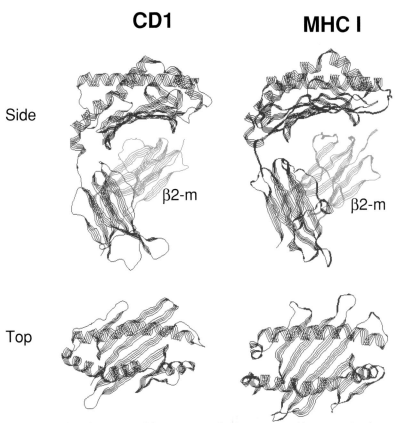

Figure 2.1. Side and top views of the structures of CD1 (represented by mouse CD1d1) and MHC I (represented by human leukocyte antigen HLA-A2).

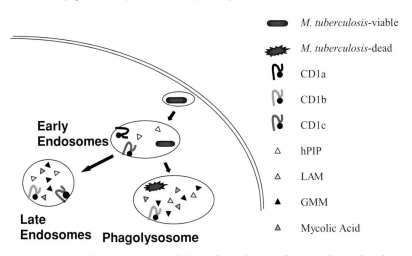

Figure 2.3. Intracellular distribution of CD1 isoforms during infection with *M. tuberculosis*.

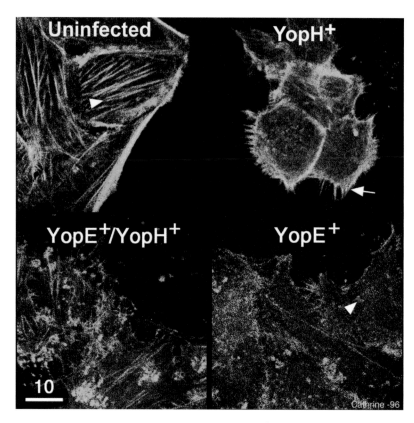

Figure 7.2. YopH and YopE have distinct effects on host cell cytoskeleton. Confocal images of HeLa cells, not infected, or infected with a *Yersinia* multiple *yop* mutant strain expressing either YopH or YopE, or infected with *Yersinia* wt. YopE fragments the F-actin cytoskeleton whereas YopH affects the integrity of focal adhesions and associated stress fibers. The combined effect is seen with the *Yersinia* wt strain (NB. this strain translocates less amount of the effectors compared to the multiple mutant strain). Cellular F-actin were visualized by staining with fluorescein-conjugated phalloidin (green); vinculin-containing focal adhesions were visualized by indirect immunofluorenscence (red). The yellow color represents colocalization of microfilaments and vinculin. Vinculin-containing focal adhesions (arrow heads) and vinculin-containing retraction fibers (arrows) are shown. All sections were scanned under identical conditions and show the basolateral side of the cells. Scale bar: 10 ?m.

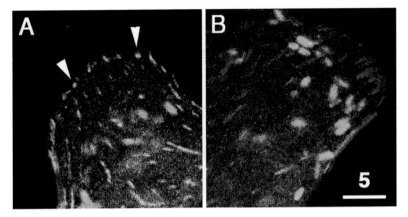

Figure 7.3. YopH contains an inherent sequence that mediates localisation to host cell peripheral focal complex structures. Confocal images of HeLa cells infected with a *Yersinia* multiple *yop* mutant strain expressing the PTPase inactive YopHC403A (A) or an in frame deletion mutant thereof, YopHC403AΔ223-226 (B) (region of importance for localisation to focal complex structures). Arrowheads indicate representative focal complexes where YopH and vinculin colocalise (yellow). NB: some bacteria are stained by the YopH antiserum. All sections were scanned under identical conditions and shows the basolateral side of cells. Scale bar: 5 ?m.

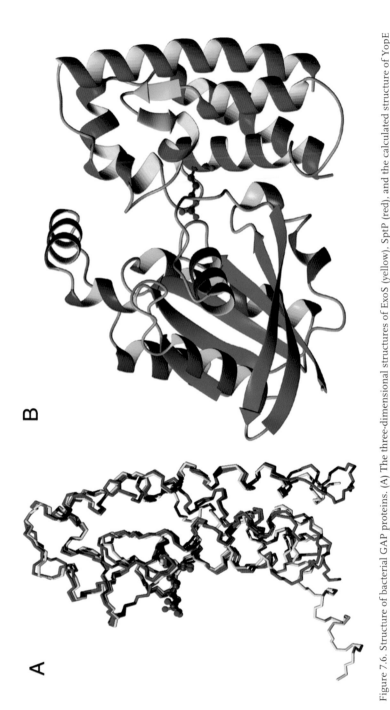

Figure 7.6. Structure of bacterial GAP proteins. (A) The three-dimensional structures of ExoS (yellow), SptP (red), and the calculated structure of YopE (blue), superimposed on top of each other. The N-terminal part of the respective protein is located to the left in the image. The arginine-finger extends out to the left in the middle of the protein structure. (B) Ribbon diagram of modeled YopE in complex with Rac. Rac1 is shown in blue-green and YopE in red-yellow. The main building block of the GAP proteins is a right-handed anti-parallel four-helix bundle with a characteristic up-down-up-down topology. Two characteristic bulges (I and II) and the neighbouring helices encompass the two most conserved regions of the bacterial GAPs (Stebbins and Galán, 2000; Wurtele *et al.*, 2001).

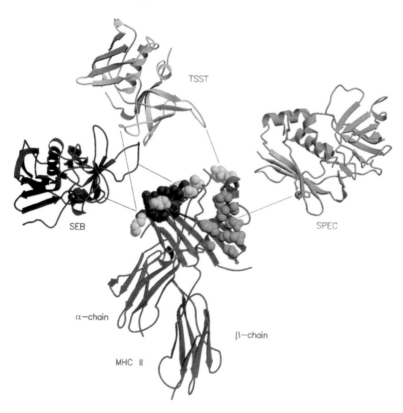

Figure 8.2. Different modes of binding to MHC-II molecules. At least three separate binding orientations have been found for bacterial SAGs. The extracellular domain of MHC-II is shown in blue and the MHC-II residues with which SEB interacts are shown as space-filling spheres in red. SEB has been moved by a simple translation away from its position on the human HLA-DR1 molecule (from the crystal structure (63)). TSST forms a complex with MHC-II using the same face of the toxin (when compared to SEB) but in a different orientation. TSST has also been moved up from its position by a simple translation from the crystal structure (62). Residues on MHC-II that interact with TSST upon complex formation include those that interact with SEB (red spheres) and, in addition, those shown as yellow spheres. Note that TSST bridges one end of the peptide binding groove and interacts with two residues of the □-chain of MHC-II. The orientation of SPE-C (grey) on MHC-II is mediated by zinc and His-81 on the MHC-II □-chain (shown as grey spheres). SPE-C binds across the peptide groove and excludes contact between MHC-class II and TcR (16). SPE-C only binds in this orientation. SEA, on the other hand, binds in both this orientation and in the SEB orientation also.

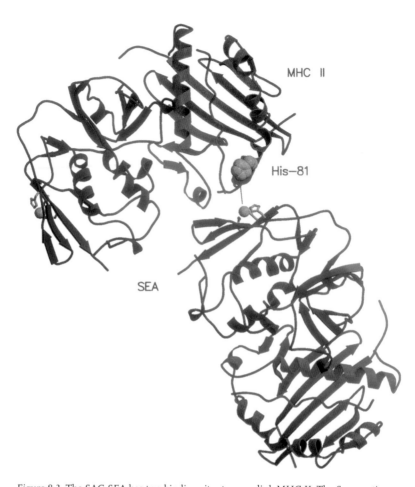

Figure 8.3. The SAG SEA has two binding sites to cross-link MHC-II. The Superantigen SEA is displayed in its dual orientation to MHC class II. One orientation is identical to the SEB site to the MHC class II □-chain whereas the other much higher affinity zinc mediated binding orientation is to the MHC class II □-chain. This picture is looking down from above with the distal domains of MHC-II □ and □-chains shown in blue. On the left of the figure, an SEA molecule (red) is bound via the generic □-chain binding site in the same fashion as SEB. On the right, another SEA molecule is bound to the same MHC-II molecule via a zinc bridge between His81 on the MHC-II □-chain and His187, His225, and Asp227 in SEA. This interaction is about 100 times stronger than the first and therefore occurs first. There is some cooperation between the two SEA molecules. Once bound, it is obvious to see that cross-linking of another MHC-II molecules can occur via both bound SEA molecules. This image is only inferred from mutational data. There is yet no crystal structure of an SEA/MHC-II complex.

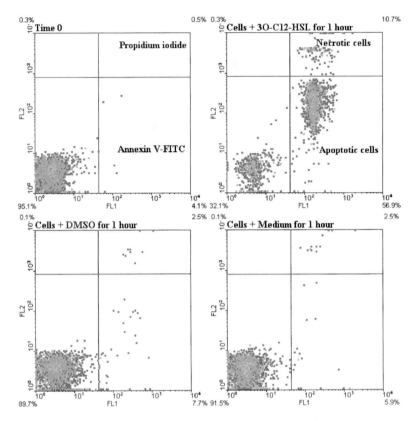

Figure 9.5. 3O-C12-HSL induces apoptosis in murine splenocytes. 10^6 murine leucocytes were incubated with 100 μM 3O-C12-HSL for 1 hour. An aliquot of 10^5 cells was removed for labelling with Annexin V-FITC, which detected apoptotic cells and propidium iodide, which stained for necrotic cells. Analysis was by flow cytometry.

Figure 9.6 (following page). 3O-C12-HSL induces apoptosis in murine peritoneal macrophages. (A) Macrophages incubated in 3O-C12-HSL exhibit several hallmarks of apoptosis. A normal cell is shown for comparison (i). Cytoplasm pinches off in a process known as blebbing (arrow-ii) until it is completely reduced (iii). Meanwhile endonucleases are activated which cleave chromatin causing it to condense and line the nuclear membrane. Nuclear morphology in (iv) is typical of a cell undergoing this process. Eventually the nucleus fragments into apoptotic bodies (arrow-v). Macrophages were incubated in 12.5, 25, 50 and 100 μM 3O-C12-HSL and 3O-C6-HSL and CTCM alone for 1, 2, 4 and 6h. Macrophages were cytospun, fixed with methanol and stained with Giemsa.

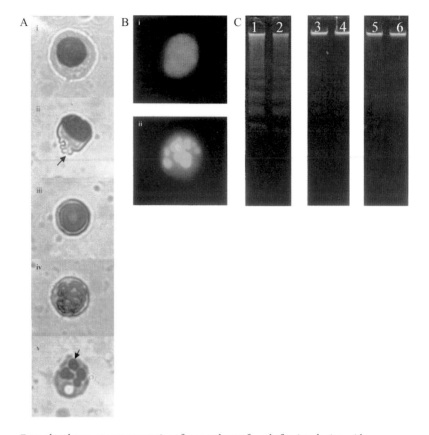

Examples shown are representative of macrophages found after incubation with 3O-C12-HSL at all concentrations and times. Control and 3O-C6-HSL incubations showed no apoptotic morphology. (B) Representative Hoechst stained cells. Hoechst 33342 binds DNA which allows the determination of nuclear morphology. DNA in the nucleus of normal cells is distributed diffusely (i) and fragmented in apoptotic cells (ii). Macrophages were incubated in 50 μM 3O-C12-HSL and 3O-C6-HSL and CTCM for 3 h, detached and stained. 3O-C12-HSL incubation induced the apoptotic morphology shown and 3O-C6-HSL and CTCM incubated cells appeared normal. (C) Incubation with 3O-C12-HSL yields the characteristic DNA ladder. DNA was extracted from macrophages incubated for 3 h in 50 μM 3O-C12-HSL (lanes 1 and 2) and 50 μM 3O-C6-HSL (lanes 3 and 4) and CTCM (lanes 5 and 6). 12 cultures of 1×10^5 macrophages were detached and pooled allowing sufficient amounts of DNA to be analysed yet ensuring that results reflect previous experiments. DNA was visualised on a 1.8% agarose gel by ethidium bromide staining.

PART I   Recognition of bacteria

CHAPTER 1

The dendritic cell in bacterial infection: Sentinel or Trojan horse?

Benjamin M. Chain and Janusz Marcinkiewicz

1.1 INTRODUCTION

Dendritic cells play a key role in the initiation and regulation of T-cell dependent immune responses. Much of their significance lies in their role as a cell linking the evolutionarily ancient innate immune system to the more complex and sophisticated adaptive immune system. Understanding their function in the context of bacterial infection, therefore, where the strands of innate and adaptive immunity are so closely interwoven, is likely to be particularly significant.

The cell biology of the dendritic cell poses a number of specific questions relating to bacterial physiology and pathophysiology. In particular, much of the literature in the field has been concerned either with understanding how dendritic cells process and present bacterial proteins in the context of a "particulate" as opposed to a "soluble" form, or with mapping the interactions between dendritic cells and bacterial cell wall components. This chapter first provides a brief overview of present understanding of the dendritic cell system and its role in immune responses, and then addresses questions relating more specifically to the interaction between dendritic cells and bacteria.

1.2 DENDRITIC CELLS AND THE IMMUNE RESPONSE

1.2.1 The dendritic cell family

T-cell recognition of antigen has a requirement for the antigen to be first *processed* and then *presented* by another cell, termed the "antigen presenting cell." This requirement, first determined empirically, can now be understood in terms of the well-established model of T-cell recognition, involving the tripartite molecular interaction between T-cell antigen receptor, antigen peptide fragment, and MHC molecule (see Chapter 2 for more details). The requirement for multiple other ligand/receptor interactions between T cell and antigen presenting cell in order to achieve full T-cell activation ("co-stimulation") adds further molecular detail to this overall recognition process. The nature of the antigen presenting cell, which is responsible for T cell activation *in vivo*, therefore becomes a question central to the understanding of T-cell immunity.

The dendritic antigen presenting cell was first identified by Steinman (Steinman, 1991), as a rare cell type found in the T-cell areas of spleen and lymph nodes of mice, which serves as a potent activator of T cells. These T-cell–associated dendritic cells must be clearly distinguished from follicular dendritic cells, found within B-cell follicles and concerned with the trapping and storage of antigen/antibody complexes for B-cell recognition. This latter cell type will not be discussed in this chapter.

The principle characteristics of the T-cell associated dendritic cells are the ability to activate both naïve and memory T cells (associated with high surface expression of both class I and class II MHC molecules), an unusual morphology showing extensive thin cytoplasmic processes or dendrites (*in vivo* these cells were sometimes described as "interdigitating cells" for the same reason), and an absence of Fc receptors and phagocytic activity. The latter features were of particular importance in distinguishing these cells from the macrophage, which had previously been believed to be the main cell type involved in the presentation of antigen to T cells. The inability of dendritic cells to phagocytose immediately raised the question of how such cells would process and present bacteria or other particulate antigens, a question which was indeed addressed in a number of early studies (Kaye et al., 1985; Guidos et al., 1984).

Although dendritic cells, as originally defined, are cells localised within the T cell areas of secondary lymphoid tissue, it is now generally accepted that this cell is closely related to antigen presenting cells found within most other tissues of the body. This relationship has been explored most thoroughly in relation to skin (Macatonia et al., 1987; Larsen et al., 1990b) where there is compelling evidence for a differentiation pathway that links skin Langerhans' cells to dendritic cells within the draining lymph nodes (Hill et al., 1990). In this model, Langerhans' cells respond to local inflammatory stimuli by migrating out of the skin, via afferent lymphatics (where they were previously identified as veiled cells, because of their extensive membrane ruffling), and then into the lymph node where they transform into interdigitating dendritic cells. These dendritic cells are quite short lived and disappear from the T-cell areas (perhaps by apoptosis) within a few days of arrival (Garside et al., 1998).

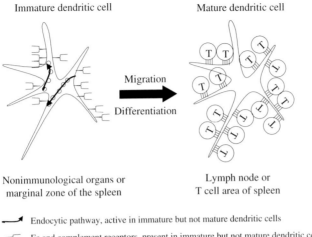

Immature dendritic cell   Mature dendritic cell

Migration

Differentiation

Nonimmunological organs or   Lymph node or
marginal zone of the spleen   T cell area of spleen

Endocytic pathway, active in immature but not mature dendritic cells

Fc and complement receptors, present in immature but not mature dendritic cells.

T   Many T cells interacting with a single mature dendritic cell via the immunological synapse, to form a cluster.

Figure 1.1. Dendritic cells exist in immature and mature forms, distinguished both by function and anatomical location.

A key feature of this model is that dendritic cells exist in two quite distinct differentiation stages, a more immature ("precursor") form found primarily outside lymphoid tissue and a mature form identical to the interdigitating cells of secondary lymphoid tissue (see Fig. 1.1). Immature dendritic cells have now been identified in many organs, including heart, liver, kidney, etc. (e.g., Larsen et al., 1990a). An immature dendritic cell type has also been described in the spleen, within the marginal zone surrounding the white pulp (Leenen et al., 1998). Appropriate stimulation induces migration and differentiation of these cells into interdigitating cells of the T-cell areas (Sousa and Germain, 1999). In addition, many *in vitro* models that mimic this two-step dendritic cell differentiation have been described (Sallusto and Lanzavecchia, 1994; Romani et al., 1989). Dendritic cell precursors differ from their mature counterparts in both quantitative and qualitative respects. In general, immature forms have higher endocytic capacity, express several Fc and complement receptors (see Chapter 4), and are phagocytic (albeit rather weakly). They are less efficient in activating resting naïve T cells and express lower levels of the various molecular determinants of antigen presentation (see below). Immature dendritic cells, therefore, may represent the "antigen capture" arm of the antigen presentation system, whereas mature dendritic cells represent the "antigen presentation" arm. The anatomical separation of antigen presentation (which takes place in lymph nodes or spleen) from the site of infection is a fundamental feature of the immune system. Indeed, the presence of differentiated dendritic cells outside lymphoid tissue is almost invariably associated with chronic inflammation and pathology.

The view of the dendritic cell system presented above has rapidly won widespread acceptance. Its most influential implication is the idea that antigen presentation is an inducible rather than a constitutive process. Under resting conditions, the flow of maturing dendritic cells from tissue to lymph node is small (although not absent; Anderson et al., 2001), and the extent of antigen presentation is limited. In the face of immune challenge, this flow dramatically increases and antigen presentation therefore also increases. The molecular signals that drive dendritic cell migration and differentiation are still being elucidated and include microbial receptors on the dendritic cells (see below), inflammatory cytokines (Cumberbatch et al., 2001), chemokines (Caux et al., 2000), and reactive oxygen species (Rutault et al., 1999) and their products (Alderman et al., manuscript in preparation). Many of these mediators are often produced by components of "innate" immunity (e.g., macrophages, neutrophils), leading some to suggest that an innate immune response is a necessary determinant of antigen presentation (Janeway, 1992). It seems more likely, however, that tissue response to injury (Ibrahim et al., 1992) rather than immune recognition is the underlying cause of dendritic cell migration/differentiation. Delivery of sterile gold beads (Porgador et al., 1998), topical sensitisers (Hill et al., 1993), and sterile allogeneic transplants (Larsen et al., 1990b) are all potent activators of dendritic cell migration and maturation.

1.2.2 Relationship between dendritic cells and macrophages

The relationship between the dendritic cell and the macrophage has been a much debated issue. The consensus is that most dendritic cells share a common precursor with the circulating monocyte and hence with the tissue macrophage. Immature dendritic cells are recruited from a bone marrow-derived blood precursor, via specific adhesion molecules on the precursor surface (Strunk et al., 1997). Recruitment is increased during an ongoing immune response, via the release of specific chemokines. Interaction with the extracellular matrix may also be important in regulating dendritic cell differentiation (Randolph et al., 1998). *In vitro*, dendritic cells can also be derived from blood monocytes, but there is little evidence that this process is important *in vivo*. There have been persistent, but often contradictory, reports that one or more other populations of nonmyeloid dendritic cells exist (e.g.,

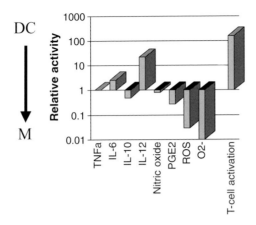

Figure 1.2. Comparative functional analysis of dendritic cells and macrophages. Bone marrow culture murine dendritic cells, and mouse peritoneal macrophages were stimulated with LPS and IFN-γ, and the production of a variety of mediators measured. T-cell activation was measured using unstimulated cells, in allogeneic antigen presentation assays. For each parameter, the ratio between dendritic cells and macrophages is given. Note that the DC are much more efficient at T cell stimulation and at producing IL-12, a T-cell regulatory cytokine. In contrast, macrophages show greater production of prostaglandins, reactive oxygen species, and nitric oxide. Experimental details are given in Marcinkiewicz et al. (1999).

lymphoid dendritic cells or plasmacytoid dendritic cells); the lineage relationships of these various other populations remain very unclear, as do their physiological importance, and they are not discussed further in this chapter.

Once within tissues, immature dendritic cells can be clearly distinguished from macrophages in terms both of morphology and cell surface phenotype. Nevertheless, immature dendritic cells do share many properties with macrophages, and differences are quantitative rather than qualitative. Immature dendritic cells, both in lymphoid tissue and outside it, share many surface markers with macrophages, including many of the myeloid lineage markers, various isotypes of Fc receptors, and complement receptors (Leenen et al., 1997; Woodhead et al., 1998; King and Katz, 1989). As discussed further below, immature dendritic cells, including skin Langerhans cells (Sousa et al., 1993), have been shown to be phagocytic, have high rates of fluid phase endocytosis, and have a well-developed phagolysosome system. Nevertheless, when compared directly to macrophages, even immature dendritic cells are only weakly phagocytic and have a much reduced lysosomal function (see Fig. 1.2).

It is best to regard dendritic cells as a distinct member of the myeloid family, sharing some molecular and functional properties with the other members of the family (both macrophages and granulocytes), but characterised by extreme specialisations that maximise the efficiency of antigen presentation.

1.2.3 The molecular cell biology of the dendritic cells

The dendritic cell is the only cell able to simulate a primary T-cell immune response (at least in the normal physiological situation). In contrast, effector T cells (both CD4 helpers and CD8 cytotoxic cells) have less stringent requirements and are activated by their targets, whether these be B cells (leading to T-cell dependent antibody production), macrophages (in T-cell dependent macrophage activation), or any cell expressing class I MHC and the antigen peptide, which is the target of the cytotoxic CD8 T cells. Memory T cells lie in between naïve T cells and effector cells, in terms of their requirement for dendritic cell presentation. Even with memory cells, however, dendritic cells provide the most efficient presentation.

The molecular features of dendritic cells responsible for their potent antigen presenting cell activities are not fully understood. The expression of high levels of MHC molecules (both class I and class II) and the expression of a panoply of "co-stimulatory" molecules involved in optimising T-cell activation are two important features. Dendritic cells are also able to interact with many T cells simultaneously, both *in vitro* and *in vivo*, to form clusters. This interaction is mediated principally by ICAM/β_2 integrin interactions (DCs express all three ICAM molecules at high level; King and Katz, 1989). Cluster formation allows T cells of different specificities to interact with each other and also stabilises T cell/dendritic cell interactions independently of antigen recognition, to allow sufficient time for the formation of the "immunological synapse," which is essential for T cell triggering. The long dendritic cell processes, which are so characteristic of this cell type, also presumably maximise opportunities of T-cell–antigen interaction (Al Alwan et al., 2001). Finally, dendritic cells within the lymph node are the major producers of IL-12, a cytokine that initiates the T helper 1 type of response that leads to macrophage activation and bacterial clearance. IL-12 production is principally regulated by the interaction of CD40 on the dendritic cell surface with CD40 ligand on activated T cells. The dendritic cell therefore acts as a bridge transmitting paracrine signals between helper and effector T cells within a cluster.

1.3 DENDRITIC CELLS AND BACTERIAL IMMUNE RESPONSES

This brief outline of the workings of the dendritic cell system provides a framework for specific questions regarding the role of the dendritic cell system in bacterial infection. It is worth noting, however, that remarkably few studies have focused specifically on this interaction, and much of what follows remains, therefore, speculative.

1.3.1 What activates dendritic cell migration/differentiation in response to bacterial infection?

Bacteria, and several bacterial components such as endotoxin, are potent activators of dendritic cell migration and differentiation (Sallusto and Lanzavecchia, 1994; Sousa and Germain, 1999). This response is primarily activated by engagement of receptor complexes (sometimes called pattern recognition receptors to distinguish them from the antigen-specific receptors of T and B cells) that recognise bacterial components. The molecular details of pattern recognition receptors, and how they transduce signals within the cell, is an area of very active research (Triantafilou et al., 2001). The family of Toll-like receptors now believed to be very important in this process are briefly discussed below, but other families of receptors may well exist such as TREM (triggering receptor expressed on myeloid cells)-1 (Bouchon et al., 2001). Immature dendritic cells, at least *in vitro*, express many Toll receptors, allowing them to respond directly to bacterial challenge. However, dendritic cell response may also be indirectly mediated by cytokines such as TNF-α and IL-1 produced by other cell types in response to bacterial invasion. Not all pattern recognition receptors on the dendritic cell stimulate migration, however. The DEC 205 lectin and the mannose receptor, for example, serve to facilitate binding and uptake of mannose-containing structures into processing compartments, but do not induce migration or differentiation (Mahnke et al., 2000).

Engagement of Toll receptors also induces IL-12 and other pro-inflammatory cytokines, suggesting that most bacterial responses are directed toward a Th1, rather than Th2, type of response. However, some bacterial toxins may interfere with Th1 priming and deviate the response toward a Th2 response (Boirivant et al., 2001; Cong et al., 2001). Such deviation is discussed in more detail in Chapter 11.

1.3.2 Toll-like receptors (TLRs) and bacterial recognition

It is now rapidly becoming established that the TLRs, cell surface proteins with a intracellular domain homologous to that of the IL-1 receptor (so called Toll/IL-1 receptor homology – TIR domain), are crucial for the recognition and discrimination of microbes. There are at least ten *tlr* genes in mammals and they can form homo-dimers (and possibly also hetero-dimers), suggesting that the range of bacterial components that can be recognised by these

Table 1.1. *Specificity of the TLRs*

TLR	Bacterial ligands binding
TLR2 and TLR6	peptidoglycan, Mycoplasma lipoprotein
TLR2 and TLR?[a]	lipoproteins, lipoarabinomannan, certain LPS molecules
TLR3	double stranded RNA (viral)
TLR4	enteric and other bacterial LPS molecules
TLR5	flagellin
TLR9	CpG DNA

[a] Nature of TLR2 binding partner not defined.

cell surface proteins may be large (Kimbrell and Beutler, 2001). The known ligands for the various TLRs is shown in Table 1.1. It is believed that the TLRs require additional proteins to form a recognition complex at the surface of myeloid cells. Among these proteins are CD14, MD2, and the β_2-integrin, Mac-1, all of which can confer increased cellular responsiveness to LPS and certain other agonists. In addition to controlling innate responses to microorganisms, by activating NF-κB through the intracellular adapter protein MyD88, it has been reported that MyD88-deficient mice have a major defect in activation of antigen-specific Th1 lymphocytes. This suggests that the TLRs may play a role in controlling adaptive immune responses (Schnare et al., 2001). The role of TLRs in the activation of NF-κB is described in Chapter 6.

1.3.3 How do dendritic cells process bacterial antigens?

The extent to which dendritic cells take up and process bacteria directly remains debatable. Many studies show that immature dendritic cells, at least *in vitro*, phagocytose bacteria and other particulates and then process and present bacterial antigens. There is also limited evidence for bacterial phagocytosis *in vivo* (Inaba et al., 1993; Paglia et al., 1998). Phagocytosis in these experiments is often measured in the presence of an enormous excess of free bacteria, which does not reflect the normal physiological situation. Even under these conditions, the phagocytic index of dendritic cells is often much smaller than that of macrophages. Furthermore, dendritic cells are ill-equipped to kill any bacterium that is internalised, because their ability to produce an oxygen burst or to synthesise nitric oxide is much less than that shown by macrophages (Fig. 1.2) (Yu et al., 1996; Marcinkiewicz et al., 1999; Bryniarski et al., 2000).

A more likely general scenario, therefore, is that dendritic cells normally act in concert with components of the innate immune system in first killing and then processing bacteria (Bryniarski et al., 2000). In early stages of infection, the neutrophil is the major phagocyte present at sites of infection. Neutrophil phagocytosis is extremely efficient and is likely to remove rapidly most free bacteria from the dendritic cell microenvironment. Both phagocytosed bacteria and any remaining extracellular bacteria can be efficiently killed by the combination of oxygen radical production and hypochlorous acid formed by neutrophil myeloperoxidase (Marcinkiewicz et al., 2000). Dendritic cell processing of dead bacteria can then occur by the action of cell surface proteinases on the dendritic cell, by the uptake of bacterial fragments via lectin or scavenger receptors, or perhaps by the uptake of apoptotic neutrophils containing internalised bacteria. The latter would be particularly important in stimulating a bacterial CD8 T-cell response (believed to be important for intracellular bacterial infection), since uptake of cell associated antigen seems to load preferentially class I MHC via the ill-defined "cross-priming" pathway (Albert et al., 1998). Bacterial fragments may alternatively enter afferent lympatics and be carried down to the draining lymph nodes to be processed and presented *in situ*.

1.3.4 Dendritic cells as a means of bacterial invasion – sentinels or Trojan horses?

A specialised case of dendritic cell phagocytosis concerns those bacteria that normally propagate within the cell. Many such bacteria target macrophages, raising the question of whether invasion of dendritic cells can also occur. These bacteria include the most important pathogenic species, such as *Mycobacteria, Listeria*, and *Salmonella*. This ability to survive within the killing machine of the macrophage is a key bacterial immune evasion strategy. Several recent studies have addressed this problem directly, although most have used *in vitro* models of dendritic cell function, which may not reflect the situation *in vivo*. *In vitro* internalisation of live bacteria into dendritic cells has been demonstrated for *Mycobacterium tuberculosis* (Gonzalez-Juarrero and Orme, 2001), *M. avium* (Mohagheghpour et al., 2000), *Listeria monocytogenes* (Kolb-Maurer et al., 2000), and *Salmonella typhimurium* (Niedergang et al., 2000). In some cases, processing and presentation of bacterial antigens by infected dendritic cells has been demonstrated (Svensson et al., 1997; Tascon et al., 2000; Paschen et al., 2000).

In many cases, the ability of dendritic cells to take up antigen is linked to the ability of the host to mount an effective immune response. Bacterial invasion of dendritic cells, however, may be neither necessary nor desirable. Even attenuated Salmonella, for example, that were unable to survive within host macrophages could survive within dendritic cells, illustrating the very limited bactericidal activity of these cells (Niedergang et al., 2000; Bryniarski et al., 2000). Uptake of live bacteria by dendritic cells, rather than being a beneficial first step in initiating adaptive immunity, may represent escape from phagocytosis and killing by neutrophils and macrophages. Furthermore, in view of the extensive migration of dendritic cells, bacteria may use this escape system to disseminate within the host (Pron et al., 2001; Niedergang et al., 2000). This model would be consistent with the observation that the absence of CD18 (Vazquez-Torres et al., 1999) blocked the spread of Salmonella to lymphoid tissue. The absence of caspase1 (Monack et al., 2000), the enzyme responsible for IL-1 and IL-18 generation, and hence Langerhans' cell migration (Antonopoulos et al., 2001), also blocked Salmonella spread. Under conditions of bacterial overload, therefore, dendritic cells may be contributing to bacterial pathogenesis – an interesting example of immune evasion.

1.4 DENDRITIC CELLS AND THE GUT FLORA – A PARADOX?

The interaction between the immune system and bacteria of the gut is a special case. Limiting immune response to beneficial bacteria is presumably as important as mounting a protective response against pathogenic bacteria (see Chapter 10 for further discussion of this important point). Many features of the gut-associated lymphoid system, which contribute to this immunostasis, will not be reviewed here. Two important recent studies, however, have looked specifically at the role of dendritic cells in gut mucosal immunity. One highlighted a population of dendritic cells found within lymphatics that drain the gut mucosa. These cells contain fragments of gut epithelial cells and may present components of gut flora to the regulatory T cells that limit immune responsiveness and inflammation (Huang et al., 2000). Another study examined the interaction between dendritic cells and the gut epithelium in the face of a strong bacterial challenge (Rescigno et al., 2001). Exposure of the epithelium to Salmonella or E. coli caused rapid recruitment of dendritic cells to the subepithelial lamina propria. The most striking finding was that in response to this bacterial stimulus, the dendritic cells extended processes among the epithelial cells (at the same time maintaining the overall integrity of the epithelial barrier) and took up bacteria from the lumen. Dendritic cell recruitment was transitory, and the cells presumably migrated to the lymph node within hours of bacterial challenge.

It seems likely that in the gut, as in other tissues, the flow of activated antigen-loaded dendritic cells into the lymphoid tissue and subsequent T-cell activation is regulated primarily by the extent of activation of innate immunity and tissue damage. Under normal conditions, insufficient bacterial components penetrate the gut barrier to initiate an inflammatory response; specific mechanisms may well exist to ensure that the threshold of this activation is higher than elsewhere. Transport of bacterial proteins via the M cells, and directly into underlying Peyer's patches, however, may result in presentation by Peyer's patch dendritic cells and subsequent localised production of IgA. In cases in which bacteria manage to penetrate the gut barrier, however, either by active invasion, or as a consequence of gut damage, innate immune activation occurs, and this in turn drives dendritic cell migration and antigen presentation.

1.5 CONCLUSION

The view of the dendritic antigen presenting cell acting at the interface among the evolutionary distinct entities of innate immunity (represented by the myeloid lineages), adaptive immunity (represented by the lymphocyte lineages), and the microbe is well exemplified by the bacterial immune response (Fig. 1.3). Dendritic cells recognise and respond to bacterial

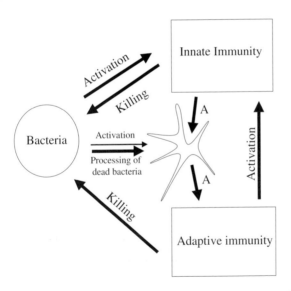

Figure 1.3. Interactions between innate and adaptive immunity in response to bacterial infection. Bacteria activate dendritic cells, either directly via pattern recognition receptors on the dendritic cell or more usually via the activation of innate immunity or tissue damage. The cells of innate immunity kill bacteria, allowing processing and presentation by dendritic cells. Activated dendritic cells present bacterial antigens to T cells, initiating adaptive immunity. In turn, adaptive immunity activates innate immunity further or kills bacteria directly.

invasion using the ancient self/nonself recognition systems of innate immunity. They process and present bacterial antigens after the bacteria have been killed by the powerful phagocytic and bactericidal systems of neutrophils and macrophages. Finally, they integrate the antigenic and inflammatory stimuli, which they experience at the site of infection, and transmit them to the lymph node, or spleen, where they deliver appropriate antigen-specific and antigen-nonspecific signals to T lymphocytes. The end result is the initiation of an adaptive immune response, which is both antigen specific and able to direct and amplify the local innate immune response, resulting, in most cases, in rapid clearance of bacterial infection and a return to the status quo. The importance of these cells in inducing antibacterial immunity suggests that they will be targets for immune evasion. Little is yet known of the antidendritic cell strategies of bacteria, as these cells have not been studied in the same detail as macrophages. The possibility that bacteria may utilise dendritic cells as a means of dispersal within the host is one novel example of immune evasion that requires further study.

REFERENCES

Al Alwan, M.M., Rowden, G., Lee, T.D., and West, K.A. (2001). The dendritic cell cytoskeleton is critical for the formation of the immunological synapse. *Journal of Immunology* **166**, 1452–1456.

Albert, M.L., Pearce, S.F., Francisco, L.M., Sauter, B., Roy, P., Silverstein, R.L., and Bhardwaj, N. (1998). Immature dendritic cells phagocytose apoptotic cells via alphavbeta5 and CD36, and cross-present antigens to cytotoxic T lymphocytes. *Journal of Experimental Medicine* **188**, 1359–1368.

Anderson, C.C., Carroll, J.M., Gallucci, S., Ridge, J.P., Cheever, A.W., and Matzinger, P. (2001). Testing time-, ignorance-, and danger-based models of tolerance. *Journal of Immunology* **166**, 3663–3671.

Antonopoulos, C., Cumberbatch, M., Dearman, R.J., Daniel, R.J., Kimber, I., and Groves, R.W. (2001). Functional caspase-1 is required for Langerhans cell migration and optimal contact sensitization in mice. *Journal of Immunology* **166**, 3672–3677.

Boirivant, M., Fuss, I.J., Ferroni, L., De Pascale, M., and Strober, W. (2001). Oral administration of recombinant cholera toxin subunit B inhibits IL-12-mediated murine experimental (trinitrobenzene sulfonic acid) colitis. *Journal of Immunology* **166**, 3522–3532.

Bouchon, A., Facchetti, F., Weigand, M.A., and Colonna, M. (2001). TREM-1 amplifies inflammation and is a crucial mediator of septic shock. *Nature* **410**, 1103–1107.

Bryniarski, K., Biedron, R., Petrovska, L., Free, P., Chain, B.M., and Marcinkiewicz, J. (2000). Phagocytosis of bacteria by mouse bone-marrow derived dendritic cells affects their ability to process a heterologous soluble antigen *in vitro. Central European Journal of Immunology* **2000**, 210–216.

Caux, C., Ait-Yahia, S., Chemin, K., de Bouteiller, O., Dieu-Nosjean, M.C., Homey, B., Massacrier, C., Vanbervliet, B., Zlotnik, A., and Vicari, A. (2000). Dendritic cell biology and regulation of dendritic cell trafficking by chemokines. *Springer Seminars in Immunopathology* **22**, 345–369.

Cong, Y., Oliver, A.O., and Elson, C.O. (2001). Effects of cholera toxin on macrophage production of co-stimulatory cytokines. *European Journal of Immunology* **31**, 64–71.

Cumberbatch, M., Dearman, R.J., Antonopoulos, C., Groves, R.W., and Kimber, I. (2001). Interleukin (IL)-18 induces Langerhans cell migration by a tumour necrosis factor-α. *Immunology* **102**, 323–330.

Garside, P., Ingulli, E., Merica, R.R., Johnson, J.G., Noelle, R.J., and Jenkins, M.K. (1998). Visualization of specific B and T lymphocyte interactions in the lymph node. *Science* **281**, 96–99.

Gonzalez-Juarrero, M. and Orme, I.M. (2001). Characterization of murine lung dendritic cells infected with *Mycobacterium tuberculosis. Infection and Immunity* **69**, 1127–1133.

Guidos, C., Wong, M., and Lee, K.C. (1984). A comparison of the stimulatory activities of lymphoid dendritic cells and macrophages in T proliferative responses to various antigens. *Journal of Immunology* **133**, 1179–1184.

Hill, S., Edwards, A.J., Kimber, I., and Knight, S.C. (1990). Systemic migration of dendritic cells during contact sensitization. *Immunology* **71**, 277–281.

Hill, S., Griffiths, S., Kimber, I., and Knight, S.C. (1993). Migration of dendritic cells during contact sensitization. *Advances in Experimental Medicine and Biology* **329**, 315–320.

Huang, F.P., Platt, N., Wykes, M., Major, J.R., Powell, T.J., Jenkins, C.D., and MacPherson, G.G. (2000). A discrete subpopulation of dendritic cells transports apoptotic intestinal epithelial cells to T cell areas of mesenteric lymph nodes [see comments]. *Journal of Experimental Medicine* **191**, 435–444.

Ibrahim, M.A., Chain, B.M., and Katz, D.R. (1992). The injured cell: the role of the dendritic cell system as a sentinel receptor pathway. *Immunology Today* **16**, 181–186.

Inaba, K., Inaba, M., Naito, M., and Steinman, R.M. (1993). Dendritic cell progenitors phagocytose particulates, including bacillus Calmette-Guerin organisms, and sensitize mice to mycobacterial antigens *in vivo. Journal of Experimental Medicine* **178**, 479–488.

Janeway, C.A.J. (1992). The immune system evolved to discriminate infectious nonself from noninfectious self. *Immunology Today* **13**, 11–16.

Kaye, P.M., Chain, B.M., and Feldmann, M. (1985). Dendritic cells can present *Mycobacterium tuberculosis* to primed T cells. *Journal of Immunology* **134**, 207–219.

Kimbrell, D.A. and Beutler, B. (2001). The evolution and genetics of innate immunity. *Nature Review of Genetics* **2**, 256–267.

King, P.D. and Katz, D.R. (1989). Human tonsillar dendritic cell-induced T cell responses: analysis of molecular mechanisms using monoclonal antibodies. *European Journal of Immunology* **19**, 581–587.

Kolb-Maurer, A., Gentschev, I., Fries, H.W., Fiedler, F., Brocker, E.B., Kampgen, E., and Goebel, W. (2000). *Listeria monocytogenes*-infected human dendritic cells: uptake and host cell response. *Infection and Immunity* **68**, 3680–3688.

Larsen, C.P., Morris, P.J., and Austyn, J.M. (1990a). Migration of dendritic leukocytes from cardiac allografts into host spleens. A novel pathway for initiation of rejection. *Journal of Experimental Medicine* **171**, 307–314.

Larsen, C.P., Steinman, R.M., Witmer Pack, M., Hankins, D.F., Morris, P.J., and Austyn, J.M. (1990b). Migration and maturation of Langerhans cells in skin transplants and explants. *Journal of Experimental Medicine* **172**, 1483–1493.

Leenen, P.J., Radosevic, K., Voerman, J.S., Salomon, B., van Rooijen, N., Klatzmann, D., and Van Ewijk, W. (1998). Heterogeneity of mouse spleen dendritic cells: *in vivo* phagocytic activity, expression of macrophage markers, and subpopulation turnover. *Journal of Immunology* **160**, 2166–2173.

Leenen, P.J., Voerman, J.S., Radosevic, K., van Rooijen, N., and Van Ewijk, W. (1997). Mouse spleen dendritic cells. Phagocytic activity and expression of macrophage markers. *Advances in Experimental Medicine and Biology* **417**, 91–95.

Macatonia, S.E., Knight, S.C., Edwards, A.J., Griffiths, S., and Fryer, P. (1987). Localization of antigen on lymph node dendritic cells after exposure to the contact sensitizer fluorescein isothiocyanate. Functional and morphological studies. *Journal of Experimental Medicine* **166**, 1654–1667.

Mahnke, K., Guo, M., Lee, S., Sepulveda, H., Swain, S.L., Nussenzweig, M., and Steinman, R.M. (2000). The dendritic cell receptor for endocytosis, DEC-205, can recycle and enhance antigen presentation via major histocompatibility complex class II-positive lysosomal compartments. *Journal of Cell Biology* **151**, 673–684.

Marcinkiewicz, J., Chain, B., Nowak, B., and Grabowska, A. (1999). Distinct mediator profile of murine dendritic cells and peritoneal macrophages. *Central European Journal of Immunology* **24**, 9–15.

Marcinkiewicz, J., Chain, B., Nowak, B., Grabowska, A., Bryniarski, K., and Baran, J. (2000). Antimicrobial and cytotoxic activity of hypochlorous acid: interactions with taurine and nitrite. *Inflammation Research* **49**, 280–289.

Mohagheghpour, N., van Vollenhoven, A., Goodman, J., and Bermudez, L.E. (2000). Interaction of *Mycobacterium avium* with human monocyte-derived dendritic cells. *Infection and Immunity* **68**, 5824–5829.

Monack, D.M., Hersh, D., Ghori, N., Bouley, D., Zychlinsky, A., and Falkow, S. (2000). Salmonella exploits caspase-1 to colonize Peyer's patches in a murine typhoid model. *Journal of Experimental Medicine* **192**, 249–258.

Niedergang, F., Sirard, J.C., Blanc, C.T., and Kraehenbuhl, J.P. (2000). Entry and survival of *Salmonella typhimurium* in dendritic cells and presentation of recombinant antigens do not require macrophage-specific virulence factors. *Proceedings of the National Academy of Sciences USA* **97**, 14,650–14,655.

Paglia, P., Medina, E., Arioli, I., Guzman, C.A., and Colombo, M.P. (1998). Gene transfer in dendritic cells, induced by oral DNA vaccination with *Salmonella typhimurium*, results in protective immunity against a murine fibrosarcoma. *Blood* **92**, 3172–3176.

Paschen, A., Dittmar, K.E., Grenningloh, R., Rohde, M., Schadendorf, D., Domann, E., Chakraborty, T., and Weiss, S. (2000). Human dendritic cells infected by *Listeria monocytogenes*: induction of maturation, requirements for phagolysosomal escape and antigen presentation capacity. *European Journal of Immunology* **30**, 3447–3456.

Porgador, A., Irvine, K.R., Iwasaki, A., Barber, B.H., Restifo, N.P., and Germain, R.N. (1998). Predominant role for directly transfected dendritic cells in antigen presentation to CD8+ T cells after gene gun immunization. *Journal of Experimental Medicine* **188**, 1075–1082.

Pron, B., Boumaila, C., Jaubert, F., Berche, P., Milon, G., Geissmann, F., and Gaillard, J.L. (2001). Dendritic cells are early cellular targets of *Listeria monocytogenes* after intestinal delivery and are involved in bacterial spread in the host. *Cell Microbiology* **3**, 331–340.

Randolph, G.J., Beaulieu, S., Lebecque, S., Steinman, R.M., and Muller, W.A. (1998). Differentiation of monocytes into dendritic cells in a model of transendothelial trafficking. *Science* **282**, 480–483.

Rescigno, M., Urbano, M., Valzasina, B., Francolini, M., Rotta, G., Bonasio, R., Granucci, F., Kraehenbuhl, J.P., and Ricciardi-Castagnoli, P. (2001).

Dendritic cells express tight junction proteins and penetrate gut epithelial monolayers to sample bacteria. *Nature Immunology* **2**, 361–367.

Romani, N.S., Koide, M., Crowley, M., Witmer-Pack, A.M., Livingstone, C., Fathman, C.G., Inaba, K., and Steinman, R.M. (1989). Presentation of exogenous protein antigens by dendritic cells to T cell clones. Intact protein is presented best by immature epidermal Langerhans' cells. *Journal of Experimental Medicine* **169**, 1169–1178.

Rutault, K., Alderman, C., Chain, B.M., and Katz, D.R. (1999). Reactive oxygen species activate human peripheral blood dendritic cells. *Free Radicals in Biology and Medicine* **26**, 232–238.

Sallusto, F. and Lanzavecchia, A. (1994). Efficient presentation of soluble antigen by cultured human dendritic cells is maintained by granulocyte/macrophage colony-stimulating factor plus interleukin 4 and downregulated by tumor necrosis factor α. *Journal of Experimental Medicine* **179**, 1109–1118.

Schnare, M., Barton, G.M., Holt A.C., Takeda, K., Akira, S., and Medzhitov, R. (2001). Toll-like receptors control activation of adaptive immune responses. *Nature Immunology* **2**, 947–950.

Sousa, C. and Germain, R.N. (1999). Analysis of adjuvant function by direct visualization of antigen presentation in vivo: endotoxin promotes accumulation of antigen-bearing dendritic cells in the T cell areas of lymphoid tissue. *Journal of Immunology* **162**, 6552–6561.

Sousa, C., Stahl, P.D., and Austyn, J.M. (1993). Phagocytosis of antigens by Langerhans' cells *in vitro*. *Journal of Experimental Medicine* **178**, 509–519.

Steinman, R.M. (1991). The dendritic cell system and its role in immunogenicity. *Annual Review of Immunology* **9**, 271–296.

Strunk, D., Egger, C., Leitner, G., Hanau, D., and Stingl, G. (1997). A skin homing molecule defines the langerhans cell progenitor in human peripheral blood. *Journal of Experimental Medicine* **185**, 1131–1136.

Svensson, M., Stockinger, B., and Wick, M.J. (1997). Bone marrow-derived dendritic cells can process bacteria for MHC-I and MHC-II presentation to T cells. *Journal of Immunology* **158**, 4229–4236.

Tascon, R.E., Soares, C.S., Ragno, S., Stavropoulos, E., Hirst, E.M., and Colston, M.J. (2000). *Mycobacterium tuberculosis*-activated dendritic cells induce protective immunity in mice. *Immunology* **99**, 473–480.

Triantafilou, K., Triantafilou, M., and Dedrick, R.L. (2001). A CD14-independent LPS receptor cluster. *Nature Immunology* **2**, 338–345.

Vazquez-Torres, A., Jones-Carson, J., Baumler, A.J., Falkow, S., Valdivia, R., Brown, W., Le, M., Berggren, R., Parks, W.T., and Fang, F.C. (1999).

Extraintestinal dissemination of Salmonella by CD18-expressing phagocytes. *Nature* **401**, 804–808.

Woodhead, V.E., Binks, M.H., Chain, B.M., and Katz, D.R. (1998). From sentinel to messenger: an extended phenotypic analysis of the monocyte to dendritic cell transition. *Immunology* **94**, 552–559.

Yu, D., Imajoh-Ohmi, S., Akagawa, K., and Kanegasaki, S. (1996). Suppression of superoxide-generating ability during differentiation of monocytes to dendritic cells. *Journal of Biochemistry* (Tokyo) **119**, 23–28.

CHAPTER 2

CD1 and nonpeptide antigen recognition systems in microbial immunity

Kayvan R. Niazi, Steven A. Porcelli, and Robert L. Modlin

2.1 INTRODUCTION

Until recently, it was generally believed that proteins encoded within the MHC (or major histocompatibility complex) locus carry out the majority of the immunologically relevant antigen presentation functions necessary to alert the immune system to pathogenic or oncogenic challenges. This notion was initially based on the identification of the MHC as the critical genetic locus involved in tissue graft rejection. Genetic and immunological studies identified the protein products of two distinct families of genes called MHC I and MHC II as being responsible for graft rejection. Additional studies demonstrated that the primary role of these proteins is to signal the immune system to respond to invading pathogens through the presentation of *peptide* fragments derived from endogenous and exogenous protein sources to different classes of T lymphocytes. Thus, two distinct pathways of *peptide* antigen presentation for the detection of invading intracellular and extracellular pathogens (or pathogens which reside in intracellular vacuoles) by MHC I and MHC II, respectively, were identified and investigated. The role of the key cell population, the dendritic cell, in antigen presentation has been described in Chapter 1. In the past ten years, however, a new paradigm in immunology has emerged whereby nonpeptide lipid antigens are presented to T cells by the MHC-related protein, CD1.

2.2 EVOLUTION OF CD1

The human CD1 (hCD1) polypeptides are members of a family of glycoproteins that are conserved throughout mammalian evolution and were initially identified as thymocyte differentiation antigens (Porcelli and Modlin, 1999). The hCD1 gene locus is located on chromosome 1 (Albertson et al., 1988) and contains five genes (*CD1A-E*) that encode five polypeptides (CD1a–e) (Porcelli and Modlin, 1999; Angenieux et al., 2000). To date, hCD1 homologues have been discovered in all species of placental mammals investigated including mice, rats, rabbits, guinea pigs, sheep, pigs, and cows (Porcelli and Modlin, 1999; Chun et al., 1999; Denham et al., 1994) as well as the bandicoot, a marsupial mammal (Cisternas and Armati, 2000). Amino acid sequence analysis of the known mammalian CD1 protein sequences has revealed the existence of two CD1 families: group I CD1 and group II CD1 (Calabi et al., 1989)). Human CD1a, -b, and -c belong to group I CD1, while CD1d belongs to group II CD1. Of the five human CD1 proteins, CD1e is the least well understood but may represent a third group of CD1 (Angenieux et al., 2000). Interestingly, the mouse genome does not contain any group I CD1 genes, making it an unsuitable model for the study of group I CD1. Because the current data regarding T-cell recognition of bacterial-derived antigens presented in the context of human CD1 predominantly involves the group I CD1 proteins, this chapter focuses on the biochemical and biological features and potential roles of CD1a, -b, and -c in human immunity to microbial infection.

2.3 TISSUE DISTRIBUTION OF CD1

Unlike MHC I, which is ubiquitously expressed, the human CD1 proteins demonstrate a unique tissue distribution particular to each isoform *in vivo* (Porcelli and Modlin, 1999). For example, CD1b is expressed on cortical thymocytes and subsets of dendritic cells. CD1a is also expressed on cortical thymocytes and dendritic cells, but is also found on Langerhans cells. CD1c expression mirrors that of CD1a as well as being present on mantle zone B cells. Expression of the group I CD1 proteins can be induced on peripheral blood mononuclear cell- (PBMC-) derived monocytes *in vitro* through the addition of GM-CSF (Sallusto and Lanzavecchia, 1994; Porcelli et al., 1992; Kasinrerk et al., 1993) suggesting that CD1 may be involved in inflammatory immune responses *in vivo*. That CD1 expression is induced by GM-CSF has proven significant with respect to the provision of readily available CD1-expressing antigen presenting cells (APCs) for the scientific dissection of the CD1 antigen presentation pathway.

2.4 STRUCTURE OF CD1

The hCD1 proteins are similar to the MHC I proteins in many respects. Like MHC I, hCD1 is composed of an approximately forty-five kDa heavy chain and twelve kDa β_2-microglobulin (Kefford et al., 1984). Although the human CD1 proteins have not been crystallized, the crystal structure of the related mouse CD1d1 protein has revealed significant structural similarity between CD1 and MHC I (Zeng et al., 1997). Because of the amino acid homology between mouse CD1d1 and the other CD1 proteins, it is generally accepted that the human members of the CD1 family are likely to adopt similar three-dimensional structures. Based on the crystal structure of mouse CD1d1, the putative antigen binding groove of CD1 is believed to be grossly similar to that of MHC I as it too comprises two antiparallel α-helices that overlay a floor composed of antiparallel β strands (Fig. 2.1). However, the CD1 antigen binding groove demonstrates two significant structural differences from that of MHC I.

First, the CD1 proteins are essentially nonpolymorphic (van Agthoven and Terhorst, 1982; Amiot et al., 1988). Thus, the anticipated antigen-binding groove of each CD1 isoform would be identical among all members of a given species. The second significant difference occurs with respect to key structural features such as charge, spatial organization, and size of the proposed CD1 antigen-binding groove with respect to those of MHC I (Zeng et al., 1997). Based on X-ray crystallographical findings and the close amino acid sequence similarities of the different CD1 proteins, it is expected that the proposed antigen-binding grooves of these proteins are electrostatically neutral. This suggests that CD1 binds antigens of a hydrophobic nature and not the charged peptides typically associated with MHC I. Also, unlike MHC I, which contains as many as six small pockets (designated A–F) in its antigen binding groove, CD1 possesses two large pockets (designated as the A' and F' pockets). Lastly, although the α-helices of CD1 are in greater proximity to those of MHC I, the depth of the proposed CD1 antigen binding groove far exceeds that of MHC I.

2.5 CD1 AND ANTIGEN PRESENTATION

These structural differences collectively cause the greatest distinction in immunological function between CD1 and the MHC proteins. Whereas MHC I presents short peptides derived from the proteolytic degradation of intracellular and extracellular proteins to T cells, respectively, the human group I CD1 proteins present bacteria-derived long-chain glycolipid and lipoglycan antigens. To date, four different mycobacteria-derived antigens

CD1 **MHC I**

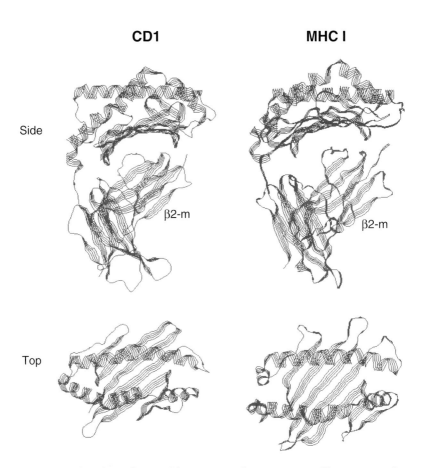

Side

β2-m

β2-m

Top

Figure 2.1. Side and top of views of the structures of CD1 (represented by mouse CD1d1) and MHC I (represented by human leukocyte antigen HLA-A2).

presented by the group I human CD1 proteins have been identified: the mycolic acids (Beckman et al., 1994), the glycosylated mycolic acid derivatives (e.g., glucose monomycolate; Moody et al., 1997), the phosphatidylinositol mannosides (which include lipoarabinomannan or LAM; Sieling et al., 1997), and the hexosyl-1-phophoisoprenoids or hPIP (which include mannosyl-1β-phosphoisoprenoid; Moody et al., 2000). All four are mycobacterial cell wall constituents. More importantly, all of these antigens, with the exception of hPIP, share the common structural motif of dual (or branched) long-chain hydrocarbon groups covalently linked via a hydrophilic cap, which is formed by either polar or charged groups (Fig. 2.2). The hydrophilic cap can be further modified by the incorporation of carbohydrate moieties. The hPIP antigen differs from the common structural motif found in the other three antigens

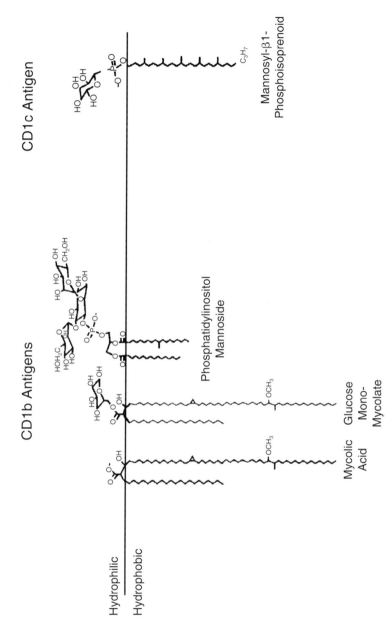

Figure 2.2. Structures of known group I CD1 antigens. The hPIP antigen is represented by mannosyl-1ß-phosphoisoprenoid.

in that it possesses a single chain hydrocarbon. It is interesting to note that CD1b presents the three antigens possessing dual acyl chains, whereas CD1c presents hPIP. These findings may imply that although CD1b and CD1c are the most closely related of the human CD1 isoforms, they bind distinct structural motifs. The identification of additional CD1c antigens may provide additional evidence in support of this hypothesis.

In addition to the identified antigens presented above, other as yet unidentified mycobacterial antigens presented by group I CD1 have also been documented (Rosat et al., 1999; Sieling et al., 2000; Stenger et al., 1997; Ochoa et al., 2001). Based on the structural similarity of the four known antigens, these antigens are likely to share similar structural motifs. Lastly, the presentation of lipid antigens purified from other bacterial species including *Haemophilus influenzae* (Fairhurst et al., 1998), *Rhodococcus equi* and *Nocardia farcinica* (P.A. Sieling, D.B. Moody, and S.A. Porcelli, personal communication), and *Staphylococcus aureus* (P.A. Sieling, personal communication) by CD1 has also been observed. These findings imply that the CD1 proteins may play an important lipid antigen-presentation function in the human immune response to a variety of bacterial infections.

2.6 CD1-RESTRICTED T LYMPHOCYTES

CD1-restricted T cells recognizing bacterial-derived lipids and glycolipids have been derived from normal human peripheral blood mononuclear cells (PBMC) (Porcelli et al., 1992; Sieling et al., 1995; Rosat et al., 1999; Sieling et al., 2000; Stenger et al., 1997), infected patient PBMC (Stenger et al., 1997), and disease lesions (Moody et al., 1997; Sieling et al., 1995; Sieling et al., 2000) suggesting that they are a ubiquitous subpopulation of circulating T cells. Group I CD1 restricted T cells can be derived from all currently defined subsets of circulating T cells such as CD4+, CD8+, and CD4−/CD8− T cells. In general, glycolipid-specific, hCD1-restricted T cells are cytotoxic and exhibit a T_H1 phenotype. In *in vitro* T cell assays, CD1-restricted T cells specifically destroy both mycobacteria-infected APCs (Jackman et al., 1998) as well as APCs that acquire exogenous mycobacterial antigens (Porcelli et al., 1992; Moody et al., 1997; Sieling et al., 1995, 2000; Moody et al., 2000; Rosat et al., 1999; Stenger et al., 1997) suggesting that they serve in controlling the extent of bacterial infection *in vivo*. Interestingly, both CD4+ (Ochoa et al., 2001) and CD8+ (Stenger et al., 1998a) CD1-restricted T cells may further limit mycobacterial infection *in vivo* through the release of the antimicrobial peptide, granulysin.

2.7 CD1 AND ANTIGEN INTERACTION

Studies focusing on CD1b and its presentation of the antigens lipoara-binomannan (LAM; Sieling et al., 1995) and glucose monomycolate (GMM; Moody et al., 1997) suggest that T cells recognize the hydrophilic cap structures of lipid antigens. LAM belongs to the phosphatidylinositol family of antigens and is composed of an arabinan head, a branched mannan core, and a phosphatidylinositol unit, which often includes tuberculostearic acid (branched C:19) and palmitic acid (C:16). T-cell recognition of LAM is dependent on the polysaccharide head group of LAM, as variant forms of LAM possessing different subsets of the carbohydrates present in the cap structure of LAM did not optimally stimulate T cells. In more detailed experiments using GMM, the alteration of the fine stereochemistry of the hydrophilic glucose head group by substitution with mannose or galactose also ablated T-cell recognition. In addition, shortening the naturally occurring C80 wax-ester mycolate to a C32 mycolate did not significantly alter T-cell activity.

Consistent with the predicted hydrophobic nature of the CD1 antigen-binding groove, parallel studies of antigen binding by CD1b have revealed that CD1b binds the hydrocarbon chains present in lipid antigens *in vitro* (Ernst et al., 1998). This was demonstrated through competition experiments between the two CD1b glycolipid antigens, LAM and PIM_2, which contain identical hydrocarbon chains but different hydrophilic cap structures. Unlike PIM_2, deacylated LAM, which varies from LAM in that it does not possess either hydrocarbon chain, was unable to inhibit CD1b/LAM binding. Lastly, CD1b/LAM interaction demonstrated a $K_d \sim 10^{-8}$ M, which approximates the affinity of the MHC proteins for antigenic peptides. Together with the T-cell antigen recognition data, these data provide a model of CD1/antigen interaction where the antigens' long-chain acyl chains are positioned deep within the hydrophobic A' and F' pockets while the hydrophilic cap structures are proximal to the membrane-distal surface of CD1 for interaction with the T-cell receptor (TCR).

To date, little is known regarding the precise molecular interactions that occur between antigen and CD1. Mutational analyses of the CD1b antigen binding groove revealed that the long-chain antigen GMM, which possesses two hydrocarbon chains 22 and 58 carbons in length, is dependent predominantly on the amino acid residues located between the A'/F' pocket intersection and the distal A' pocket for optimal presentation (Niazi et al., 2001). Simultaneous analysis of the presentation of synthetic GMM (or sGMM), which differs from GMM in that it contains two shorter hydrocarbon chains of 18 carbons and 14 carbons in length, was more dependent on the residues located at the A′/F′ pocket intersection and the distal F′ pocket. Thus, although the presentation of both antigens was dependent on many common residues, these two antigens that vary in hydrocarbon chain length utilize different subsets of hydrophobic residues located in both the A′ and F′ pockets in antigen presentation. Whether or not both pockets are required for the CD1 presentation of all lipid antigens remains unknown but may be elucidated through additional mutational studies and CD1/antigen co-crystallization studies.

Mutational analysis of the CD1b residues recognized by the TCR of antigen-specific T cells has further defined the area of interaction between the membrane distal surface of CD1b and the TCR (Melian et al., 2000). These studies demonstrated that CD1b interacts with TCR at a different angle (almost perpendicular to the groove axis) than observed for MHC I and TCR ($\sim 45°$ in relation to the groove axis). These differences are likely to be due to both differences in the structures of CD1 and MHC I as well as differences in the structures of the antigens presented by these two molecules. Additional studies of the TCRs cloned from various CD1b restricted T cells recognizing different antigens suggest that the recognition of different CD1b/antigen complexes occurs as a result of genetic variability in the CDR3 region of the TCR, similar to TCRs that recognize MHC/peptide complexes (Grant et al., 1999).

2.8 REQUIREMENTS FOR CD1 ANTIGEN PRESENTATION

Because of the availability of the CD1-restricted T cells, antibodies, and purified antigens, the CD1 antigen presentation pathway is in the process of being dissected. Unlike MHC I and II, the presentation of lipid antigens by CD1 is independent of TAP1/2 or HLA-DM, respectively (Porcelli et al., 1992; Beckman et al., 1996). Whether CD1 requires accessory proteins for antigen presentation remains unknown although the CD1b presentation of LAM is dependent on the expression of the macrophage mannose receptor (MMR) (Prigozy et al., 1997). Furthermore, in the case of CD1b, the presentation of many antigens (e.g., mycolic acid (Porcelli et al., 1992), GMM (Sugita et al., 1996), and LAM (Sieling et al., 1995) is dependent on endosomal acidification as presentation can be blocked through the addition of endosomal acidification inhibitors.

2.9 INTRACELLULAR TRAFFICKING OF CD1

That CD1b requires a low pH environment is not surprising given the nature of the intracellular compartments through which CD1b traffics to acquire antigen. Intracellular immunofluoresence studies of CD1b traffic have revealed that CD1b predominantly resides in late endosomal/lysosomal/MIIC compartments where the pH is ~4.5 (Prigozy et al., 1997; Sugita et al., 1996). Previous studies of MHC II antigen acquisition (which is likely to occur in the same compartments as CD1b) have revealed that endosomal acidification may be important for two reasons. First, a low pH environment activates the proteases that provide suitable peptides for MHC II binding (Authier et al., 1996). Second, this environment produces conformational effects on MHC II structure (Lee et al., 1992) that may facilitate peptide sampling in the presence of HLA-DM (Vogt et al., 1999). Although CD1b is unlikely to depend on similar protease activities to provide its supply of lipid antigens, it too adopts different three-dimensional conformations that better enable it to bind antigen at low pH through the exposure of buried hydrophobic amino acid residues (Ernst et al., 1998).

In 1996, it was demonstrated that the traffic of CD1b to late endosomes/lysosomes is dependent on the presence of the CD1b cytoplasmic tail (Sugita et al., 1996). The CD1b cytoplasmic tail contains a tyrosine-containing amino acid motif (YXXZ) where Y = tyrosine, X = any amino acid, and Z = bulky hydrophobic amino acid. This motif is also found in HLA-DM, a protein that resides predominantly in the late endosome/MIIC compartments. Thus, deletion or substitution of this motif with other sequences abrogates CD1b traffic to the late endosomal/lysosomal compartments (Jackman et al., 1998; Sugita et al., 1996). Antigen presentation studies of cells expressing wild-type versus CD1b possessing an altered cytoplasmic tail strongly suggest that CD1b traffic to these compartments is required for the presentation of both exogenously (Jackman et al., 1998; Geho et al., 2000) and endogenously (i.e., through live bacterial infection) acquired antigens (Jackman et al., 1998).

Comparison of the cytoplasmic tails of CD1a, CD1b, and CD1c has revealed the presence of a similar YXXZ motif in CD1c but not CD1a. Not surprisingly, CD1a does not traffic to the later endosomal compartments but rather predominantly localizes to ARF-6+ early recycling endosomes (Sugita et al., 1999). Consistent with its absence from low pH-late endosomes, CD1a antigen presentation does not depend on endosomal acidification.

Interestingly, immunofluorescence studies of CD1c demonstrate that CD1c also traffics through early endosomes (Sugita et al., 2000; Briken et al., 2000). Functional studies of CD1c antigen presentation provide indirect evidence for the binding of antigen by CD1c in these early compartments. First, acidification inhibitors do not inhibit CD1c antigen presentation as observed for CD1b, which traffics into more mature endosomal compartments

(Sugita et al., 2000; Briken et al., 2000). Furthermore, the CD1c presentation of antigens studied thus far is independent of the presence of the YXXZ targeting motif within the CD1c cytoplasmic tail (Geho et al., 2000; Briken et al., 2000). Experiments involving various forms of the antigen hPIP containing hydrocarbons of increasing length may provide additional indirect evidence for antigen acquisition by CD1c in early endosomal compartments. It is generally believed that lipid antigens containing longer hydrocarbon tails are trafficked deeper into the endosomal system than those with shorter tails. Consistent with this hypothesis, the presentation efficiency of hPIP by CD1c is inversely proportional to its hydrocarbon chain length (Moody et al., 2000). Kinetic studies of hPIP presentation by CD1c also suggest that CD1c binds antigen in early endosomal compartments (Briken et al., 2000). In addition to early endosomes, the presence of CD1c in late endosomes/lysosomes/MIICs has also been reported suggesting that CD1c may also bind antigen in more mature endosomal compartments as observed for CD1b (Sugita et al., 2000). Together with the unusual single-hydrocarbon chain motif of hPIP discussed earlier, differences in intracellular traffic between CD1b and CD1c may reflect different antigen-binding capabilities between these proteins.

2.10 CD1 AND MYCOBACTERIAL INFECTIONS

The evolutionary selective pressure behind the existence of such diverse intracellular trafficking patterns for the group I CD1 is initially difficult to appreciate. However, analyses of CD1 distribution in mycobacteria-infected antigen-presenting cells may provide some insight into this matter. Upon infection, viable mycobacteria reside within phagosomes, which avoid fusion with late endosomes/lysosomes through inhibition of the vacuolar H+ATPase (Sturgill-Koszycki et al., 1994). Simultaneous analyses of the intracellular distribution of the group I CD1 isoforms and mycobacteria in infected dendritic cells demonstrate that CD1 distribution is relatively unchanged by infection (Schaible et al., 2000). Thus, CD1a and CD1c intersect with mycobacteria in early endosomes. In contrast, CD1b (and to a lesser extent CD1c) intersect with mycobacteria in mature phagolysosomes where the bacilli are likely to be nonviable. Of equal importance, mycobacteria-secreted glycolipids such as LAM, PIM, and other unidentified lipids localize to late endosomes where they are likely to bind CD1b and CD1c (Schaible et al., 2000; Xu et al., 1994). Thus, the combined efforts of the group I CD1 proteins apparently ensure the surveillance of the intracellular compartments containing either live mycobacteria and/or secreted glycolipids for immune recognition. Based on the results discussed here and on the data discussed

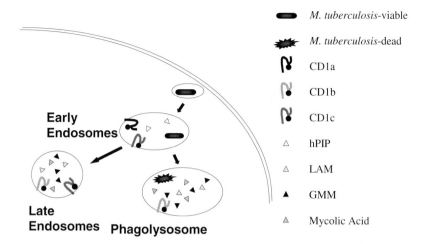

Figure 2.3. Intracellular distribution of CD1 isoforms during infection with *M. tuberculosis*.

earlier regarding inhibition of CD1 antigen presentation by acidification inhibitors, a model of the intracellular traffic of the group I CD1 isoforms in relation to *M. tuberculosis* bacilli and shed glycolipids is presented in Figure 2.3.

The notion that human CD1 antigen presentation is relevant to host defense against microbial pathogens is supported by experiments focusing on CD1 expression in leprosy where the level of CD1 expression correlates with cell-mediated immunity and the ability to limit infection. The etiologic agent of leprosy is *Mycobacterium leprae*. In humans, leprosy demonstrates a spectrum of disease severity. Thus, in the milder, immunologically responsive form of the disease (i.e., tuberculoid leprosy), CD1 expression is significantly greater than the expression detected in the more severe, immunologically unresponsive form of leprosy (i.e., lepromatous leprosy; Sieling et al., 1999). Whether or not disease severity is caused by the level of CD1 expression locally, however, is unlikely as the overall number of dendritic cells present in disease lesions differs greatly between tuberculoid and lepromatous samples. More recent studies of T cells collected from *M. tuberculosis*-infected patients revealed an increase in the frequency of antigen-specific, CD1-restricted T cells in comparison to uninfected controls, further suggesting that CD1 plays a direct role in the host response to bacterial infection (Moody et al., 2000).

The exact mechanisms behind the differences in CD1 expression in the various forms of leprosy remain unknown. Studies of monocyte-derived dendritic cells infected by *M. tuberculosis* revealed that cell-surface expression of the group I CD1 isoforms are specifically down-regulated while levels of MHC I and MHC II remain unchanged by infection (Stenger et al., 1998b). CD1 down-regulation occurred only in the presence of live mycobacteria and not heat-killed bacilli, suggesting that it is an active process. Further analyses of the messenger RNAs (mRNAs) encoding the group I CD1 isoforms demonstrated that the mechanism of cell-surface down-regulation is likely to occur at the mRNA level. That pathogens can alter MHC I and MHC II antigen presentation pathways to evade the host immune response has been documented in various viral, bacterial, and parasite infections (Zhong et al., 1999, 2000; Alcami and Koszinowski, 2000; Luder et al., 1998). The down-regulation of CD1 expression by *M. tuberculosis* is the first report of the deployment of a similar immune evasion strategy by bacteria that targets the CD1 antigen presentation pathway.

2.11 CONCLUSION

In this chapter, we examined the group I CD1 proteins, which together with MHC I and MHC II, represent a third family of antigen presentation molecules. Although great effort has been put forth to delineate a role for CD1 and other nonclassical MHC and MHC-like proteins in adaptive immunity, their role in host defense is currently a matter of debate. It was recently suggested that CD1 serves solely as a housekeeping protein that functions to monitor lipid synthesis in normal host cells (Shinkai and Locksley, 2000). Two points in support of this hypothesis include: (1) that mice, which lack the group I CD1 genes (represented by hCD1a, b, and c), are resistant to infection by *M. tuberculosis* and (2) that CD1 evolved millions of years prior to the appearance of *M. tuberculosis* as a human pathogen. While it may be unlikely that CD1 evolved to combat *M. tuberculosis* infection specifically, it is altogether possible that the group 1 CD1 proteins originally evolved to aid in the resistance against more ancient mycobacteria (or other bacteria) that have since become categorized as nonpathogenic. Furthermore, the absence of the group 1 CD1 genes in the murine system does not provide significant evidence against the role of these proteins in the human antimycobacterial defense as mycobacteria and related pathogens are not pathogens for mice.

On the contrary, the group I CD1 proteins demonstrate several characteristics that make their classification as housekeeping proteins and not antigen presentation proteins difficult. As stated earlier, group I CD1 is mainly expressed on professional antigen presenting cells, which serve as the primary means of notifying the immune system of the existence of pathogenic challenges. In addition, group I CD1 expression correlates with disease severity in humans providing further support for a role in antigen presentation. The CD1 proteins demonstrate unique intracellular trafficking capabilities that correlate with the intracellular compartments, which contain live and/or nonviable mycobacteria and their secreted lipid cell wall constituents in infection. Within these compartments, CD1b (and possibly CD1c) can undergo conformational changes that allow binding of antigen. Also, CD1 expression is specifically down-regulated on antigen-presenting cells by infection with *M. tuberculosis* suggesting that the presence of group I CD1 may inhibit mycobacterial survival in immunocompetent hosts. Lastly and most importantly, the CD1 proteins present lipid antigens from a variety of different bacterial species to T cells to stimulate cytotoxicity, the release of inflammatory cytokines, and the release of antimicrobial peptides such as granulysin to destroy intracellular bacteria and limit infection. These data taken together suggest that the CD1 proteins are likely to be involved in promoting the immune response to bacterial infection.

Although the current data does not argue against the CD1 proteins originally serving as a means to monitor cellular health through the surveillance of intracellular lipid pools, the existing evidence is strongly consistent with the group I CD1 serving as an MHC-independent antigen presentation pathway. As such, the CD1 proteins represent exciting targets for the design of antimicrobial vaccines. Indeed, synthetic analogues of known CD1 antigens have been successfully created and utilized to elicit T cell cytotoxicity *in vitro* (Moody et al., 1997). Further insights into the CD1 antigen presentation pathway should provide novel strategies to control and prevent bacterial infection *in vivo*.

ACKNOWLEDGMENTS

The authors thank P.A. Sieling for his helpful discussion in the preparation of this chapter and D. Branch Moody for his assistance in preparing Figure 2.2.

REFERENCES

Albertson, D.G., Fishpool, R., Sherrington, P., Nacheva, E., and Milstein, C. (1988). Sensitive and high resolution in situ hybridization to human chromosomes using biotin labelled probes: assignment of the human thymocyte CD1 antigen genes to chromosome 1. *European Molecular Biology Organisation Journal* 7, 2801–2805.

Alcami, A. and Koszinowski, U.H. (2000). Viral mechanisms of immune evasion. *Trends in Microbiology* 8, 410–418.

K. R. NIAZI, S. A. PORCELLI, AND R. L. MODLIN

Amiot, M., Dastot, H., Fabbi, M., Degos, L., Bernard, A., and Boumsell, L. (1988). Intermolecular complexes between three human CD1 molecules on normal thymus cells. *Immunogenetics* **27**, 187–195.

Angenieux, C., Salamero, J., Fricker, D., Cazenave, J.P., Goud, B., Hanau, D., and de La Salle, H. (2000). Characterization of CD1e, a third type of CD1 molecule expressed in dendritic cells. *Journal of Biological Chemistry* **275**, 37,757–37,764.

Authier, F., Posner, B.I., and Bergeron, J.J. (1996). Endosomal proteolysis of internalized proteins. *FEBS Letters* **389**, 55–60.

Beckman, E.M., Porcelli, S.A., Morita, C.T., Behar, S.M., Furlong, S.T., and Brenner, M.B. (1994). Recognition of a lipid antigen by CD1-restricted alpha beta+ T cells [see comments]. *Nature* **372**, 691–694.

Beckman, E.M., Melian, A., Behar, S.M., Sieling, P.A., Chatterjee, D., Furlong, S.T., Matsumoto, R., Rosat, J.P., Modlin, R.L., and Porcelli, S.A. (1996). CD1c restricts responses of mycobacteria-specific T cells. Evidence for antigen presentation by a second member of the human CD1 family. *Journal of Immunology* **157**, 2795–2803.

Briken, V., Jackman, R.M., Watts, G.F., Rogers, R.A., and Porcelli, S.A. (2000). Human CD1b and CD1c isoforms survey different intracellular compartments for the presentation of microbial lipid antigens. *Journal of Experimental Medicine* **192**, 281–288.

Calabi, F., Jarvis, J.M., Martin, L., and Milstein, C. (1989). Two classes of CD1 genes. *European Journal of Immunology* **19**, 285–292.

Chun, T., Wang, K., Zuckermann, F.A., and Gaskins, H.R. (1999). Molecular cloning and characterization of a novel CD1 gene from the pig. *Journal of Immunology* **162**, 6562–6571.

Cisternas, P.A. and Armati, P.J. (2000). Immune system cell markers in the northern brown bandicoot, *Isoodon macrourus. Developmental and Comparative Immunology* **24**, 771–782.

Denham, S., Shimizu, M., Bianchi, A.T., Zwart, R.J., Carr, M.M., and Parkhouse, R.M. (1994). Monoclonal antibodies recognising differentiation antigens on porcine B cells. *Veterinary Immunology and Immunopathology* **43**, 259–267.

Ernst, W.A., Maher, J., Cho, S., Niazi, K.R., Chatterjee, D., Moody, D.B., Besra, G.S., Watanabe, Y., Jensen, P.E., Porcelli, S.A., Kronenberg, M., and Modlin, R.L. (1998). Molecular interaction of CD1b with lipoglycan antigens. *Immunity* **8**, 331–340.

Fairhurst, R.M., Wang, C.X., Sieling, P.A., Modlin, R.L., and Braun, J. (1998). CD1 presents antigens from a gram-negative bacterium, *Haemophilus influenzae* type B. *Infection and Immunity* **66**, 3523–3526.

Geho, D.H., Fayen, J.D., Jackman, R.M., Moody, D.B., Porcelli, S.A., and Tykocinski, M.L. (2000). Glycosyl-phosphatidylinositol reanchoring unmasks distinct antigen-presenting pathways for CD1b and CD1c. *Journal of Immunology* **165**, 1272–1277.

Grant, E.P., Degano, M., Rosat, J.P., Stenger, S., Modlin, R.L., Wilson, I.A., Porcelli, S.A., and Brenner, M.B. (1999). Molecular recognition of lipid antigens by T cell receptors. *Journal of Experimental Medicine* **189**, 195–205.

Jackman, R.M., Stenger, S., Lee, A., Moody, D.B., Rogers, R.A., Niazi, K.R., Sugita, M., Modlin, R.L., Peters, P.J., and Porcelli, S.A. (1998). The tyrosine-containing cytoplasmic tail of CD1b is essential for its efficient presentation of bacterial lipid antigens. *Immunity* **8**, 341–351.

Kasinrerk, W., Baumruker, T., Majdic, O., Knapp, W., and Stockinger, H. (1993). CD1 molecule expression on human monocytes induced by granulocyte-macrophage colony-stimulating factor. *Journal of Immunology* **150**, 579–584.

Kefford, R.F., Calabi, F., Fearnley, I.M., Burrone, O.R., and Milstein, C. (1984). Serum beta 2-microglobulin binds to a T-cell differentiation antigen and increases its expression. *Nature* **308**, 641–642.

Lee, J.M., Kay, C.M., and Watts, T.H. (1992). Conformational changes in mouse MHC class II proteins at acidic pH. *International Immunology* **4**, 889–897.

Luder, C.G., Lang, T., Beuerle, B., and Gross, U. (1998). Down-regulation of MHC class II molecules and inability to up-regulate class I molecules in murine macrophages after infection with *Toxoplasma gondii*. *Clinical Experimental Immunology* **112**, 308–316.

Melian, A., Watts, G.F., Shamshiev, A., De Libero, G., Clatworthy, A., Vincent, M., Brenner, M.B., Behar, S., Niazi, K., Modlin, R.L., Almo, S., Ostrov, D., Nathenson, S.G., and Porcelli, S.A. (2000). Molecular recognition of human CD1b antigen complexes: evidence for a common pattern of interaction with alpha beta TCRs. *Journal of Immunology* **165**, 4494–4504.

Modlin, R.L. (1998). An antimicrobial activity of cytolytic T cells mediated by granulysin. *Science* **282**, 121–125.

Moody, D.B., Reinhold, B.B., Guy, M.R., Beckman, E.M., Frederique, D.E., Furlong, S.T., Ye, S., Reinhold, V.N., Sieling, P.A., Modlin, R.L., Besra, G.S., and Porcelli, S.A. (1997). Structural requirements for glycolipid antigen recognition by CD1b-restricted T cells. *Science* **278**, 283–286.

Moody, D.B., Reinhold, B.B., Reinhold, V.N., Besra, G.S., and Porcelli, S.A. (1999). Uptake and processing of glycosylated mycolates for presentation to CD1b-restricted T cells. *Immunology Letters* **65**, 85–91.

Moody, D.B., Ulrichs, T., Muhlecker, W., Young, D.C., Gurcha, S.S., Grant, E., Rosat, J.P., Brenner, M.B., Costello, C.E., Besra, G.S., and Porcelli,

S.A. (2000). CD1c-mediated T-cell recognition of isoprenoid glycolipids in *Mycobacterium tuberculosis* infection. *Nature* **404**, 884–888.

Niazi, K.R., Chiu, M.W., Mendoza, R.M., Degano, M., Khurana, S., Moody, D.B., Melian, A., Wilson, I.A., Kronenberg, M., Porcelli, S.A., and Modlin, R.L. (2001). The A' and F' pockets of human CD1b are both required for optimal presentation of lipid antigens to T cells. *Journal of Immunology* **166**, 2562–2570.

Ochoa, M.T., Stenger, S., Sieling, P.A., Thoma-Uszynski, S., Sabet, S., Cho, S., Krensky, A.M., Rollinghoff, M., Nunes, S.E., Burdick, A.E., Rea, T.H., and Modlin, R.L. (2001). T-cell release of granulysin contributes to host defense in leprosy. *Nature Medicine* **7**, 174–179.

Porcelli, S., Morita, C.T., and Brenner, M.B. (1992). CD1b restricts the response of human CD4-8-T lymphocytes to a microbial antigen. *Nature* **360**, 593–597.

Porcelli, S.A. and Modlin, R.L. (1999). The CD1 system: antigen-presenting molecules for T cell recognition of lipids and glycolipids. *Annual Review of Immunology* **17**, 297–329.

Prigozy, T.I., Sieling, P.A., Clemens, D., Stewart, P.L., Behar, S.M., Porcelli, S.A., Brenner, M.B., Modlin, R.L., and Kronenberg, M. (1997). The mannose receptor delivers lipoglycan antigens to endosomes for presentation to T cells by CD1b molecules. *Immunity* **6**, 187–197.

Rosat, J.P., Grant, E.P., Beckman, E.M., Dascher, C.C., Sieling, P.A., Frederique, D., Modlin, R.L., Porcelli, S.A., Furlong, S.T., and Brenner, M.B. (1999). CD1-restricted microbial lipid antigen-specific recognition found in the CD8+ alpha beta T cell pool. *Journal of Immunology* **162**, 366–371.

Sallusto, F. and Lanzavecchia, A. (1994). Efficient presentation of soluble antigen by cultured human dendritic cells is maintained by granulocyte/macrophage colony-stimulating factor plus interleukin 4 and downregulated by tumor necrosis factor alpha. *Journal of Experimental Medicine* **179**, 1109–1118.

Schaible, U.E., Hagens, K., Fischer, K., Collins, H.L., and Kaufmann, S.H. (2000). Intersection of group I CD1 molecules and mycobacteria in different intracellular compartments of dendritic cells. *Journal of Immunology* **164**, 4843–4852.

Sieling, P.A., Chatterjee, D., Porcelli, S.A., Prigozy, T.I., Mazzaccaro, R.J., Soriano, T., Bloom, B.R., Brenner, M.B., Kronenberg, M., Brennan, P.J., and Modlin, R.L. (1995). CD1-restricted T cell recognition of microbial lipoglycan antigens. *Science* **269**, 227–230.

Sieling, P.A., Jullien, D., Dahlem, M., Tedder, T.F., Rea, T.H., Modlin, R.L., and Porcelli, S.A. (1999): CD1 expression by dendritic cells in human leprosy lesions: correlation with effective host immunity. *Journal of Immunology* **162**, 1851–1858.

Sieling, P.A., Ochoa, M.T., Jullien, D., Leslie, D.S., Sabet, S., Rosat, J.P., Burdick, A.E., Rea, T.H., Brenner, M.B., Porcelli, S.A., and Modlin, R.L. (2000). Evidence for human CD4+ T cells in the CD1-restricted repertoire: derivation of mycobacteria-reactive T cells from leprosy lesions. *Journal of Immunology* **164**, 4790–4796.

Shinkai, K. and Locksley, R.M. (2000). CD1, tuberculosis, and the evolution of major histocompatibility complex molecules. *Journal of Experimental Medicine* **191**, 907–914.

Stenger, S., Mazzaccaro, R.J., Uyemura, K., Cho, S., Barnes, P.F., Rosat, J.P., Sette, A., Brenner, M.B., Porcelli, S.A., Bloom, B.R., and Modlin, R.L. (1997). Differential effects of cytolytic T cell subsets on intracellular infection. *Science* **276**, 1684–1687.

Stenger, S., Hanson, D.A., Teitelbaum, R., Dewan, P., Niazi, K.R., Froelich, C.J., Ganz, T., Thoma-Uszynski, S., Melian, A., Bogdan, C., Porcelli, S.A., Bloom, B.R., Krensky, A.M., and Modlin, R.L. (1998a). An antimicrobial activity of cytolytic T cells mediated by granulysin. *Science* **282**, 121–125.

Stenger, S., Niazi, K.R., and Modlin, R.L. (1998b). Down-regulation of CD1 on antigen-presenting cells by infection with *Mycobacterium tuberculosis*. *Journal of Immunology* **161**, 3582–3588.

Sturgill-Koszycki, S., Schlesinger, P.H., Chakraborty, P., Haddix, P.L., Collins, H.L., Fok, A.K., Allen, R.D., Gluck, S.L., Heuser, J., and Russell, D.G. (1994). Lack of acidification in Mycobacterium phagosomes produced by exclusion of the vesicular proton-ATPase [see comments] [published erratum appears in *Science* 1994 Mar 11; 263 (5152): 1359]. *Science* **263**, 678–681.

Sugita, M., Jackman, R.M., van Donselaar, E., Behar, S.M., Rogers, R.A., Peters, P.J., Brenner, M.B., and Porcelli, S.A. (1996). Cytoplasmic tail-dependent localization of CD1b antigen-presenting molecules to MIICs. *Science* **273**, 349–352.

Sugita, M., Grant, E.P., van Donselaar, E., Hsu, V.W., Rogers, R.A., Peters, P.J., and Brenner, M.B. (1999). Separate pathways for antigen presentation by CD1 molecules. *Immunity* **11**, 743–752.

Sugita, M., van Der, W., Rogers, R.A., Peters, P.J., and Brenner, M.B. (2000). CD1c molecules broadly survey the endocytic system. *Proceedings of the National Academy of Sciences USA* **97**, 8445–8450.

van Agthoven, A. and Terhorst, C. (1982). Further biochemical characterization of the human thymocyte differentiation antigen T6. *Journal of Immunology* **128**, 426–432.

Vogt, A.B., Arndt, S.O., Hammerling, G.J., and Kropshofer, H. (1999). Quality control of MHC class II associated peptides by HLA-DM/H2-M. *Seminars in Immunopathology* **11**, 391–403.

Xu, S., Cooper, A., Sturgill-Koszycki, S., van Heyningen, T., Chatterjee, D., Orme, I., Allen, P., and Russell, D.G. (1994). Intracellular trafficking in *Mycobacterium tuberculosis* and *Mycobacterium avium*-infected macrophages. *Journal of Immunology* **153**, 2568–2578.

Zeng, Z., Castano, A.R., Segelke, B.W., Stura, E.A., Peterson, P.A., and Wilson, I.A. (1997). Crystal structure of mouse CD1: An MHC-like fold with a large hydrophobic binding groove [see comments]. *Science* **277**, 339–345.

Zhong, G., Fan, T., and Liu, L. (1999). Chlamydia inhibits interferon gamma-inducible major histocompatibility complex class II expression by degradation of upstream stimulatory factor 1. *Journal of Experimental Medicine* **189**, 1931–1938.

Zhong, G., Liu, L., Fan, T., Fan, P., and Ji, H. (2000). Degradation of transcription factor RFX5 during the inhibition of both constitutive and interferon gamma-inducible major histocompatibility complex class I expression in chlamydia-infected cells. *Journal of Experimental Medicine* **191**, 1525–1534.

CHAPTER 3

The NRAMP family: co-evolution of a host/pathogen defence system

Richard Bellamy

3.1 INTRODUCTION

Iron is essential for the growth of a wide range of microorganisms. Successful pathogens have developed several strategies to obtain host iron including use of siderophores to remove iron from transferrin, erythrocyte lysis and haemoglobin digestion, extracting iron at the cell surface, and procurement of host intracellular iron (Weinberg, 1999). Intracellular parasites such as mycobacteria have evolved complex mechanisms to acquire iron from their host cell, the macrophage (Wheeler and Ratledge, 1994). One such mechanism is a divalent cation transporter, known as Mramp (Agranoff et al., 1999). It is hypothesised that Mramp competes with its mammalian homologue Nramp1 for intraphagosomal iron and that this is important in determining the host's susceptibility to mycobacterial infection. This review will discuss the research that has led up to the elucidation of the Nramp1–Mramp story a fascinating example of the coevolution of prokaryotic/eukaryotic mutual evasion systems.

3.2 EARLY WORK ON *Bcg*

In 1981, Gros et al. recognised that among inbred strains of mice, innate susceptibility to infection with *Mycobacterium bovis* (BCG) was determined by a host genetic factor, which they named *Bcg* (Gros et al., 1981). Mice that possessed the dominant resistance gene, *Bcg^r*, had 100- to 1000-fold fewer splenic colony forming units after intravenous injection of BCG-Montreal compared to the *Bcg^s* susceptible strains (Forget et al., 1981). The *Bcg* gene was mapped to mouse chromosome 1 (Skamene et al., 1982), in the same position as the murine susceptibility genes to *Salmonella typhimurium*, designated *Ity* (Plant and Glynn, 1976, 1979) and *Leishmania donovani*, designated *Lsh* (Bradley et al., 1977, 1979). It was soon realised that the same gene was likely to be determining susceptibility to these three antigenically unrelated pathogens (Plant et al., 1982; Skamene et al., 1982). Susceptibility to *Mycobacterium lepraemurium* and *Mycobacterium intracellulare* also appeared to be due to the *Lsh/Ity/Bcg* gene (Brown et al., 1982; Skamene et al., 1984; Goto et al., 1989). Macrophages from *Bcg*[s] strains have decreased the ability to restrict the growth of mycobacteria, salmonella, and leishmania *in vitro* (Lissner et al., 1983; Stach et al., 1984; Crocker et al., 1984; Denis et al., 1990). *In vivo Bcg* only appeared to influence the initial phase of infection and did not affect the acquired specific immune response. There has been much speculation as to what the function of *Bcg* might be (Blackwell, 1989). However the *Bcg* gene was not identified by functional studies, but by a laborious reverse genetics approach by Philippe Gros' group in Montreal (Gros and Malo, 1989).

3.3 REVERSE GENETICS APPROACH TO CLONING *Nramp1*

Linkage studies on mouse chromosome 1 identified a marker that was tightly linked to the putative *Bcg* gene in the region corresponding to human chromosome 2q (Schurr et al., 1989, 1990). A high-resolution genetic linkage map of this region was then developed in the mouse, pinpointing the *Bcg* locus to a 0.3 centi-Morgan region (cM) (Malo et al., 1993a). 1cM approximately corresponds to one megabase (Mb) of DNA, therefore the region containing *Bcg* was of a size that could be cloned and physically mapped. This resulted in the identification of a 400 kilobase (kb) bacteriophage and cosmid contig containing the region surrounding the putative *Bcg* gene (Malo et al., 1993b). Several potential candidate genes for *Bcg* were identified from this region using a technique called exon trapping (Vidal et al., 1993). One gene showed several characteristics suggesting it was the *Bcg* gene being sought: (1) Its chromosome position was in the region of interest on mouse chromosome 1; (2) it was expressed exclusively in macrophage populations from reticuloendothelial organs and a macrophage cell line; (3) it encoded a polypeptide with characteristics of a transport protein; and (4) a single nonconservative base change at position 169 of glycine to aspartic acid (G169D) was found in *Bcg*[s] and not in *Bcg*[r] strains (Vidal et al., 1993). This gene was designated *Nramp* – natural resistance-associated macrophage protein. When other genes belonging to this family were described, the original *Nramp* became N*ramp1*. *Bcg*[s] mice possess the homozygous $Nramp1^{D169/D169}$ genotype and Bcg^r mice are homozyogus for the $Nramp1^{G169}$ allele.

3.4 DEMONSTRATION THAT *Nramp1* IS *Bcg*

Nramp1 is expressed in macrophages and consists of fifteen exons spanning 11.5 kb of genomic DNA encoding a 90–100 kiloDalton (kDa) membrane protein (Vidal et al., 1993; Barton et al., 1994; Govoni et al., 1995). The Nramp1 protein contains twelve hydrophobic transmembrane domains, a predicted extracellular glycosylated loop, several phosphorylation sites, and a putative Src homology 3-binding domain (Vidal et al., 1993; Barton et al., 1994). Nramp1 protein was not detectable in macrophages from Bcg^s mice, suggesting the G169D mutation prevents proper maturation or membrane integration of the protein (Vidal et al., 1996).

The Montreal group performed three experiments to prove that *Nramp1* is the *Bcg* mycobacteria-susceptibility gene. First *Nramp1* genotypes were determined for twenty-seven different mouse strains, seven of which were Bcg^s and twenty of which were Bcg^r. Inbred strains of mice can be divided into those that are phenotypically mycobacteria-resistant, Bcg^r, and those which are mycobacteria-susceptible, Bcg^s, when inoculated intravenously with BCG-Montreal. The twenty Bcg^r strains were all found to be $Nramp1^{G169/G169}$ homozygotes and the seven Bcg^s strains were found to be $Nramp1^{D169/D169}$ homozygotes (Malo et al., 1994). The seven Bcg^s strains shared a conserved haplotype of 2.2 Mb surrounding the *Nramp1* region, indicating they had descended from a single common ancestor. This did not conclusively prove that *Nramp1* was the *Bcg* gene as the G169D mutation could have been in linkage disequilibrium with some other genetic variant.

In the second experiment an *Nramp1* gene-disrupted, "knock-out" mouse was produced and designated $Nramp1^{-/-}$. The $Nramp1^{-/-}$ mouse was found to be susceptible to BCG, *L. donovani*, and *S. typhimurium* (Vidal et al., 1995a). $Nramp1^{-/-}$ mice were then mated with $Nramp1^{D169/D169}$ mice, producing some heterozygous $Nramp1^{D169/-}$ offspring. The $Nramp1^{D169/-}$ mice had lost the capacity to control BCG growth compared to their $Nramp1^{D169/+}$ littermates. $Nramp1^{-/-}$, $Nramp1^{D169/-}$, and $Nramp1^{D169/D169}$ mice were found to have indistinguishable phenotypic resistance to BCG. In particular, $Nramp1^{-/-}$ mice were able to develop an acquired immune response controlling BCG growth in the late stage of infection as effectively as $Nramp1^{D169/D169}$ mice (Vidal et al., 1995a). This suggests that the $Nramp1^{D169}$ allele is a null allele producing no functional Nramp1 protein.

In the third experiment to conclusively prove that the G169D mutation was responsible for *Nramp1* lack of function, the Montreal group attempted to restore the BCG resistance phenotype by transferring the normal G169 allele onto the background of a homozygous $Nramp1^{D169/D169}$ mouse (Govoni et al., 1996). This transgenic mouse was resistant to BCG and *S. typhimurium*. In addition, macrophages from the transgenic mouse expressed the mature Nramp1 protein, indicating the gene's function had been restored (Govoni et al., 1996). Taken together these experiments conclusively proved that *Nramp1* is the *Lsh/Ity/Bcg* gene.

3.5 EARLY STUDIES OF *Nramp1/Bcg* FUNCTION

Numerous functional differences have been found between *Bcg^r*- and *Bcg^s*-derived macrophages including greater production of toxic oxygen radicals (Denis et al., 1988a), increased production of interleukin 1β and tumour necrosis factor-α (Roach et al., 1994), greater major histocompatibility complex (MHC) Ia surface antigen expression (Zwilling et al., 1987; Denis et al., 1988b), and increased MHC I-Aβ gene transcription (Barrera et al., 1997a).

When the amino acid sequence of Nramp1 was determined, several theories of its function were proposed on the basis of its structure. Nramp1 was noted to have some structural similarity to CrnA, a nitrite/nitrate transporter in the fungus *Aspergillus nidulans* (Unkles et al., 1991). It was therefore suggested that Nramp1 may be involved in the transport of simple nitrogen compounds (Vidal et al., 1993). It was later hypothesised that the Nramp1 SH3-binding domain could regulate transport of L-arginine, the substrate for nitric oxide (NO) synthesis (Barton et al., 1995). This was an intriguing theory because NO is a potent inhibitor of mycobacteria (Chan et al., 1992), and BCG-infected, *Bcg^r*-derived macrophages produce significantly more NO than macrophages from *Bcg^s* mice (Barrera et al., 1997b). The superior inhibition of *M. tuberculosis* by *Bcg^r*-derived macrophages compared to *Bcg^s*-derived macrophages correlates with NO production (Arias et al., 1997). However, these may not represent the primary difference between *Bcg^r* and *Bcg^s* strains and may simply be secondary to differences in host–parasite interactions. Recently, further insight into Nramp1 function has been derived from studying other proteins with similar structure.

3.6 THE *Nramp* FAMILY

Since the discovery of Nramp1, it has become apparent that there is a family of Nramp-related proteins characterised by a hydrophobic core of ten to twelve transmembrane domains. The human homologue of the mouse *Nramp1* gene has been cloned and mapped to chromosome 2q35 and has been designated *NRAMP1* (Cellier et al., 1994). The human *NRAMP1* gene contains at least fifteen exons and spans 12 kb of genomic DNA (Cellier et al., 1994; Blackwell et al., 1995). It encodes a polypeptide of 550 amino acids with 85% sequence identity to murine Nramp1 (Blackwell et al., 1995). In the mouse, two additional *Nramp*-related genes have been identified on chromosomes 15 and 17 and respectively named *Nramp2* and *Nramp-rs* (Dosik et al., 1994; Gruenheid et al., 1995). The human homologue of *Nramp2*, designated *NRAMP2*, has been cloned and mapped to chromosome 12q13 (Vidal et al., 1995b; Kishi and Tabuchi, 1998). Members of the Nramp family have been identified in several phylogenetically diverse organisms including the fly *Drosophila melanogaster*, the nematode *Caenorhabditis elegans*, the plant *Oryza sativa*, the yeast *Saccharomyces cerevisiae*, and the bacteria *Mycobacterium leprae* and *M. tuberculosis* (Cellier et al., 1996). These proteins show between 39% and 94% homology with the hydrophobic core of the human NRAMP1 protein (Cellier et al., 1996). The finding of structurally conserved Nramp homologues in prokaryotic and eukaryotic species suggests that the Nramp family has existed for more than one billion years and that the proteins have a common function. Recent work suggests that the Nramp family are metal iron transporters, which has provided insight into how *Nramp1* may be involved in host resistance to antigenically unrelated intracellular pathogens.

3.7 *Nramp* PROTEINS AND METAL CATION TRANSPORT

Nramp1 has structural similarity to SMF1, a manganese ion transporter found in *S. cerevisiae*. This observation led Supek and colleagues to suggest that Nramp1 may transport metal cations from the extracellular environment into macrophage cytoplasm (Supek et al., 1996). More clues to the function of Nramp1 came from the study of murine Nramp2 and its homologue in rats. Unlike Nramp1, which is exclusively expressed in reticuloendothelial cells, Nramp2 is ubiquitous. In 1997, a divalent cation transporter was isolated from rats, which was identified as Nramp2 (Gunshin et al., 1997). In the same year, the microcytic anaemia mouse, *mk*, which suffers from defective iron absorption and erythroid iron utilisation, was found to have a missense mutation in the *Nramp2* gene (Fleming et al., 1997). The Belgrade rat, which also suffers from hereditary microcytic anaemia, was then found to have an *Nramp2* missense mutation (Fleming et al., 1998). It was surprising to find that the mutation in the two species was identical, a glycine to arginine

R. BELLAMY

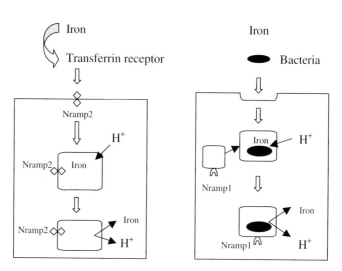

Figure 3.1. Comparison of Nramp1 and Nramp2 function as hypothesised by Gruenheid et al. (1999). The ubiquitous Nramp2 is internalised to endosomes with iron. Acidification of the endosome activates Nramp2 and iron and protons are transported into the cytoplasm. Nramp1 is recruited to the macrophage phagosome after bacteria and iron are ingested. Following acidification iron and protons are transported out of the phagosome.

substitution at position 185 (G185R) in transmembrane domain 4. This may be a highly conserved region as the *Nramp1* mutation G169D also affects transmembrane domain 4.

It has been speculated that Nramp1 is also an iron transporter, perhaps with a highly specialised function, specific to reticuloendothelial cells (Gruenheid et al., 1999). Murine Nramp2 and human NRAMP2 are integral membrane glycoproteins, which are found at the plasma membrane and in recycling endosomes (Gruenheid et al., 1999; Tabuchi et al., 2000). Nramp2 is believed to transport iron together with a proton across the endosomal membrane into the cytoplasm (Fig. 3.1) (Gruenheid et al., 1999).

In contrast, Nramp1 is localised to the late endosomal compartment of resting macrophages and is recruited to the phagosome on phagocytosis (Fig. 3.1) (Gruenheid et al., 1997). Iron overload has been found to be a risk factor for the development of tuberculosis in humans (Gordeuk et al., 1996). This raises the question of whether Nramp1 could affect the growth of mycobacteria by pumping intra-phagosomal iron into the cytoplasm. Nramp1 protein levels in microglial cells are influenced by iron availability, suggesting a link between *Nramp1* and iron (Atkinson et al., 1997). Increased iron concentrations increase *M. avium* growth within macrophages (Gomes and Appelberg, 1998). In conditions of low iron concentration, resistant $Nramp1^{G169/G169}$ mice limit *M. avium* replication more effectively than susceptible $Nramp1^{D169/D169}$ mice. However, $Nramp1^{G169/G169}$ mice are not able to control *M. avium* replication better than $Nramp1^{D169/D169}$ mice when excess iron is present (Gomes and Appelberg, 1998). This suggests that the Nramp1 iron transport mechanism may become saturated at high iron concentration. Macrophages from $Nramp1^{-/-}$ gene-disrupted mice have an impaired ability to acidify phagosomes following infection with live *M. bovis* (Hackham et al., 1998). It has therefore been suggested that phagosome acidification may be needed for co-transport of iron and protons out of the phagosome (Gruenheid et al., 1999).

M. leprae and M. tuberculosis both possess members of the Nramp family (Cellier et al., 1996; Agranoff et al., 1999). The *M. tuberculosis* Nramp homologue, called Mramp, is a pH-dependent transporter of iron and a broad range of other divalent metallic cations (Agranoff et al., 1999). Mramp function is maximal between pH 5.5 and 6.5, equivalent to that which exists within phagosomes (Agranoff et al., 1999). It seems likely that Mramp and Nramp1 are competing for iron within the phagosome and that the outcome of this competition determines whether the mycobacteria can proliferate. This is an astonishing example of how host and pathogen can evolve a similar mechanism to compete for the same substrate.

3.8 *Nramp* AND HUMAN SUSCEPTIBILITY TO INTRACELLULAR PATHOGENS

NRAMP1 is a strong candidate gene for human susceptibility to several infectious diseases caused by intracellular pathogens including tuberculosis, leprosy, leishmaniasis, and typhoid. Several polymorphisms have been identified in the human *NRAMP1* gene enabling studies on genetic susceptibility in human populations to be carried out (Liu et al., 1995). Two different strategies have been used: linkage-based family studies and association-based case-control studies.

The advantage of linkage studies is that they are not susceptible to the effects of confounding factors. However, linkage studies have relatively low power and in multifactorial diseases they can only detect extremely large genetic effects (Risch and Merikangas, 1996). Weak evidence of linkage was found between *NRAMP1* polymorphisms and leprosy in twenty Vietnamese families ($P < 0.02$) (Abel et al., 1998) and between *NRAMP1* and tuberculosis in ninety-eight Brazilian families ($P = 0.025$) (Shaw et al., 1997). In another study in seven families from French Polynesia, leprosy was not found to be linked to *NRAMP1* markers (Roger et al., 1997). In a genome-wide screen for tuberculosis susceptibility genes in 173 African sibling pairs, the nearest marker to *NRAMP1* showed only weak evidence of linkage to tuberculosis ($P = 0.06$) (Bellamy et al., 2000). Overall these linkage results are only weakly positive, indicating that susceptibility to mycobacteria in humans is not solely due to *NRAMP1* variants. This is probably similar to the situation in mice where it has also been found that *Nramp1* is not the single major gene determining susceptibility to *M. tuberculosis* (Medina and North, 1996).

Association studies have much greater power than linkage studies to explore the effects of candidate genes that exert moderate effects on disease risk (Risch and Merikangas, 1996). In a large case-control study of over 800 subjects in The Gambia, a highly statistically significant association was found between *NRAMP1* polymorphisms and tuberculosis (Bellamy et al., 1998). Individuals with tuberculosis had approximately four times the odds of possessing the high-risk *NRAMP1* genotype compared to healthy controls (Bellamy et al., 1998). This demonstrated that *NRAMP1* is an important tuberculosis-susceptibility gene in humans, but it does not account for the majority of the familial clustering effect in this disease (Bellamy et al., 2000). Therefore, there are likely to be many more human genes involved in susceptibility to tuberculosis and other mycobacterial diseases.

3.9 CONCLUSION

Nramp1, the natural resistance-associated macrophage protein gene, has been shown to influence susceptibility to several antigenically unrelated, intracellular pathogens in mice. The Nramp family includes proteins expressed in a diverse range of prokaryotic and eukaryotic species. The ubiquitously expressed *Nramp2* encodes a divalent metal ion transporter that regulates iron uptake in the gut. It has been hypothesised that Nramp1 is also a metal ion transporter, perhaps restricting the availability of iron for mycobacteria within phagosomes. *Mycobacterium tuberculosis* has its own member of the Nramp family, named Mramp, and this is also an iron transporter. It is intriguing that host and pathogen appear to have evolved highly similar proteins to compete for the availability of a single nutrient.

ACKNOWLEDGMENTS

I am grateful to Professor AVS Hill for helpful discussions and to the Wellcome Trust for funding.

REFERENCES

Abel, L., Sanchez, F.O., Oberti, J., Thuc, N.V., Hoa, L.V., Lap, V.L., Skamene, E., Lagrange P.H., and Schurr, E. (1998). Susceptibility to leprosy is linked to the human *NRAMP1* gene. *Journal of Infectious Diseases* **177**, 133–145.

Agranoff, D., Monahan, I.M., Mangan, J.A., Butcher, P.D., and Krishna, S. (1999). *Mycobacterium tuberculosis* expresses a novel pH-dependent divalent cation transporter belonging to the Nramp family. *Journal of Experimental Medicine* **190**, 717–724.

Arias, M., Rojas, M., Zabaleta, J., Rodriguez, J.I., Paris, S.C., Barrera, L.F., and Garcia, L.F. (1997). Inhibition of virulent *Mycobacterium tuberculosis* by *Bcg*^{r} and *Bcg*^{s} macrophages correlates with nitric oxide production. *Journal of Infectious Diseases* **176**, 1552–1558.

Atkinson, P.G.P., Blackwell, J.M., and Barton, C.H. (1997). *Nramp1* locus encodes a 65kDa interferon-γ-inducible protein in murine macrophages. *Biochemical Journal* **325**, 779–786.

Barrera, L.F., Kramnik, I., Skamene, E., and Radzioch, D. (1997a). I-Aβ gene expression regulation in macrophages derived from mice susceptible or resistant to infecton with *M. bovis* BCG. *Molecular Immunology* **34**, 343–355.

Barrera, L.F., Kramnik, I., Skamene, E., and Radzioch, D. (1997b). Nitrite production by macrophages derived from BCG-resistant and -susceptible congenic mouse strains in response to IFN-γ and infection with BCG. *Immunology* **82**, 457–464.

Barton, C.H., White, J.K., Roach, T.I.A., and Blackwell, J.M. (1994). NH$_2$-terminal sequence of macrophage-expressed natural resistance-associated macrophage protein (*Nramp*) encodes a proline/serine-rich putative *Src* homology 3-binding domain. *Journal of Experimental Medicine* **179**, 1683–1687.

Barton, C.H., Whitehead, S.H., and Blackwell, J.M. (1995). *Nramp* transfection transfers *Ity/Lsh/Bcg*-related pleiotropic effects on macrophage activation: influence on oxidative burst and nitric oxide pathways. *Molecular Medicine* **1**, 267–279.

Bellamy, R., Ruwende, C., Corrah, T., McAdam, K.P.W.J., Whittle, H.C., and Hill, A.V.S. (1998). Variation in the human *NRAMP1* gene and human tuberculosis in an African population. *New England Journal of Medicine* **338**, 640–644.

Bellamy, R.J., Beyers, N., McAdam, K.P.W.J., Ruwende, C., Gie, R., Samaai, P., Bester, D., Meyer, M., Corrah, T., Collin, M., Camidge, D.R., Wilkinson, D., Hoal-van Helden, E., Whittle, H.C., Amos, W., van Helden, P., and Hill, A.V.S.H. (2000). Genetic susceptibility to tuberculosis in Africans: a genome-wide scan. *Proceedings of the National Academy of Sciences, USA* **97**, 8005–8009.

Blackwell, J.M. (1989). The macrophage resistance gene *Lsh/Ity/Bcg*. 27th Forum in Immunology. *Research in Immunology* **140**, 767–828.

Blackwell, J.M., Barton, H., White, J.K., Searle, S., Baker, A.-M., Williams, H., and Shaw, M.-A. (1995). Genomic organization and sequence of the human *NRAMP* gene: identification and mapping of a promoter region polymorphism. *Molecular Medicine* **1**, 194–205.

Bradley, D.J. (1977). Genetic control of *Leishmania* populations within the host. II. Genetic control of acute susceptibility of mice to *L. donovani* infection. *Clinical Experimental Immunology* **30**, 130–140.

Bradley, D.J., Taylor, B.A., Blackwell, J.M., Evans, E.P., and Freeman, J. (1979). Regulation of *Leishmania* populations within the host. -III. Mapping of the locus controlling susceptibility to visceral leishmaniasis in the mouse. *Clinical Experimental Immunology* **37**, 7–14.

Brown, I.N., Glynn, A.A., and Plant, J.E. (1982). Inbred mouse strain resistance to *Mycobacterium lepraemurium* follows the *Ity/Lsh* pattern. *Immunology* **47**, 149–156.

Cellier, M., Govoni, G., Vidal, S., Kwan, T., Groulx, N., Liu, J., Sanchez, F., Skamene, E., Schurr, E., and Gros, P. (1994). Human natural resistance-associated macrophage protein: cDNA cloning, chromosomal mapping, genomic organization, and tissue-specific expression. *Journal of Experimental Medicine* **180**, 1741–1752.

Cellier, M., Belouchi, A., and Gros, P. (1996). Resistance to intracellular infections: comparative genomic analysis of *Nramp*. *Trends in Genetics* **12**, 201–205.

Chan, J., Xing, Y., Magliozzo, R.S., and Bloom, B.R. (1992). Killing of virulent *Mycobacterium tuberculosis* by reactive nitrogen intermediates produced by activated murine macrophages. *Journal of Experimental Medicine* **175**, 1111–1122.

Crocker, P.R., Blackwell, J.M., and Bradley, D.J. (1984). Expression of the natural resistance gene *Lsh* in resident liver macrophages. *Infection and Immunity* **43**, 1033–1040.

Denis, M., Forget, A., Pelletier, M., and Skamene, E. (1988a). Pleiotropic effects of the *Bcg* gene: III. Respiratory burst in *Bcg*-congenic macrophages. *Clinical Experimental Immunology* **73**, 370–375.

Denis, M., Buschman, E., Forget, A., Pelletier, M., and Skamene, E. (1988b). Pleiotropic effects of the *Bcg* gene. II. Genetic restriction of responses to mitogens and allogeneic targets. *Journal of Immunology* **141**, 3988–3993.

Denis, M., Forget, A., Pelletier, M., Gervais, F., and Skamene, E. (1990). Killing of *Mycobacterium smegmatis* by macrophages from genetically susceptible and resistant mice. *Journal of Leukocyte Biology* **47**, 25–30.

Dosik, J.K., Barton, C.H., Holiday, D.L., Krall, M.M., Blackwell, J.M., and Mock, B.A. (1994). An *Nramp*-related sequence maps to mouse chromosome 17. *Mammalian Genome* **5**, 458–460.

Fleming, M.D., Trenor, C.C. 3rd, Su, M.A., Foernzler, D., Beier, D.R., Dietrich, W.F., and Andrews, N.C. (1997). Microcytic anaemia mice have a mutation in *Nramp2*, a candidate iron transporter gene. *Nature Genetics* **16**, 383–386.

Fleming, M.D., Romano, M.A., Su, M.A., Garrick, L.M., Garrick, M.D., and Andrews, N.C. (1998). *Nramp2* is mutated in the anaemic Belgrade (*b*) rat: evidence of a role for Nramp2 in endosomal iron transport. *Proceedings of the National Academy of Sciences, USA* **95**, 1148–1153.

Forget, A., Skamene, E., Gros, P., Miailhe, A.-E., and Turcott, R. (1981). Differences in response among inbred mouse strains to infection with small doses of *Mycobacterium bovis* BCG. *Infection and Immunity* **32**, 42–47.

Gomes, M.S. and Appelberg, R. (1998). Evidence for a link between iron metabolism and *Nramp1* gene function in innate resistance against *Mycobacterium avium*. *Immunology* **95**, 165–168.

Gordeuk, V.R., McLaren, C.E., MacPhail, A.P., Deichsel, G., and Bothwell, T.H. (1996). Associations of iron overload in Africa with hepatocellular carcinoma and tuberculosis: Strachan's 1929 thesis revisited. *Blood* **87**, 3470–3476.

Goto, Y., Buschman, E., and Skamene, E. (1989). Regulation of host resistance to *Mycobacterium intracellulare in vivo* and *in vitro* by the *Bcg* gene. *Immunogenetics* **30**, 218–221.

Govoni, G., Vidal, S., Cellier, M., Lepage, P., Malo, D., and Gros, P. (1995). Genomic structure, promoter sequence, and induction of expression of the mouse *Nramp1* gene in macrophages. *Genomics* **27**, 9–19.

Govoni, G., Vidal, S., Gauthier, S., Skamene, E., Malo, D., and Gros, P. (1996). The *Bcg/Ity/Lsh* locus: genetic transfer of resistance to infections in C57BL/6J mice transgenic for the *Nramp1*^{G169} allele. *Infection and Immunity* **64**, 2923–2929.

Gros, P. and Malo, D. (1989). A reverse genetics approach to *Bcg/Ity/Lsh* gene cloning. *Research in Immunology* **140**, 774–777.

Gros, P., Skamene, E., and Forget, A. (1981). Genetic control of natural resistance to *Mycobacterium bovis* BCG. *Journal of Immunology* **127**, 417–421.

Gruenheid, S., Cellier, M., Vidal, S., and Gros, P. (1995). Identification and characterization of a second mouse *Nramp* gene. *Genomics* **25**, 514–525.

Gruenheid, S., Pinner, E., Desjardins, M., and Gros, P. (1997). Natural resistance to infection with intracellular pathogens: the Nramp1 protein is recruited to the membrane of the phagosome. *Journal of Experimental Medicine* **185**, 717–730.

Gruenheid, S., Canonne-Hergaux, F., Gauthier, S., Hackam, D.J., Grinstein, S., and Gros, P. (1999). The iron transport protein Nramp2 is an integral membrane glycoprotein that colocalizes with transferrin in recycling endosomes. *Journal of Experimental Medicine* **189**, 831–841.

Gunshin, H., Mackenzie, B., Berger, U.V., Gunshin, Y., Romero, M.F., Boron, W.F., Nussberger, S., Goilan, J.L., and Hediger, M.A. (1997). Cloning and characterization of a mammalian proton-coupled metal-ion transporter. *Nature* **388**, 482–488.

Hackham, D.J., Rotstein, O.D., Zhang, W.-J., Gruenheid, S., Gros, P., and Grinstein, S. (1998). Host resistance to intracellular infection: mutation of natural resistance-associated macrophage protein 1 (*Nramp1*) impairs phagosome acidification. *Journal of Experimental Medicine* **188**, 351–364.

Kishi, F. and Tabuchi, M. (1998). Human natural resistance-associated macrophage protein 2: gene cloning and protein identification. *Biochemical and Biophysical Research Communications* **251**, 775–783.

Lissner, C.R., Swanson, R.N., and O'Brien, A.D. (1983). Genetic control of innate resistance of mice to *Salmonella typhimurium*: expression of the *Ity* gene in peritoneal and splenic macrophages isolated *in vitro*. *Journal of Immunology* **131**, 3006–3013.

Liu, J., Fujiwara, M., Buu, N.T., Sanchez, F.O., Cellier, M., Paradis, A.J., Frappier, D., Skamene, E., Gros, P., Morgan, K., and Schurr, E. (1995). Identification of polymorphisms and sequence variants in the human homologue of the mouse natural resistance-associated macrophage protein gene. *American Journal of Human Genetics* **56**, 845–853.

Malo, D., Vidal, S.M., Hu, J., Skamene, E., and Gros, P. (1993a). High resolution linkage map in the vicinity of the host resistance locus *Bcg*. *Genomics* **16**, 655–663.

Malo, D., Vidal, S., Lieman, J.H., Ward, D.C., and Gros, P. (1993b). Physical delineation of the minimal chromosomal segment encompassing the murine host resistance locus *Bcg*. *Genomics* **17**, 667–675.

Malo, D., Vogan, K., Vidal, S., Hu, J., Cellier, M., Schurr, E., Fuks, A., Bumstead, N., Morgan, K., and Gros, P. (1994). Haplotype mapping and sequence analysis of the mouse *Nramp* gene predict susceptibility to infection with intracellular parasites. *Genomics* **23**, 51–61.

Medina, E. and North, R.J. (1996). Evidence inconsistent with a role for the *Bcg* gene (*Nramp1*) in resistance of mice to infection with virulent *Mycobacterium tuberculosis*. *Journal of Experimental Medicine* **183**, 1045–1051.

Plant, J.E. and Glynn, A.A. (1976). Genetics of resistance to infection with *Salmonella typhimurium* in mice. *Journal of Infectious Diseases* **133**, 72–78.

Plant, J.E. and Glynn, A.A. (1979). Locating *Salmonella* resistance gene on mouse chromosome 1. *Clinical Experimental Immunology* **37**, 1–6.

Plant, J.E., Blackwell, J.M., O'Brien, A.D., Bradley, D.J., and Glynn, A.A. (1982). Are the *Lsh* and *Ity* disease resistance genes at one locus on mouse chromosome 1? *Nature* **297**, 510–511.

Risch, N. and Merikangas, K. (1996). The future of genetic studies of complex human diseases. *Science* **273**, 1516–1517.

Roach, T.I.A., Chatterjee, D., and Blackwell, J.M. (1994). Induction of early-response genes KC and JE by mycobacterial lipoarabinomannans: regulation of KC expression in murine macrophages by *Lsh/Ity/Bcg* (candidate *Nramp*). *Infection and Immunity* **62**, 1176–1184.

Roger, M., Levee, G., Chanteau, S., Gicquel, B., and Schurr, E. (1997). No evidence for linkage between leprosy susceptibility and the human natural resistance-associated macrophage protein 1 (*NRAMP1*) gene in French Polynesia. *International Journal of Leprosy* **65**, 197–202.

Schurr, E., Skamene, E., Forget, A., and Gros, P. (1989). Linkage analysis of the *Bcg* gene on mouse chromosome 1: identification of a tightly linked marker. *Journal of Immunology* **141**, 4507–4513.

Schurr, E., Skamene, E., Morgan, K., Chu, M.L., and Gros, P. (1990). Mapping of Col3a1 and Col6a3 to proximal murine chromosome 1 identifies conserved linkage of structural protein genes between murine chromosome 1 and human chromosome 2q. *Genomics* **8**, 477–486.

Shaw, M.A., Collins, A., Peacock, C.S., Miller, E.N., Black, G.F., Sibthorpe, D., Lins-Lainson, Z., Shaw, J.J., Ramos, F., Silveira, F., and Blackwell, J.M. (1997). Evidence that genetic susceptibility to *Mycobacterium tuberculosis* in a Brazilian population is under oligogenic control: linkage study of the candidate genes *NRAMP1* and *TNFA*. *Tubercle and Lung Disease* **78**, 35–45.

Skamene, E., Gros, P., Forget, A., Kongshavn, P.A.L., St.-Charles, C., and Taylor, B.A. (1982). Genetic regulation of resistance to intracellular pathogens. *Nature* **297**, 506–509.

Skamene, E., Gros, P., Forget, A., Patel, P.J., and Nesbitt, M. (1984). Regulation of resistance to leprosy by chromosome 1 locus in the mouse. *Immunogenetics* **19**, 117–120.

Stach, J.L., Gros, P., Forget, A., and Skamene, E. (1984). Phenotypic expression of genetically-controlled natural resistance to *Mycobacterium bovis* (BCG). *Journal of Immunology* **132**, 888–892.

Supek, F., Supekova, L., Nelson, H., and Nelson, N. (1996). A yeast manganese transporter related to the macrophage protein involved in conferring resistance to mycobacteria. *Proceedings of the National Academy of Sciences, USA* **93**, 5105–5110.

Tabuchi, M., Yoshimori, T., Yamaguchi, K., Yoshida, T., and Kishi, F. (2000). Human NRAMP2/DMT1, which mediates iron transport across membranes, is localized to late endosomes and lysosomes in Hep-2 cells. *Journal of Biological Chemistry* **275**, 22,220–22,228.

Unkles, S.E., Hawker, K.L., Grieve, C., Campbell, E.I., Montagne, P., and Kinghorn, J.R. (1991). *crnA* encodes a nitrate transporter in *Aspergillus nidulans. Proceedings of the National Academy of Sciences, USA* **88**, 204–208.

Vidal, S.M., Malo, D., Vogan, K., Skamene, E., and Gros, P. (1993). Natural resistance to infection with intracellular parasites: isolation of a candidate gene for *Bcg. Cell* **73**, 469–485.

Vidal, S., Tremblay, M.L., Govoni, G., Gauthier, S., Sebastiani, G., Malo, D., Skamene, E., Olivier, M., Jothy, S., and Gros, P. (1995a). The *Ity/Lsh/Bcg* locus: natural resistance to infection with intracellular parasites is abrogated by disruption of the *Nramp1* gene. *Journal of Experimental Medicine* **182**, 655–666.

Vidal, S., Belouchi, A.-M., Cellier, M., Beatty, B., and Gross, P. (1995b). Cloning and characterization of a second human *NRAMP* gene on chromosome 12q13. *Mammalian Genome* **6**, 224–230.

Vidal, S.M., Pinner, E., Lepage, P., Gauthier, S., and Gros, P. (1996). Natural resistance to intracellular infections. *Nramp1* encodes a membrane phosphoglycoprotein absent in macrophages from susceptible (*Nramp1^{D169}*) mouse strains. *Journal of Immunology* **157**, 3559–3568.

Weinberg, E.D. (1999). Iron loading and disease surveillance. *Emerging Infectious Diseases* **5**, 346–352.

Wheeler, P.R. and Ratledge, C. (1994). Metabolism of *Mycobacterium tuberculosis.* In *Tuberculosis: pathogenesis, protection and control,* ed. B.R. Bloom, pp. 353–385. Washington DC: ASM Press.

Zwilling, B.S., Vespa, L., and Massie, M. (1987). Regulation of I-A expression by murine peritoneal macrophages: differences linked to the *Bcg* gene. *Journal of Immunology* **138**, 1372–1376.

PART II

Evasion of humoral immunity

CHAPTER 4

Evasion of complement system pathways by bacteria

Michael A. Kerr and Brian Henderson

4.1 INTRODUCTION

Paul Ehrlich, better known for his work on chemotherapeutics, coined the term "complement" in the 1890s to denote the activity in sera that could "complement" the lysis of bacteria induced by specific antibody. By the early 1900s, complement was recognised as composed of two components, and by the 1920s it was believed that at least four serum factors were involved. However, it was not until the 1960s that analytical biochemistry was sufficiently rigorous to allow the identification of the majority of the known complement pathway components. Individual components were named as they were discovered, which accounts for the still confusing nature of the nomenclature for describing the complement pathways (for comprehensive reviews of complement, see Law and Reid, 1995; Fearon, 1998; Crawford and Alper, 2000; Kirschfink, 2001; Walport, 2001a, 2001b).

Three pathways of complement activation have now been described (Fig. 4.1). The classical pathway, first to be discovered, is generally considered to require immune complexes for activation. A second pathway, termed, naturally enough, the alternative pathway, was first proposed by Pillemer in the late 1950s but was not taken seriously until the late 1960s when sufficient evidence had accrued. This pathway is now generally considered to be activated by cell surfaces that are not protected by host-derived complement inhibitors (see Lindahl et al., 2000). A third pathway was elucidated in the late 1980s–early 1990s. This has been termed the lectin pathway and is activated by the collectin (i.e., collagen-like lectin), mannose-binding lectin (MBL) (Gadjeva et al., 2001) and by ficolins (proteins containing both a collagen-like and a fibrinogen-like domain; Matsushita and Fujita, 2001). These serum proteins can opsonise bacteria and then interact with proteinases homologous to

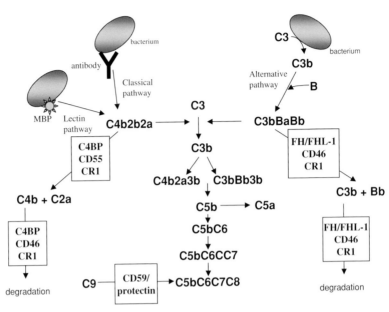

Figure 4.1. The complement pathways and their controlling proteins. The three pathways (classical, lectin, and alternative) produce C3 convertases – C4bC2b2a in the case of the classical and lectin pathways and C3bBaBb in the case of the alternative pathway. These enzyme complexes are very labile; they are formed when the C4bC2 complex is cleaved by C1s or MASP or when the C3bB complex is cleaved by D. The decay of the activity is the result of the loss of the enzymic subunits, C2a and Bb, from the complexes. The smaller subunits C2b and Ba contain the binding sites of C2 and B for C4b and C3b, respectively. The convertases dissociate even more quickly in the presence of appropriate RCA proteins (boxed). The association of C9 with the rest of the lytic complex is inhibited by the GPI-anchored protein, CD59 (protectin).

C1r and C1s, known as mannose-binding lectin associated serine proteinases (MASPs). The activated MASPs, in turn, cause the antibody-independent activation of the classical pathway. These three pathways overlap in terms of their activators and activity and must not be thought of as being totally discrete.

Because the alternative complement pathway is spontaneously and continuously activated, and could thus cause tissue damage, a number of genes encoding proteins termed regulators of complement activation (RCAs) have evolved (Fig. 4.1). These include membrane bound proteins such as CR1 (CD35), CD46 (membrane cofactor protein – MCP), and CD55 (decay accelerator factor – DAF) and soluble proteins such as C4 binding protein

Table 4.1. *Bacterial components (or host responses to bacteria) associated with activation of the three pathways of complement activation*

Complement pathway	Bacterial component or host response
Classical Pathway	Natural antibody (IgM, IgG) via C1q
	Direct binding via C1q
	Lipid A and LPS (*Klebsiella, Escherichia, Shigella, Salmonella*)
	Lipoteichoic acid (group B streptococci)
	Capsular polysaccharide (*H. influenza*)
	OMPs (*Proteus mirabilis, Sal. minnesota, Klebsiella pneumoniae*)
	C1q binding via C-reactive protein (CRP) (*Strep. pneumoniae*)
Lectin Pathway	Mannose-binding lectin
	Other collectins?
	Ficolins
Alternative Pathway	Bacterial cell wall components (LPS, peptidoglycan, teichoic acid)

(C4BP), factor H (FH), and factor H-like protein 1 (FHL-1), also known as reconectin and factor H-related proteins 1–4. These proteins are encoded by closely linked genes on human chromosome 1 and are composed almost entirely of domains of approximately sixty residues known as short consensus repeats (SCRs) or complement control protein repeats (CCPs). The RCAs are major targets for bacterial and viral evasion mechanisms (Lindahl et al., 2000). A further complement inhibitory protein is protectin (CD59), a GPI anchored protein that inhibits the C5b-8 catalysed insertion of C9 into cell membranes (Davies et al., 1989). The total number of proteins involved in complement activation must be approaching forty and for this reason we will refer to complement as the complement system throughout this chapter. Examples of the constituents of bacteria able to trigger the three complement activation pathways are listed in Table 4.1.

4.2 BIOLOGICAL FUNCTIONS OF THE COMPLEMENT SYSTEM

As is highlighted in other chapters, multicellular organisms have evolved protective mechanisms, which can be grouped under the umbrella term

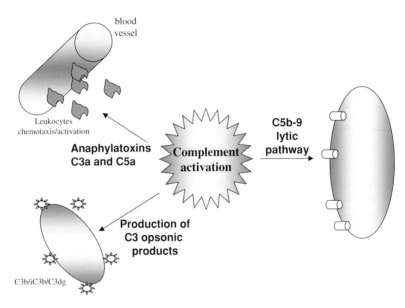

blood
vessel

Leukocytes
chemotaxis/activation

Anaphylatoxins
C3a and C5a

Complement
activation

C5b-9
lytic
pathway

Production of
C3 opsonic
products

C3b/iC3b/C3dg

Figure 4.2. Main actions of the complement system. The major antibacterial components are the breakdown products of C3 (C3b, iC3b and C3dg) that covalently bond to the bacterial surface and enhance the process of phagocytosis and bacterial killing. The anaphylatoxins (C3a and C5a) are chemotactic and activate recruited leukocytes. The lytic pathway of complement forms a complex with the ability to insert into the bacterial membrane and form a damaging pore.

inflammation. The complement system is one of the major effector arms of both the innate and adaptive immune responses and is recognised to be involved in: (i) the killing of microorganisms; (ii) the solubilisation and clearing of immune complexes, and (iii) the enhancement of B lymphocyte responses (the latter effect is reviewed by Fearon and Locksley, 1996; Carroll, 1998; Carroll, 2000). There is also growing evidence for the ability of the complement system to act to control the key regulatory cytokine, IL-12 (Karp and Wills-Karp, 2001).

The activation of the complement system provides three sets of antimicrobial proteins: (i) opsonins (principally C3b and its products and also C4b) to bind to bacteria and enhance bacterial phagocytosis and antibody formation, (ii) anaphylotoxins (C3a, C5a) to enhance inflammatory events and cause leukocyte activation/chemoattraction, and (iii) a lytic complex to kill bacteria (Fig. 4.2). C3 is present at a concentration of 1.3 mg ml^{-1} in serum and is the key participant in the antimicrobial actions of the complement system. Cleavage of C3 by one of the C3 convertases (C4b2a of the classical pathway or

C3bBb of the alternative pathway) produces C3a and C3b. The former is a potent anaphylatoxin. The cleavage of C3 results in the exposure of a previously inaccessible internal thioester, which is extremely reactive with nucleophiles such as molecules bearing hydroxyl (e.g., water) or amino groups. This allows C3b to covalently bond to the surface of bacteria. The very large amount of water in the extracellular fluid prevents significant diffusion of C3b from its site of production. Bound C3b can be further cleaved on the microbial cell surface (generally by reactions involving the complement regulatory proteins) to produce smaller bound fragments – iC3b and C3dg. These proteins, along with C4b, are recognised by the complement receptors types I to IV (CR1 to CR4). CR1 (CD35) is present on a range of cells including B lymphocytes, neutrophils, monocytes/macrophages, follicular dendritic cells (see Chapter 1), glomerular epithelial cells, and erythrocytes and binds to C3b and C4b. CR2 (CD21) is found on B lymphocytes, follicular dendritic cells, and cervical and nasopharyngeal epithelial cells and recognises C3b, iC3b and C3dg. CR3 (CD18/CD11b), a member of the β_2-integrin family, is found on neutrophils, monocytes/macrophages, follicular dendritic cells, and NK cells and recognises iC3b. Transendothelial migration increases substantially the levels of CR3 on human neutrophils and their capacity to phagocytose *Escherichia coli* (Hofman et al., 2000). The final receptor CR4 (CD18/CD11c), also a member of the β_2-integrin family, is present on neutrophils, monocytes, and tissue macrophages and recognises iC3b. Binding of bacteria (opsonised with these various breakdown products of activated C3) to phagocytes bearing one or other of the complement receptors enhances binding to the phagocyte surface and uptake into a phagolysosome with concomitant killing.

A major problem in inflammation is that the leukocytes required to ingest and kill invading bacteria are sequestered in the blood vessels. Chemotactic proteins are required to cause leukocytes to traffic into infected sites. C5a was the first host chemotactic protein discovered. It binds to its cognate receptor (C5a-R), a G-protein coupled receptor, on a range of myeloid cells acting both as a potent chemotactic factor and as a leukocyte activator. C5a causes neutrophil degranulation and activation of the respiratory burst. It is now recognised that the terminal lytic pathway (C5b-9) may also contribute to the activation of leukocytes found during complement activation.

The most visually compelling image of the complement system is the presence of membrane lesions, normally photographed using sheep or rabbit erythrocytes, formed by the action of the terminal components (C5 to C9) of the complement system. These proteins form what is termed the membrane-attack complex (MAC). The initiation of this complex begins with the cleavage of C5 into C5a and C5b. This latter protein binds to C6 (forming C5b6) and this complex, in turn, binds C7 (forming C5b67). The interaction of C7 results in the complex having the ability to insert itself into lipid bilayers. C8 then binds to this membrane-associated complex (C5b678), which can then associate with as many as fourteen C9 monomers to form a membrane pore. It is thought that this pore has antibacterial actions (Joiner et al., 1985) and, as will be discussed, there is increasing evidence for this. However, opsonophagocytosis still seems to be the major antibacterial defence mechanism of the complement system.

4.3 THE INVOLVEMENT OF THE COMPLEMENT SYSTEM IN ANTI-BACTERIAL DEFENCES

Now that we have briefly described the salient features of the complement system, what is the evidence for its role in protection against bacterial infections? The main evidence supporting the role of the complement system in antibacterial protection comes from individuals with deficiencies in individual complement genes (Table 4.2). Such genetic deficiencies can be broadly divided into seven categories: (i) classical pathway genes, (ii) mannose-binding lectin, (iii) alternative pathway genes, (iv) C3, (v) genes encoding the MAC, (vi) regulatory protein genes, and (vii) complement receptors. Further evidence is now emerging from the generation of complement gene transgenics (Mold, 1999). Deficiencies in the components of the classical pathway result largely in individuals with the symptoms of SLE or immune complex disease. These individuals can also suffer recurrent infections. Low levels of serum mannose-binding lectin are associated with recurrent infections in young children, but not in adults. Deficiency in the gene encoding the alternative pathway protein, D, results in recurrent upper respiratory tract infection while deficiency in alternative pathway protein, P (properdin), results in an enhanced susceptibility to fatal fulminant meningococcal infections. The importance of C3 is demonstrated by the finding that individuals deficient in this gene have recurrent infections. In contrast, deficiencies in the genes involved in construction of the MAC are manifest by an enhanced susceptibility to recurrent infections with *Neisseria* spp. Individuals deficient in C9 are generally healthy suggesting that the C5-8 complex is sufficient to cause damage to bacterial cell walls. However, C9 deficiency can be associated with recurrent Neisserial infection. Even in Japan where C9 deficiency is rather common (1 in 1,000), usually without symptoms, patients with recurrent meningococcal meningitis are most likely to be C9 deficient (Ngata et al., 1989). Deficiency in complement control proteins, such as factor I, can result in recurrent infections. Complement receptor protein deficiencies are

Table 4.2. *Human infections associated with genetic deficiencies of complement components*

Complement component deficient	Infection or condition
C1q	Chronic bacterial infections with encapsulated organisms (22%)
C2, C4	Primarily immune complex disease. Some infections (20%)
H, I	Recurrent infections with pyogenic organisms (40–100%)
Properdin, D	Meningococcal meningitis (70%)
C3	Recurrent infection with pyogenic organisms (*Strep. pneumoniae, Strep. pyogenes, H. influenzae, Staph. aureus*) (80%)
CR3	Recurrent infection with pyogenic organisms (*Strep. pneumoniae, Strep. pyogenes, Pseudomonas* spp, *Staph. aureus*) (100%)
C5	Recurrent meningococcal and gonococcal infections and recurrent infections with staphylococci, streptococci, proteus, pseudomonas, and enterobacter (60%)
C6, C7 or C8	Recurrent meningococcal and gonococcal infections (75%)
C9	Recurrent meningococcal and gonococcal infections (8%)

Note: Table indicates the most frequently observed infections associated with complement deficiencies together with an indication of the frequency of those infections that have been observed in individual cases. For example, individuals with C9 deficiency almost always suffer from Neisserial infection but only 8% of such people seem to suffer from recurrent infection.

also associated with pathology (see reviews by Colten and Rosen, 1992; Mold, 1999; Walport 2001a, 2001b).

Over the past decade, transgenic mice have been produced in which a small number of complement genes have been inactivated (Table 4.3). The response of these mice to bacterial infection or endotoxin challenge is generally deficient. Surprisingly, the lack of C3 appeared to have no influence

Table 4.3. *Response to infection by mice with complement gene knockouts*

Gene inactivated	Response	Reference
C1q	increased mortality/SLE	Botto 1998
C3/C4	reducedLD$_{50}$ to GBS[a]	Wessels et al. 1995
C3/C4	enhanced response to endotoxin	Fischer et al. 1997
C3/C4	increased lethality to CLP[b]	Prodeus et al. 1997
C3/C4	enhanced *E. coli* colonisation	Springall et al. 2001
CR3	no influence on *Mycobacterium tuberculosis* infection	Hu et al. 2000
C3a receptor	enhanced lethality to endotoxin	Kildsgaard et al. 2000
C5	decreased clearance of *Pseudomonas aeruginosa*	Cerquetti et al. 1986
C5	hypersusceptibility to *M. tuberculosis* infection	Jagannath et al. 2000
C5	deficit in granulomatous response to *M. tuberculosis*	Actor et al. 2001
C5a receptor	decreased mucosal clearance of *Ps. aeruginosa*	Hopken et al. 1996
Urokinase	decreased clearance of *Ps. aeruginosa*	Gyetko et al. 2000

[a]GBS: Group B streptococci.
[b]CLP: Caecal ligation and puncture-induced peritonitis.

on infection by *Mycobacterium tuberculosis* (Hu et al., 2000) although the lack of C5 decreased the host protective response to this organism. These findings generally support the clinical information available from the investigation of natural complement deficiencies in *Homo sapiens*.

4.4 BACTERIAL EVASION OF THE COMPLEMENT SYSTEM

With such a plethora of mechanisms for the activation of complement, it is clear that in any infection the susceptibility of the microorganism will depend on which mechanisms are activated and to what extent. This will also change during the course of an infection as the balance of innate and adaptive immunity changes. Many aspects of the interaction of an organism with the immune system might indirectly or directly affect the amount of complement activation, e.g., the type and amount of antibody produced, or the intensity of the acute phase response. That bacteria can directly evade the complex defences of the complement system has also been recognised for some time

Table 4.4. *Bacterial strategies to evade the complement system*

Evasion strategy	Bacterium	Molecules involved
Bacterial capsule	GAS	hyaluronic acid-containing capsule
	Group B streptococci	type III capsular polysaccharide and sialic acid
	Neisseria spp	capsule and capsule containing sialic acid/LPS
	Staph. aureus	
	Haemophilus spp	
	E. coli	lipopolysaccharide
	Salmonella spp	lipopolysaccharide
	Meningococci	lipopolysaccharide
Proteinases	*P. gingivalis*	gingipain
	GAS	C5a peptidase
	Ps. aeruginosa	elastase
Chemical inactivation	*H. pylori*	Urea/ammonia
	Ps. aeruginosa	
Binding to RCA proteins	GAS	M protein family
		Protein H
	Strep. pneumoniae	Hic
	Bord. pertussis	filamentous haemagglutinin
	N. gonorrhoeae	Por1A/Por1B
	B. burgdorferi	many (CRASP/OspE etc.)
	Y. enterocolitica	YadA
Ibhibition of lytic pathway	GAS	Streptococcal inhibitor of complement (SIC)
	Y. enterocolitica	Ail
	S. typhimurium	Rck and Trat
	E. coli	Trat and binding protectin
	H. pylori	binding protectin
	Moraxella catarrhalis	?

but only in the last decade or so has the range of mechanisms that bacteria utilise to achieve this begun to be defined (Mold, 1999; Rautemaa and Meri, 1999; Lindahl et al., 2000; Table 4.4). Viruses have also been shown to be able to evade complement-mediated attack by, for example, encoding proteins with homology to host complement control proteins such as protectin (Albrecht et al., 1992).

One mechanism to avoid activation of the classical pathway system is to prevent the production or the action of antibacterial antibodies. A number of mechanisms to inhibit the antibody response to bacteria are now known to exist and include antigenic variation, immunoglobulin-binding proteins, and specific proteases that cleave and inactivate immunoglobulins. Streptococci have developed a range of mechanisms to evade complement-mediated attack including a recently reported endoglycosidase that can hydrolyse the chitobiose core of the asparagine-linked glycan on human IgG, rendering this protein more sensitive to the bacterial cysteine proteinase SpeB (Collin and Olsen, 2001). The ability of bacteria to target immunoglobulins in order to prevent antibody-dependent killing will be discussed in detail by Mogens Kilian in Chapter 7 and will not be discussed further in this chapter.

While we are probably only scratching the surface of the potential mechanisms that bacteria have evolved to evade complement-mediated killing, it is possible to divide these mechanisms into distinct types, the aim of most of them being the prevention of effective deposition of C3 (Table 4.4). The best known mechanism is the use of outer membrane structures – capsule or O-antigenic side chains – to: (i) prevent activation of complement, (ii) inhibit the binding of complement components (C3b/C4b) to bacteria, or (iii) sterically hinder the recognition of these bound proteins. A second mechanism utilises proteases to interfere with the normal proteolytic control of the complement cascade. Related to this is the specific cleavage and inactivation of the anaphylatoxin, C5a and the receptor for C5a. The recent finding of a staphylococcal protein that, in some way, blocks the C5a receptor is a modification of this theme (Veldkamp et al., 2000). The third mechanism, and one that appears to have been evolved by a growing number of bacteria, is the ability to utilise the various host RCA proteins to inhibit the action of the classical and alternative complement pathway C3 convertases. Trypanosomes and some viruses also possess genes encoding proteins with homology to the RCA genes (reviewed by Wurzner, 1999). Whereas this functional mimicry appears to be common among viruses, there are only a few examples reported of bacterial mimicry of complement components. The fourth mechanism of complement evasion involves interference with the lytic process by interference with C7 or C8/C9 insertion into membranes. Thus, it is now clear that bacteria have evolved to defeat the three antibacterial mechanisms of the complement pathway: opsonophagocytosis, proinflammatory actions of anaphylatoxins, and membrane lysis via the MAC (Fig. 4.2). The means by which bacteria achieve this will now be described in more detail.

4.5 BACTERIAL CELL SURFACES AND COMPLEMENT ACTIVATION

The alternative complement pathway is continuously being activated by opsonising surfaces that lack RCA proteins (CR1, CD55, and CD46 – see Fig. 4.1). These proteins are synthesised by host cells but not by bacteria. Thus, all bacteria should be opsonised and subject to the multifarious actions of the complement defence system. However, if the active complement components cannot bind to the bacterial membrane, then they will be relatively ineffective as opsonins or as membrane pore formers. The biochemistry of the surface of a microorganism, particularly its carbohydrate composition, will have a profound effect on its ability to activate complement. The classic example of a bacterial cell surface component that negates the complement system is the capsule.

The capsule of Group A streptococci (GAS) has been most intensively studied. These organisms have a hyaluronic acid capsule produced by a three gene operon: *hasA, hasB, hasC* (Wessels, 2000). This capsule is essentially nonimmunogenic and provides resistance to phagocytosis, although it does not influence the amount of C3b deposited on the bacterial surface (Dale et al., 1996). It is believed that the capsule acts to prevent leukocyte receptors interacting with C3 products deposited on the bacterium (Dale et al., 1996). Acapsular strains have decreased virulence and colonisation capacities (Moses et al., 1997). In contrast to GAS, group B streptococci have capsules composed of type III capsular polysaccharide containing in sialic acid, which is likely to promote the binding of soluble factor H and thus downregulates complement activation and prevents deposition of C3b (Marques et al., 1992). Similarly, capsules of group B meningococci are rich in sialic acid, which diminishes complement attack for, one assumes, the same reasons (Craven et al., 1980). The end result is therefore similar to that of bacteria that have evolved proteins able to bind the soluble RCA proteins. The capsule is also a major virulence factor for staphylococci (Wilkinson et al., 1979; Thakker et al., 1998), pneumococci (Winkelstein et al., 1980) and *E. coli* (Taylor and Robinson, 1980).

Gram-negative bacteria are coated in lipopolysaccharide (LPS). This amphiphilic molecule (see Henderson et al., 1998) is one of the most potent proinflammatory molecules produced by bacteria. C1q can bind directly to lipid A in LPS, and activate the classical pathway, but it is also recognised that LPS molecules with long O-specific side chains can prevent the binding of C1q to the bacterial membrane (presumably by steric hindrance), and this appears to be the major cause of serum resistance in salmonellae

(Joiner et al., 1982). The role of sialic acid in capsules has been mentioned. Sialic acid can be incorporated into the terminal sugar in the lipooligosaccharides of meningococci and gonococci (Mandrell et al., 1990) and can induce serum resistance by binding factor H (Ram et al., 1998).

As has been described, the thio-ester of C3 (and C4) is inactivated by strong nucleophiles (Hostetter and Gordon, 1987) and therefore organisms that secrete ureases (e.g., *H. pylori*) may utilise the ammonia that they produce to destroy C3 (Rokita et al., 1998). The ammonia-inactivated C3 is rapidly cleaved by the host's own complement-regulating proteinases. This mechanism can be classified as chemical inactivation and is included here because it works at the bacterial surface.

4.6 BACTERIAL PROTEINASES AND ANAPHYLATOXINS

The complement system is a proteinase-controlled mechanism for generating macromolecular complexes and mediators. It is therefore possible to modulate the activity of this system with external proteinases. Many bacteria secrete proteinases. However, some secrete more proteinases or more powerful proteinases, than others. A good example is the oral Gram-negative pathogen, *Porphyromonas gingivalis* (Potempa et al., 2000), which is a causative organism of periodontal disease – a chronic inflammatory and destructive disease of the tissues supporting the teeth. This organism produces a number of powerful proteinases, including cysteine proteinases known as gingipains, that can cleave both C3 and C5 (Chen et al., 1992). C3 is cleaved into C3a-like and C3b-like fragments but, of importance, it has been reported that the C3b is not bound to the bacterial surface (Schenkein, 1989). The cleavage of C5, however, does generate active C5a (Wingrove et al., 1992). In addition to cleaving C3 and C5, *P. gingivalis* proteinases have also been reported to inactivate the leukocyte C5a receptor (Jagels et al., 1996).

Other bacteria-producing complement-regulating proteinases include *Pseudomonas aeruginosa*, which produces an elastase that can inactivate complement components (Schultz and Miller, 1974) and GAS, which produce a proteinase with specificity for cleaving C5a. This so-called C5a peptidase is a 130kDa serine proteinase anchored to the surface of GAS that cleaves C5a at its receptor binding site (Cleary et al., 1992) and can inhibit the recruitment of phagocytes to sites of infection (Ji et al., 1996). Knockout of the gene encoding the C5a peptidase resulted in a decreased ability to colonise the throat and induce pneumonia in mice (Husman et al., 1997).

Another organism that has recently been reported to target the C5a receptor is *Staphylococcus aureus*, which secretes a protein able to downregulate the

Table 4.5. *Bacteria binding to the soluble (fluid phase) RCA proteins*

RCA protein	Bacterium	Bacterial ligand
C4BP	*Strep. pyogenes*	some M proteins
	Bord. pertussis	filamentous haemagglutinin
	N. gonorrhoeae	porin
FH	*Strep. pyogenes*	some M proteins
	Strep. pyogenes	protein H
	Strep. pneumoniae	H-binding inhibitor (Hic)
	N. gonorrhoeae	porin
	Borrelia burgdorferi	CRASP-2[a]
	Yersinia enterocolitica	YadA
FHL-1	*Strep. pyogenes*	some M proteins
	Strep. pneumoniae	protein H
	B. burgdorferi	many (e.g., CRASP-1, OspE etc.)

[a] CRASP: complement regulators acquiring surface protein.

expression of C5a on human neutrophils. This protein is not a proteinase and its mechanism of action is still under investigation (Veldkamp et al., 2000).

4.7 EVASION USING RCA PROTEINS

The Achilles heel of the complement system is the host's requirement to control inappropriate activation. As briefly described, this is controlled by a range of membrane-associated and soluble proteins most of which are of the RCA protein family. The membrane-bound RCA proteins are receptors for the entry of various viruses into cells: (i) HIV (CR1), (ii) Epstein-Barr virus and HIV (CR2), (iii) measles virus (CD46), and (iv) various Picornaviruses (CD55) (see Walport, 2001a). In addition, CD46 is the host pilus receptor for pathogenic *Neisseria* spp. (Kallstrom et al., 1997) and CD55 is a receptor for many *E. coli* strains (Nowicki et al., 1993).

The major RCA proteins targeted by bacteria to evade complement-mediated killing are the soluble ones, namely: C4BP, FH, and FHL-1 (Zipfel and Skerka, 1999; Lindahl et al., 2000; see Table 4.5). C4BP inhibits the classical pathway C3 convertase (C4b2a) whereas FH and FHL-1 inhibit the alternative pathway C3 convertase (C3bBb). The role that sialic acid plays in the binding of RCA proteins to bacteria has been described. The prototypic RCA-binding proteins are the M proteins of GAS, which are highly variable proteins bound to the bacterial surface by an LPSTGE motif and which extend from the cell surface as an α-helical coiled-coil dimer (see Cunningham, 2000 for review). The N-terminal region of the M proteins confers serotype specificity on GAS and is highly variable. Gene deletion and complementation experiments have established that M proteins inhibit phagocytosis (Perez-Casal et al., 1992). The M proteins bind both C4BP, FH, and FHL-1. Binding to C4BP is via the highly variable N-terminus of the M protein (Johnsson et al., 1996) and the binding site on C4BP for the M protein overlaps with that used by C4b (Blom et al., 2000). *Bordetella pertussis*, the causative agent of whooping cough, also binds C4BP and this binding involves the major virulence factor of this organism – the filamentous haemagglutinin (Berggard et al., 1997). *Neisseria gonorrhoeae* also binds to C4BP. In this organism, the cell surface protein binding to this RCA component is the major outer membrane protein, porin (Por). This protein is allelic consisting of two main isoforms, Por1A and Por1B. Analysis of the binding of these two proteins to C4BP has demonstrated that Por1A interacts with C4BP by hydrophobic interactions while Por1B associates through ionic forces (Ram et al., 2001). This binding interaction between the porin and C4BP may explain the phenomenon of stable serum resistance in *Neisseria* spp.

A number of bacteria also encode proteins that bind to FH and FHL-1. These two proteins are the alternative splice products of a single factor H gene, with FH containing twenty CCP domains and FHL-1 seven such domains. Both proteins share the ability to control C3 activation. In addition to binding to C4BP, streptococcal M proteins also bind to FH (Horstmann et al., 1988) and to FHL-1 and binding to the latter protein may be the more important interaction in terms of streptococcal inhibition of the alternative complement pathway (Johnsson et al., 1998; Kotarsky et al., 1998). The M protein binds to the SCR7 repeat of FH and FHL-1 (Kotarsky et al., 1998). *Strep. pyogenes* also produces proteins structurally related to the M proteins. One of these, protein H, also binds to FH and FHL-1. It has been shown, using isogenic mutants, that for survival in blood either M protein or protein H is sufficient. However, inactivation of both genes resulted in rapid killing (Kihlberg et al., 1999).

There are a number of other bacteria that are reported to be able to bind FH or FHL-1 on their surfaces although the mechanisms are not quite so well defined as are the M proteins. *Yersinia enterocolitica* utilises YadA, a protein involved in the adherence of this bacterium, to bind FH (China et al., 1993). *Strep. pneumoniae* has recently been shown to encode a protein – termed factor H-binding inhibitor of complement (Hic) – the gene being found in the *pspC* locus. Hic contains a LPXTGX motif and is therefore cell-anchored. Inactivation of the gene encoding this protein prevented binding of FH to bacteria. Recombinant Hic was found to bind to FH with high affinity (Kd = 2.3×10^8 M^{-1}) (Janulczyk et al., 2000). The binding of *N. gonorrhoeae* porins to C4BP has been described. These porins also bind to FH (Ram et al., 1999, 2001). Another organism that has recently been shown to bind FH and FHL-1 is the causative agent of Lyme disease, *Borrelia burgdorferi*. A number of proteins appear to be involved. In one report it was claimed that binding is due to two surface-associated proteins of 28 and 21 kDa which have been named complement regulators acquiring surface proteins (CRASP). The 28 kDa protein (CRASP-1) preferentially binds to FHL-1 and CRASP-2, the 21 kDa protein, preferentially binds to FH. The former protein was found in all isolates of this bacterium whereas CRASP-2 was only detected in a proportion of isolates (Kraiczy et al., 2001). Meri and coworkers have identified a 35 kDa protein in *B. burgdorferi* that binds to FH and FHL-1 (Alitalo et al., 2001) and have also reported that the 19 kDa outer surface protein of this organism, OspE, binds to FH (Hellwage et al., 2001). It is possible that OspE is equivalent to CRASP-2, however, the other two proteins, CRASP-1 and the 35 kDa protein, seem to be different, suggesting that *B. burgdorferi* expresses three RCA-binding proteins.

4.8 TARGETING THE LYTIC PATHWAY

The clinical evidence from deficiencies in the common terminal components of the complement pathways suggests that the only increase in susceptibility to infection is with the Gram-negative *N. meningitidis* (Mold, 1999; Walport, 2001a). However, some Gram-negative bacteria are susceptible to human serum, suggesting that the lytic pathway is important in antibacterial defence. It is still not entirely clear how the C5b-9 complex kills cells, and there is controversy as to the nature of the lytic complex. Some workers have reported that C9-deficient serum is lytic for serum-sensitive *E. coli*, suggesting that C5b-8 is sufficient to cause cell lysis (Harriman et al., 1981, Pramoonjago et al., 1992a). Others have reported that C9 is necessary for killing *E. coli* (Joiner et al., 1985). C9 alone has now been shown to be directly lytic to *E. coli* and to require processing in the bacterial periplasm to convert it from a protoxin to a toxin. This conversion requires the formation of disulphide bonds catalysed by bacterial disulphide bond-forming proteins such as DsbA and DsbB (Wang et al., 2000).

The strongest evidence that the lytic pathway of complement activation is important in the colonisation and growth of bacteria is the growing number of reports that bacteria have evolved mechanisms to evade the lytic mechanism. Two main mechanisms appear to be utilised. The first is the inhibition of the formation of the lytic complex on the cell membrane by independently evolved bacterial proteins. The second is analogous to the binding of the RCA proteins. In this case, the bacteria bind CD59 (protectin). For example, the *Y. enterocolitica* outer membrane protein Ail (attachment invasion locus), an adhesin like YadA, has been found to block complement-mediated lysis when expressed in *E. coli* (Rossi et al., 1998). A homologous protein in *Salmonella typhimurium*, Rck, inhibited the formation of the MAC (Heffernan et al., 1994). *Salmonella* spp. and *E. coli* express the Trat lipoprotein, which is thought to interfere with the binding of C5b6 (Pramoonjago et al., 1992b). *Moraxella catarrhalis*, a commensal of the upper respiratory tract, also seems to have a way of inhibiting MAC formation (Verduin et al., 1994). Gram-negative bacteria have been assumed to be much more sensitive to the lytic pathway of complement attack than Gram-positive bacteria. Therefore, it is somewhat surprising to find that GAS produce a protein known as streptococcal inhibitor of complement (SIC) (Akkesson et al., 1996). SIC is a secreted protein containing a large number of proline residues. The *sic* gene is extremely variable (Stockbauer et al., 1998). Analysis of the mechanism of recombinant SIC in preventing complement-mediated lysis of guinea pig erythrocytes suggests that it blocks the uptake of C5b-7 complexes onto the cell membrane. In this respect, SIC has a similar action to clusterin, a protein found in serum (Fernie-King et al., 2001).

Protectin (CD59) is a GPI-anchored host protein (like CD14), which acts to block cell lysis by preventing the C5b-8 complex catalysed insertion and polymerisation of C9 (Meri et al., 1990). Both *E. coli* (Rautemaa et al., 1998) and *Helicobacter pylori* (Rautemaa et al., 2001) have been found to incorporate CD59 onto their cell surfaces where it protects against complement-mediated lysis. Binding of protectin required the GPI phospholipid moiety as protectin lacking this lipid anchor was unable to prevent complement-mediated lysis.

4.9 UTILISATION OF THE COMPLEMENT SYSTEM

In addition to evading complement-mediated killing, some bacteria actually utilise the complement system for their own advantage. *Mycobacterium tuberculosis* can live within macrophages but needs to enter these cells. The normal mechanism involves binding of the bacterium to macrophage complement receptors through C3 fragments. It has been found that mycobacteria produce on their cell wall a protein functionally analogous to C4b that can bind with C2b to form a surface bound C3 convertase, thus coating the organism in

C3b and allowing uptake into the macrophage (Schorey et al., 1997). Another mycobacterial protein, heparin-binding haemagglutinin, which binds C3, is also thought to enhance the uptake of bacteria into macrophages (Mueller-Ortiz et al., 2001).

4.10 CONCLUSION

The complement system is the major bacterial killing mechanism employed by mammalian hosts and is therefore a key target for the evolution of evasion mechanisms. It is now emerging that bacteria have one or more systems for evading complement-mediated killing. In addition, as described in other chapters, many organisms can also avoid phagocytosis or killing within phagocytes. The key to survival is the avoidance of the generation of C3b and/or the binding of this key component to the bacterial cell wall. A fascinating evolutionary development involves the bacterial genes encoding proteins that can bind RCA proteins and block activation of the C3 convertases. The finding that a growing number of bacteria, both Gram-positive and Gram-negative, have evolved mechanisms to avoid deposition of the membrane attack complex is revealing the obvious and unexpected role of this lytic pathway in antibacterial defences. It is certain that many more evasion mechanisms will be elucidated in the near future. The finding that *Staph. aureus* secretes a protein able to downregulate the C5a-R is an unexpected strategy. A key question that need to be addressed is what role does the complement system play in controlling the enormous numbers of bacteria, our commensal microbiota, that live on our mucosal surfaces.

REFERENCES

Actor, J.K., Breij, E., Wetsel, R.A., Hoffmann, H., Hunter, R.L, and Jagannath, C. (2001). A role for complement C5 in organism containment and granulomatous response during murine tuberculosis. *Scandinavian Journal of Immunology* **53**, 464–474.

Akesson, P., Sjoholm, A.G., and Bjorck, L. (1996). Protein SIC, a novel extracellular protein of *Streptococcus pyogenes* interfering with complement function. *Journal of Biological Chemistry* **371**, 1081–1088.

Albrecht, J.C., Nicholas, J., Cameron, K.R., Newman, C., Fleckenstein, B., and Honess, R.W. (1992). Herpesvirus saimiri has a gene specifying a homologue of the cellular membrane glycoprotein CD59. *Virology* **190**, 527–530.

Alitalo, A., Meri, T., Ramo, L., Jokiranta, T.S., Heikkila, T., Seppala, I.J.T., Oksi, J., Viljanen, M., and Meri, S. (2001). Complement evasion by *Borrelia burgdorferi*:

serum-resistant strains promote C3b inactivation. *Infection and Immunity* 69, 3685–3691.

Berggard, K., Johnsson, E., Mooi, F.R., and Lindahl, G. (1997). *Bordetella pertussis* binds to the human complement regulator C4BP: role of filamentous haemagglutinin. *Infection and Immunity* 65, 3638–3643.

Blom, A.M., Berggard, K., Webb, J.H., Lindahl, G., Villoutreix, B.O., and Dahlback, B. (2000). Human C4b-binding protein has overlapping, but not identical, binding sites for C4b and streptococcal M proteins. *Journal of Immunology* 164, 5238–5336.

Botto, M. (1998). C1q knock-out mice for the study of complement deficiency in autoimmune disease. *Experimental and Clinical Immunogenetics* 15, 231–234.

Carroll, M.C. (1998). The role of complement and complement receptors in induction and regulation of immunity. *Annual Reviews of Immunology* 16, 545–568.

Carroll, M.C. (2000). The role of complement in B cell activation and tolerance. *Advances in Immunology* 74, 61–88.

Cerquetti, M.C., Sordelli, D.O., Bellanti, J.A., and Hooke, A.M. (1986). Lung defences against *Pseudomonas aeruginosa* in C5-deficient mice with different genetic backgrounds. *Infection and Immunity* 52, 853–857.

Chen, Z., Potempa, J., Polanowski, A., Wikstrom, M., and Travis, J. (1992). Purification and characterisation of a 50-kDa cysteine proteinase (gingipain) from *Porphyromonas gingivalis*. *Journal of Biological Chemistry* 267, 18,896–18,901.

China, B., Sory, M.-P., N'guyen, B.T., De Bruyere, M., and Cornelis, G.R. (1993). Role of the YadA protein in prevention of opsonisation of *Yersinia enterocolitica* by C3b molecules. *Infection and Immunity* 61, 3129–3136.

Cleary, P., Prabu, U., Dale, J., Wexler, D., and Handley, J. (1992). Streptococcal C5a peptidase is a highly specific endopeptidase. *Infection and Immunity* 60, 5219–5223.

Collin, M. and Olsen, A. (2001). Effect of SpeB and EndoS from *Streptococcus pyogenes* on human immunoglobulins. *Infection and Immunology* 69, 7187–7189.

Colten, H.R. and Rosen, F.S. (1992). Complement deficiencies. *Annual Reviews of Immunology* 10, 809–834.

Craven, D.E., Peppler, M.S., Frasch, C.E., Mocca, L.F., McGrath, P.P., and Washington, G. (1980). Adherence of isolates of *Neisseria meningitidis* from patients and carriers to human buccal epithelial cells. *Journal of Infectious Diseases* 142, 556–568.

Crawford, K. and Alper, C.A. (2000). Genetics of the complement system. *Review in Immunogenetics* 2, 323–338.

Cunninghame, M.W. (2000). Pathogenesis of group A streptococcal infections. *Clinical Microbiology Reviews* **13**, 470–511.

Dale, J.B., Washburn, R.G., Marques, M.B., and Wessels, M.R. (1996). Hyaluronate capsule and surface M protein in resistance to opsonisation of group A streptococci. *Infection and Immunity* **64**, 1495–1501.

Davies, A., Simmons, D.L., Hale, G., Harrison, R.A., Tighe, H., Lachmann, P., and Waldmann, H. (1989). CD59, an LY6-like protein expressed in human lymphoid cells, regulates the action of the complement membrane attack complex on homologous cells. *Journal of Experimental Medicine* **170**, 637–654.

Fearon, D.T. and Locksley, R.M. (1996). The instructive role of innate immunity in the acquired immune response. *Science* **272**, 50–53.

Fearon, D.T. (1998). The complement system and adaptive immunity. *Seminars in Immunology* **10**, 355–361.

Fernie-King, B.A., Seilly, D.J., Willers, C., Wurzner, R., Davies A., and Lachmann, P.J. (2001). Streptococcal inhibitor of complement (SIC) inhibits the membrane attack complex by preventing uptake of C567 onto cell membranes. *Immunology* **103**, 390–398.

Fischer, M.B., Prodeus, A.P., Nicholson-Weller, A., Ma, M., Murrow, J., Reid, R.R., Warren, H.B., Lage, A.L., Moore, F.D., Rosen, F.S., and Carroll, M.C. (1997). Increased susceptibility to endotoxin shock in complement C3- and C4-deficient mice is corrected by C1 inhibitor replacement. *Journal of Immunology* **159**, 976–982.

Gadjeva, M., Thiel, S., and Jensenius, J.C. (2001). The mannan-binding-lectin pathway of the innate immune response. *Current Opinions in Immunology* **13**, 74–78.

Gyetko, M.R., Sud, S., Kendall, T., Fuller, J.A., Newstead, M.W., and Standiford, T.J. (2000). Urokinase receptor-deficient mice have impaired neutrophil recruitment in response to pulmonary *Pseudomonas aeruginosa* infection. *Journal of Immunology* **165**, 1513–1519.

Harriman, G.R., Esser, A.F.. Podack, E.R., Wunderlich, A.C., Braude, A.I., Lian, T.F., and Curd, J.G. (1981). The role of C9 in the complement-mediated killing of *Neisseria*. *Journal of Immunology* **127**, 2386–2390.

Heffernan, E.J., Wu, L., Louie, J., Okamoto, S., Fierer, J., and Guiney, D.G. (1994). Specificity of the complement resistance and cell association phenotypes encoded by the outer membrane protein genes *rck* from *Salmonella typhimurium* and *ail* from *Yersinia enterocolitica*. *Infection and Immunity* **62**, 5183–5186.

Hellwage, J., Meri, T., Heikkila, T., Alitalo, A., Panelius, J., Lahdenne, P., Seppala, I.J.T., and Meri, S. (2001). The complement regulator factor H binds to the surface protein OspE of *Borrelia burgdorferi*. *Journal of Biological Chemistry* **276**, 8427–8435.

Henderson, B., Poole, S., and Wilson, M. (1998). *Bacteria-Cytokine Interactions. In Health and Disease*. Portland Press: London.

Hofman, P., Piche, M., Farahi Far, D., Le Negrate, G., Selva, E., Landraud, L., Alliana-Schmid, A., Boquet, P., and Rossi, B. (2000). Increased *Escherichia coli* phagocytosis in neutrophils that have transmigrated across a cultured intestinal epithelium. *Infection and Immunity* **68**, 449–455.

Hopken, U.E., Lu, B., Gerard, N.P., and Gerard, C. (1996). The C5a chemoattractant receptor mediates mucosal defence to infection. *Nature* **383**, 86–89.

Horstmann, R.D., Sievertsen, H.J., Knobloch, J., and Fischetti, V.A. (1988). Antiphagocytic activity of streptococcal M protein: selective binding of complement control protein factor H. *Proceedings of the National Academy of Sciences USA* **85**, 1657–1661.

Hostetter, M.K. and Gordon, D.L. (1987). Biochemistry of C3 and related thioester proteins in infection and inflammation. *Reviews of Infections Diseases* **9**, 97–109.

Hu, C., Mayadas-Norton, T., Tanaka, K., Chan, J., and Salgame, P. (2000). *Mycobacterium tuberculosis* infection in complement receptor 3-deficient mice. *Journal of Immunology* **165**, 2596–2602.

Husman, L.K., Yung, D.-L., Hollingshead, S.K., and Scott, J.R. (1997). Role of putative virulence factors of *Streptococcus pyogenes* in mouse models of long-term throat colonisation and pneumonia. *Infection and Immunity* **65**, 1422–1430.

Jagannath, C., Hoffmann, H., Sepulveda, E., Actor, J.K., Wetsel, R.A., and Hunter, R.L. (2000). Hypersusceptibility of A/J mice to tuberculosis is in part due to a deficiency of the fifth complement component (C5). *Scandinavian Journal of Immunology* **52**, 369–379.

Jagels, M.A., Travis, J., Potempa, J., Pike, R., and Hugli, T.E. (1996). Proteolytic inactivation of the leukocyte C5a receptor by proteinases derived from *Porphyromonas gingivalis*. *Infection and Immunity* **64**, 1984–1991.

Janulczyk, R., Iannelli, F., Sjoholm, S., Pozzi, G., and Bjorck, L. (2000). Hic, a novel surface protein of *Streptococcus pneumoniae* that interferes with complement function. *Journal of Biological Chemistry* **275**, 37,257–37,263.

Ji, Y., Schnitzler, N., DeMaster, E., and Cleary, P. (1996). C5a peptidase alters clearance and trafficking of group A streptococci by infected mice. *Infection and Immunity* **64**, 503–510.

Joiner, K.A., Hammer, C.H., Brown, E.J., Cole, R.J., and Frank, M.M. (1982). Studies on the mechanism of bacterial resistance to complement-mediated killing. I. Terminal complement components are deposited and released from *Salmonella minnesota* S218 without causing bacterial death. *Journal of Experimental Medicine* **155**, 797–808.

Joiner, K.A., Schmetz, M.A., Sanders, M.E., Murray, T.G., Hammer, C.F., Dourmashkin, R., and Frank, M.M. (1985). Multimeric complement component C9 is necessary for killing of *Escherichia coli* J5 by terminal attack complex C5b-C9. *Proceedings of the National Academy of Sciences, USA* **82**, 4808–4812.

Johnsson, E., Thern, A., Dahlback, B., Heden, L-O., Wikstrom, M., and Lindahl, G. (1996). A highly variable region in members of the streptococcal M protein family binds the human complement regulator C4BP. *Journal of Immunology* **157**, 3021–3029.

Johnsson, E., Berggard, K., Kotarsky, H., Hellwage, J., Zipfel, P.F., Sjobring, U., and Lindahl, G. (1998). Role of the hypervariable region in streptococcal M proteins: binding of a human complement inhibitor. *Journal of Immunology* **161**, 4894–4901.

Kallstrom, H., Liszewski, M.J., Atkinson, J.P., and Jonsson, A.-B. (1997). Membrane cofactor protein (MCP or CD46) is a cellular pilus receptor for pathogenic *Neisseria*. *Molecular Microbiology* **25**, 639–647.

Karp, C.L. and Wills-Karp, M. (2001). Complement and IL-12: yin and yang. *Microbes and Infection* **3**, 109–119.

Kihlberg, B.-M., Collin, M., Olsen, A., and Bjorck, L. (1999). Protein H, an antiphagocytic surface protein in *Streptococcus pyogenes*. *Infection and Immunity* **67**, 1708–1714.

Kildsgaard, J., Hollmann, T.J., Matthews, K.W., Bian, K., Murad, F., and Wetsel, R.A. (2000). Cutting edge: targeted disruption of the C3a receptor gene demonstrates a novel protective anti-inflammatory role for C3a in endotoxin shock. *Journal of Immunology* **165**, 5406–5409.

Kirschfink, M. (2001). Targeting complement in therapy. *Immunology Reviews* **180**, 177–189.

Kotarsky, H., Hellwage, J., Johnsson, E., Skerka, C., Svensson, H.G., Lindahl, G., Sjobring, U., and Zipfel, P.F. (1998). Identification of a domain in human factor H and factor H-like protein-1 required for the interaction with streptococcal M proteins. *Journal of Immunology* **160**, 3349–3354.

Kraiczy, P., Skerka, C., Kirschfink, M., Brade, V., and Zipfel, P.F. (2001). Immune evasion of *Borrelia burgdorferi* by acquisition of human complement regulators FHL-1/reconectin and factor H. *European Journal of Immunology* **31**, 1674–1684.

Law, S.K.A. and Reid, K.B.M. (1995). Complement 2nd Edition. IRL Press at Oxford University Press.

Lindahl, G., Sjobring, U., and Johnsson, E. (2000). Human complement regulators: a major target for pathogenic microorganisms. *Current Opinions in Immunology* **12**, 44–51.

Mandrell, R.E., Lesse, A.J., Sugai, J.V., Shero, M., Griffiss, J.M., Cole, J.A., Parsons, N.J., Smith, H., Morse, S.A., and Apicella M.A. (1990). *In vitro* and *in vivo* modifications of *Neisseria gonorrhoeae* lipooligosaccharide epitope structure by sialylation. *Journal of Experimental Medicine* **171**, 1649–1664.

Marques, M.B., Kasper, D.L., Pangburn, M.K., and Wessels, M.R. (1992). Prevention of C3 deposition is a virulence mechanism of type III group B *Streptococcus* capsular polysaccharide. *Infection and Immunity* **60**, 3986–3993.

Matsushita, M. and Fujita, T. (2001). Ficolins and the lectin complement pathway. *Immunological Reviews* **180**, 78–85.

Meri, S., Morgan, B.P., Davies, A., Daniels, R.H., Olavesen, M.G., Waldmann, H., and Lachmann, P.J. (1990). Human protectin (CD59), an 18,000–20,000 MW complement lysis restricting factor, inhibits C5b-8 catalysed insertion of C9 into lipid bilayers. *Immunology* **71**, 1–9.

Mold, C. (1999). Role of complement in host defence against bacterial infection. *Microbes and Infection* **1**, 633–638.

Moses, A.E., Wessels, M.R., Zalcman, K., Alberti, S., Natanson-Yaron, S., Menes, T., and Hanski, E. (1997). Relative contributions of hyaluronic acid capsule and M protein to virulence in a mucoid strain of group A *Streptococcus*. *Infection and Immunity* **65**, 64–71.

Mueller-Ortiz, S.L., Wanger, A.R., and Norris, S.J. (2001). Mycobacterial protein HbhA binds human complement component C3. *Infection and Immunity* **69**, 7501–7611.

Ngata, M., Hara, T., Aoki, T., et al. (1989). Inherited deficiency of ninth component of complement: an increased risk of meningococcal meningitis. *Journal of Pediatrics* **114**, 260–264.

Nowicki, B., Hart, A., Coyne, K.E., Lublin, D.M., and Nowicki, S. (1993). Short consensus repeat-3 domain of recombinant decay-accelerating factor is recognised by *Escherichia coli* recombinant Dr adhesin in a model of cell–cell interaction. *Journal of Experimental Medicine* **178**, 2115–2121.

Perez-Casal, J., Caparon, M.G., and Scott, J.R. (1992). Introduction of the *emm6* gene into an *emm*-deleted strain of *Streptococcus pyogenes* restores its ability to resist phagocytosis. *Research Microbiology* **143**, 549–558.

Potempa, J., Banbula, A., and Travis, J. (2000). Role of bacterial proteinases in matrix destruction and modulation of host responses. *Periodontology 2000* **24**, 153–192.

Pramoonjago, P., Kinoshita, T., Hong, K., Takata-Kozono, Y., Kozono, H., Inagi, R., and Inoue, K. (1992a). Bactericidal activity of C9-deficient human serum. *Journal of Immunology* **148**, 837–843.

Pramoonjago, P., Kaneko, M., Kinoshita, T., Ohtsubo, E., Takeda, J., Hong, K.S., Inagi, R., and Inoue, K. (1992b). Role of Trat protein, an anticomplementary protein produced in *Escherichia coli* by R100 factor, in serum resistance. *Journal of Immunology* **148**, 827–836.

Prodeus, A.P., Zhou, X., Maurer, M., Galli, S.J., and Carroll, M.C. (1997). Impaired mast cell-dependent natural immunity in C3-deficient mice. *Nature* **390**, 172–175.

Ram, S., Sharma, A.K., Simpson, S.D., Gulati, S., McQuillen, D.P., Pangburn, M.K., and Rice, P.A. (1998). A novel sialic acid binding site on factor H mediates serum resistance of sialylated *Neisseria gonorrhoeae*. *Journal of Experimental Medicine* **187**, 743–752.

Ram, S., Mackinnon, F.G., Gulati, S., McQuillen, D.P., Vogel, U., Frosch, M., Elkins, C., Guttormsen, H.-K., Wetzler, L.M., Oppermann, M., Pangburn, M.K., and Rice, P.A. (1999). The contrasting mechanisms of serum resistance of *Neisseria gonorrhoeae* and group B *Neisseria meningitidis*. *Molecular Immunology* **36**, 915–928.

Ram, S., Cullinane, M., Blom, A.M., Gulati, S., McQuillen, D.P., Monks, B.G., O'Connell, C., Boden, R., Elkins, C., Pangburn, M.K., Dahlbeck, B., and Rice, P.A. (2001). Binding of C4b-binding protein to porin: A molecular mechanism of serum resistance of *Neisseria gonorrhoeae*. *Journal of Experimental Medicine* **193**, 281–295.

Rautemaa, R., Jarvis, G.A., Marnila, P., and Meri, S. (1998). Acquired resistance of *Escherichia coli* to complement lysis by binding of glycophosphoinositol-anchored protectin (CD59). *Infection and Immunity* **66**, 1928–1933.

Rautemaa, R. and Meri, S. (1999). Complement-resistance mechanisms of bacteria. *Microbes and Infection* **1**, 785–794.

Rautemaa, R., Rautelin, H., Puolakkainen, P., Kokkola, A., Karkkainen, P., and Meri, S. (2001). Survival of *Helicobacter pylori* from complement lysis by binding of GPI-anchored protectin (CD59). *Gastroenterology* **120**, 470–479.

Rokita, E.M, Makristathis, A., Presterl, E., Rotter, M L., and Hirschl, A.M. (1998). *Helicobacter pylori* urease significantly reduces opsonisation by human complement. *Journal of Infectious Diseases* **178**, 1521–1525.

Rossi, V., Tatar, L.D., Beer, K.B., Miller, V.L., and Esser, A.F. (1998). Structural studies of Ail-mediated resistance of *E. coli* to killing by complement. *Molecular Immunology* **35**, 397.

Schenkein, H.A. (1989). Failure of *Bacteroides gingivalis* W83 to accumulate bound C3 following opsonisation with serum. *Journal of Periodontal Research* **24**, 20–27.

Schorey, J.S., Carroll, M.C., and Brown, E.J. (1997). A macrophage invasion mechanism of pathogenic mycobacteria. *Science* **277**, 1091–1093.

Schultz, D.R. and Miller, K.D. (1974). Elastase of *Pseudomonas aeruginosa*: inactivation of complement components and complement-derived chemotactic and phagocytic factors. *Infection and Immunity* **10**, 128–135.

Springall, T., Sheerin, N.S., Abe, K., Holers, V.M., Wan, H., and Sacks, S.H., (2001). Epithelial secretion of C3 promotes colonisation of the upper urinary tract by *Escherichia coli*. *Nature Medicine* **7**, 801–806.

Stockbauer, K.E., Grigsby, D., Pan, X., Fu, Y.X., Mejia, L.M., Cravioto, A., and Musser, J.M. (1998). Hypervariability generated by natural selection in an extracellular complement-inhibiting protein of serotype M1 strains of group A Streptococcus. *Proceedings of the National Academy of Sciences USA* **95**, 3128–3133.

Taylor, P.W. and Robinson, M.K. (1980). Determinants that increase the serum resistance of *Escherichia coli*. *Infection and Immunity* **29**, 278–280.

Thakker, M., Park, J.S., Carey, V., and Lee, J.C. (1998). *Staphylococcus aureus* serotype 5 capsular polysaccharide is antiphagocytic and enhances bacterial virulence in a murine bacteremia model. *Infection and Immunity* **66**, 5183–5189.

Veldkamp, K.E., Heezius, H.C.J.M., Verhoef, J., van Strijp, J.A.G., and van Kessel, K.P.M. (2000). Modulation of neutrophil chemokine receptors by *Staphylococcus aureus* supernate. *Infection and Immunity* **68**, 5908–5913.

Verduin, C.M., Jansze, M., Hol, C., Mollnes, T.E., Verhoef, J., and VanDijk, H. (1994). Differences in complement activation between complement-resistant and complement-sensitive *Moraxella (Branhamella) catarrhalis* strains occur at the level of membrane attack complex formation. *Infection and Immunity* **62**, 589–595.

Walport, M.J. (2001a). Complement. First of two parts. *New England Journal of Medicine* **344**, 1058–1066.

Walport, M.J. (2001b). Complement. Second of two parts. *New England Journal of Medicine* **344**, 1140–1144.

Wang, Y., Bjes, E.S., and Esser, A.F. (2000). Molecular aspects of complement-mediated bacterial killing. *Journal of Biological Chemistry* **275**, 4687–4692.

Wessels, M.R., Butko, P., Ma, M., Warren, H.B., Lage, A.L., and Carroll, M.C. (1995). Studies of group B streptococcal infection in mice deficient in complement component C3 or C4 demonstrate an essential role for complement in both innate and acquired immunity. *Proceedings of the National Academy of Sciences USA* **92**, 11,490–11,494.

Wessels, M.R. (2000). Capsular polysaccharide of Group A *Streptococcus*. In *Gram-positive Pathogens*, ed. V. Fischetti et al. pp. 34–42. Washington, ASM Press.

Wilkinson, B.J., Sissons, S.P., Kim, Y., and Peterson, P.K. (1979). Localisation of the third component of complement on the cell wall of encapsulated *Staphylococcus aureus* M: implications for the mechanism of resistance to phagocytosis. *Infection and Immunity* **26**, 1159–1163.

Wingrove, J.A., DiScipio, R.G., Chen, Z., Potempa, J., Travis, J., and Hugli, T.E. (1992). Activation of complement components C3 and C5 by a cysteine proteinase (gingipain-1) from *Porphyromonas (Bacteroides) gingivalis. Journal of Biological Chemistry* **267**, 18,902–18,907.

Winkelstein, J.A., Abramovitz, A.S., and Tomasz, A. (1980). Activation of C3 via the alternative complement pathway results in fixation of C3b to the pneumococcal cell wall. *Journal of Immunology* **124**, 2502–2506.

Wurzner, R. (1999). Mini Review – Evasion of pathogens by avoiding recognition or eradication by complement, in part via molecular mimicry. *Molecular Immunology* **36**, 249–260.

Zipfel, P.F. and Skerka, C. (1999). FHL-1/reconectin: a human complement and immune regulator with cell-adhesive function. *Immunology Today* **20**, 135–140.

CHAPTER 5

Bacterial immunoglobulin-evading mechanisms: Ig-degrading and Ig-binding proteins

Mogens Kilian

5.1 INTRODUCTION

The mucosae are the largest surface areas of the body directly exposed to the environment and, in consequence, throughout life these surfaces are constantly challenged by microbes. Indeed, most infectious diseases take place or are initiated on these surfaces. The approximately 400 m^2 of the mucosae in the respiratory, gastrointestinal, and genito-urinary tract of humans are protected by a multitude of innate and adaptive immune mechanisms (Ogra et al., 1999). To successfully colonise, all microorganisms, whether they are pathogens or commensals, must be able to evade local defence mechanisms. The dominant adaptive immune factor in the mucosal surfaces is secretory IgA (S-IgA), which is actively transported to the surface by a receptor-mediated mechanism. This chapter reviews the bacterial strategies to evade the functions of immunoglobulins (Ig), that involve proteolytic/glycolytic degradation or Ig-binding proteins, with a particular focus on IgA.

5.2 IgA PROTEASE-PRODUCING BACTERIA

IgA proteases constitute a diverse group of bacterial proteinases that share several unique enzymatic properties including the ability to cleave human IgA in the hinge region. They are post-proline endopeptidases and, in most cases, attack the target polypeptide immediately adjacent to a carbohydrate side chain. With a single exception, the IgA proteases cleave only IgA1. The first examples of IgA1 proteases were demonstrated in *Streptococcus sanguis*, *Neisseria meningitidis*, and *Neisseria gonorrhoeae* by Plaut and coworkers in the mid-1970s (Plaut et al., 1974; Plaut et al., 1975). The

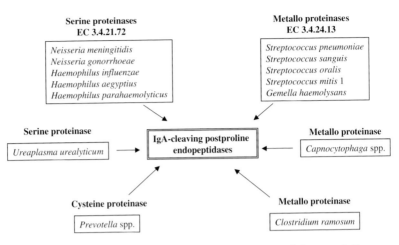

Figure 5.1. Primary structure of the hinge region of human IgA subclasses and allotypes. The arrows indicate the individual peptide bonds in the IgA1 hinge cleaved by bacterial IgA1 proteases and the bond cleaved in IgA1 and IgA2 allotype A2m(1) by the C. ramsum IgA protease. The small boxes indicate glycosylation sites in the IgA1 hinge region, those indicated by hatched boxes being variably present in IgA1 molecules.

potential biological significance of IgA1 proteases was highlighted by the subsequent demonstration that all three principal causes of bacterial meningitis: *Haemophilus influenzae, Streptococcus pneumoniae*, and *N. meningitidis* possess such enzymes (Kilian et al., 1979; Male, 1979). Subsequent comprehensive screenings of bacteria, fungi, and viruses revealed IgA1 protease activity in a limited number of bacterial species that exclusively colonise or infect the mucosal membranes of humans (Fig. 5.1). In addition to the three principal causes of bacterial meningitis, these species include important causes of respiratory tract infections, two urogenital pathogens, four species of commensal Gram-positive cocci that belong in the pharynx and oral cavity, and all species of the genera *Capnocytophaga* and *Prevotella* associated with man. Some of the latter have been implicated in the pathogenesis of periodontal diseases (for references to the individual findings, see Kilian and Russell, 1999). Apart from occasional mutants, all members of these species, including noncapsulated strains of *H. influenzae*, constitutively express IgA1 proteases. The only exceptions are *Streptococcus mitis*, in which IgA1 protease activity is restricted to approximately half of the strains of biovar 1 (Hohwy et al., 2001) and *Clostridium ramosum* in which IgA protease is produced by less than 10% of isolates examined from human faeces (Senda et al., 1985).

All IgA1 proteases that have been identified cleave one of the several prolyl-seryl or prolyl-threonyl peptide bonds that are present in the elongated hinge region in the human (Fig. 5.1) and hominoid primates (chimpanzees, gorillas, and orangutans), although the hinge region sequence of the latter show minor differences (Qiu et al., 1996). The phylogenetically recent, cleavage-susceptible insert of thirteen amino acids in the human IgA1 hinge, relative to human IgA2 and IgA of all other animal species, allows the Fab regions of IgA1 antibodies increased conformational freedom. This advantage is achieved at the expense of introducing an open stretch that, like other interdomain areas of Ig molecules, is potentially more susceptible to attack by proteolytic enzymes. In the IgA1 hinge region, this problem is counteracted by the unusual amino acid sequence, combined with extensive O-glycosylation of several serine and threonine residues (Mattu et al., 1998; Fig. 5.1). Although these properties render the IgA1 hinge resistant to cleavage by traditional proteolytic enzymes of host or microbial origin, the IgA1 proteases possess this unique cleavage specificity.

Because the hinge region of IgA2 lacks the 13-amino acid insert, this subclass is not susceptible to cleavage. The only exception to this pattern is the protease identified in some strains of *C. ramosum*. The protease of *C. ramosum* cleaves both IgA1 and the A2m(1) allotype of IgA2 at the prolyl–valyl bond just before the hinge within a sequence shared by these two humans α-chains (Fujiyama et al., 1986), but lacking in the two other IgA2 allotypes (Fig. 5.1). Apart from hominoid primates, IgA from no animal species tested is susceptible to cleavage by IgA1 proteases.

5.3 CONVERGENT EVOLUTION OF IgA PROTEASES

IgA1 proteases represent a striking example of convergent evolution of a highly specialised enzymatic activity, which by itself provides strong evidence for their biological significance. Detailed characterisation of the IgA1 proteases at gene or enzyme levels reveals that IgA1 protease activity has evolved along at least five independent lines of evolutionary events (Fig. 5.2) (for details see Kilian and Russell, 1999). Although the proteases display very similar enzymatic activity, they do so by at least three different catalytic mechanisms and, accordingly, belong to three of the major groups of proteinases: serine proteinases, metalloproteinases, and thiol proteinases (Fig. 5.2). Within each group of bacteria in the boxes in Fig. 5.2, the IgA1 protease genes (*iga*) show extensive sequence homology (Lomholt et al., 1995; Poulsen et al., 1997) indicating that they share a common ancestor or have been spread by horizontal gene transfer after diversification of the individual

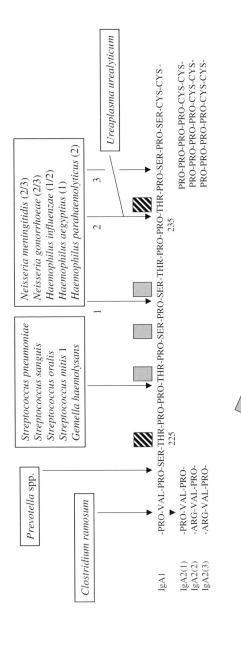

Figure 5.2. Convergent evolution of IgA proteases. Genes encoding IgA1 proteases in the bacteria boxed together are mutually related.

species. Interestingly, the recently finished genome sequence of the IgA1 protease-producing *Ureaplasma urealyticum* reveals no sequence homology to any of the known *iga* genes, and the IgA protease gene has not yet been identified.

According to results of recent studies, the IgA1/2 protease of *C. ramosum* belongs to a novel type of metalloproteinases (K. Kosowska, J. Reinholdt, L. Kjær, J. Potempa, M. Kilian, and K. Poulsen, in preparation).

5.4 SPECIFIC ADAPTATION TO HOST AND HABITAT

In accordance with the substrate specificity of IgA1 proteases, the bacteria that produce these enzymes are exclusively associated with humans and hominoid primates, although *S. pneumoniae* may cause infections in occasional animal species. Furthermore, the specificity of IgA proteases for a particular subclass of IgA is in complete agreement with the predominance of that subclass in the normal habitat of the bacteria. Thus, IgA1 accounts for 70–95% of S-IgA in secretions at mucosal membranes in the upper respiratory tract, which harbour most of the IgA1 protease-producing species. Likewise, secretions in the normal habitat of *C. ramosum*, the human gut, contain an even distribution of the two subclasses in accordance with the specificity of the IgA1/2 protease (Kett et al., 1986; Mestecky and Russell, 1986; Kirkeby et al., 2000). Whether the occurrence of *C. ramosum* in the population reflects the distribution of the IgA2 allotypes is not known. One possible exception to the apparent adaptation to IgA subclass distribution is *N. gonorrhoeae*, which in its primary habitat encounters IgA1 and IgA2 in approximately equal proportions. A possible explanation for this seeming lack of optimal adaptation is that the gonococcus is a recent offshoot from meningococci as indicated by phylogenetic studies. The selection pressure on gonococci for further adaptation may have been limited because of an abundance of other immune escape mechanisms employed by this species (Meyer et al., 1994).

The strict specificity of IgA1 proteases for human IgA1 precludes studies of the biological significance of IgA1 proteases in traditional experimental animal models. In attempts to identify potential animal models, numerous animal pathogens have been examined for their ability to cleave the IgA of their respective hosts. Early studies revealed that *Actinobacillus* (formerly *Haemophilus*) *pleuropneumoniae*, which causes fatal lower respiratory tract infections in pigs, induces degradation of porcine IgA, but the pattern of cleavage, and hence the potential consequences, are not similar to those observed with specific IgA1 proteases (Kilian et al., 1979). In contrast, screening of a large number of animal pathogenic (*Mycoplasma* and *Ureaplasma*) species has identified a protease that is capable of cleaving the hinge region of canine IgA in *Ureaplasma* strains associated with infections in dogs (Kapatais-Zoumbos et al., 1985). So far, this potential model has not been employed in studies of the biological significance of IgA proteases.

5.5 *IN VIVO* ACTIVITY

In contrast to many other microbial proteinases, IgA1 proteases are not inhibited by physiological protease inhibitors such as α2-macroglobulin and α1-protease inhibitor, a property explainable in part by the substrate specificity which, for example, precludes activation of α2-macroglobulin. However, human milk lactoferrin is reported to inactivate IgA1 protease and the related Hap protein (see below) in *H. influenzae* by extracting these auto-transporter proteins from the bacterial cell membrane by a yet unknown mechanism (Qiu et al., 1998). The full significance of this interaction in the protection against infections caused by IgA1 protease-producing bacteria is still unclear.

Both colonisation and infection with IgA1 protease-producing pathogens induce significant titres of enzyme-neutralising antibodies in serum and secretions (Gilbert et al., 1983; Brooks et al., 1992). Accordingly, neutralising antibodies against IgA1 proteases, but not the IgA1/2 protease of *C. ramosum*, have been demonstrated in human milk (Gilbert et al., 1983; Kobayashi et al., 1987). Such antibodies may play a role in the protection of the infant against IgA1 protease-producing bacteria. Besides inhibiting IgA1 protease activity, neutralizing antibodies block the release of the enzyme from *H. influenzae*, meningococci, and gonococci, because of inhibition of the autocatalytic cleavage responsible for this process (Pohlner et al., 1987). As a result, antibodies against IgA1 proteases may agglutinate the bacteria through interaction with the surface-bound protease, which constitutes a novel surface antigen (Plaut et al., 1992).

The significance of enzyme-neutralising antibodies as a selection pressure in the host–parasite relationship is strongly suggested by the fact that IgA1 proteases of most of the pathogens show considerable antigenic diversity because of accumulation of mutations combined with inter-strain and possibly inter-species recombination affecting particular regions in *iga* genes (Halter et al., 1989; Poulsen et al., 1992; Morelli et al., 1994; Lomholt et al., 1995). The most extensive antigenic diversity has been detected in IgA1 proteases of *H. influenzae* and *S. pneumoniae* (Kilian and Thomsen, 1983; Lomholt, 1995).

Production of IgA1 protease during infection is evident by the presence of active enzyme in cerebrospinal fluid of patients with *H. influenzae* meningitis and in vaginal washes of women with culture-proven gonorrhoea (for review see Kilian and Russell, 1999). In an immunochemical study of secretions collecting in the nasopharynx of ninety-seven children (age <5 years) during surgery and general anaesthesia, extensive cleavage of S-IgA1 into Fab and $Fc_2 \cdot SC$ fragments were detected in eighteen children (18.6%). A significantly increased prevalence of cleavage was observed in children with a history of atopic diseases (61.5% versus 11.9%; $p < 0.001$), which led to the hypothesis that IgA1 protease-induced abrogation of the immune barrier in the respiratory tract may be a factor in the development of allergic disease (Sørensen and Kilian, 1984). A longitudinal follow-up study of the nasopharyngeal microbiota in infants developing atopic diseases as compared to healthy infants revealed that the former were indeed colonised by higher proportions of IgA1 protease-producing bacteria (Kilian et al., 1995).

Characteristic fragments of IgA were detected on the surface of bacteria collected from dental plaque (Ahl and Reinholdt, 1991). Further indirect evidence for *in vivo* cleavage of IgA1 induced by oral *Prevotella* and *Capnocytophaga* species has been obtained by the detection of serum antibodies to a neoepitope exposed on Fabα fragments released from IgA1 upon cleavage with these proteases (Frandsen et al., 1995). However, the failure to demonstrate characteristic cleavage fragments of IgA in vaginal secretions of women with gonococcal infection (Hedges et al., 1998) and in nasal rinses of healthy human adults colonised with IgA1 protease-producing bacteria (Kirkeby et al., 2000) suggests a subtle relationship, and that cleavage, in most cases, may be restricted to the microenvironment of the individual bacterial cells. It is conceivable that the extent of cleavage *in vivo* is dependent on the number of bacteria producing IgA1 protease, the amount of enzyme secreted, the presence of protease-neutralising antibodies, and the time allowed for interaction of the IgA1 protease with its substrate. Unless bound to a bacterial surface by antigen–antibody interactions, S-IgA molecules are rapidly cleared from mucosal surfaces.

Significant inter- and intra-species differences have been detected in the IgA1 protease activity released by bacteria *in vitro* (Reinholdt and Kilian, 1997). Although there is no overall relationship between species pathogenicity and the amount of activity released, meningococci generally show activity that may be several hundred times higher than that of some of the oral streptococci. However, significant variations may be seen also among strains of *S. mitis* biovar 1 (Reinholdt and Kilian, 1997). Recently reported comparative examinations of the IgA1 protease activity of large numbers of isolates of

N. meningitidis indicate significantly enhanced activity in isolates from invasive infections compared to isolates from healthy carriers (Vitovski et al., 1999).

5.6 BIOLOGICAL SIGNIFICANCE OF IgA CLEAVAGE

Cleavage of IgA by any of the IgA proteases results in separation of antigen-binding, monomeric Fab fragments from the secondary effector mechanisms residing in the remaining part of the antibody molecule (Fc or $Fc_2 \cdot SC$). In this way, the bacteria potentially evade both the noninflammatory protective mechanisms of S-IgA on mucosal surfaces and the IgA-mediated uptake in phagocytes via the $Fc\alpha$ receptor (CD89) and other immune complex elimination mechanisms associated with IgA antibodies (Russell et al., 1999; van Egmond et al., 2001).

Studies involving gonococci and oral streptococci demonstrate that IgA1 protease activity is able to reverse the inhibitory effect of S-IgA on *in vitro* adherence of these bacteria, or purified bacterial adhesins, to epithelial cells and saliva-coated hydroxyapatite, respectively (Mulks et al., 1980; Reinholdt and Kilian, 1997; Hajishengallis et al., 1992). Longitudinal studies of the clonal turn-over of noncapsulated *H. influenzae* in the pharynx of healthy children provide further indirect support for the hypothesis that an active IgA1 protease is important for successful colonisation by these bacteria. Thus, successive clones of *H. influenzae* colonizing the same individual produce antigenically different IgA1 proteases that are not inhibited by antibodies induced by previously colonising clones (Lomholt et al., 1993).

Interactions between mucus and S-IgA-containing immune complexes are believed to contribute to the barrier function of mucous membranes. Although S-IgA diffuses freely through mucus, multiple S-IgA antibodies bound on the surface of microorganisms may mediate trapping and ensuing prevention of penetration of the surface (Biesbrock et al., 1991; Saltzman et al., 1994). Studies of this phenomenon performed with human sperm suggest that IgA1 proteases can interfere with this protective mechanism. Although sperm coated with S-IgA antibodies are unable to penetrate a layer of cervical mucus, the ability may be restored by selective elimination of the $Fc_2 \cdot SC$ moieties by IgA1 protease treatment (Bronson et al., 1987). If these results can be extrapolated to apply to IgA1 protease-producing bacteria, it may explain, in part, their ability to penetrate mucus in the presence of S-IgA antibodies.

In the event of successful penetration, loss of the Fc portion of IgA1 antibodies bound to the surface of the pathogen will preclude its elimination by poly-immunoglobulin receptor-mediated retransport through the surface epithelia and CD89-mediated phagocytosis (Russell et al., 1999; van Egmond et al., 2001). Furthermore, *in vitro* studies indicate that Fabα fragments remaining on the bacterial surface can interfere with complement activation and complement-mediated lysis in the presence of IgM and IgG antibodies (Russell et al., 1989; Jarvis and Griffiss, 1991). See Chapter 4 for more details on the complement system and of the mechanisms of bacterial evasion of complement. However, the key to the understanding of the biological significance of IgA1 proteases in invasive infections may be the relationship between cleavage-relevant IgA1 antibodies and IgA1 protease-neutralising antibodies. When colonised by IgA1 protease-producing bacteria such as *H. influenzae*, *N. meningitidis*, or *S. pneumoniae*, an IgA antibody response to bacterial surface epitopes and their IgA1 protease are likely to occur concurrently. Thus, once a human host has responded to an acquired IgA1 protease-producing bacterium with IgA1 antibodies, the cleavage of which would be beneficial to the bacterium, it is likely that neutralising antibodies prevent the IgA1 protease from functioning. This is likely to be the situation in the majority of individuals who develop immunity to these bacteria rather than disease.

According to a proposed hypothetical model (Kilian and Reinholdt, 1987), IgA1 protease activity may have particular consequences if a potential pathogen colonises a human host who already has IgA1 antibodies to surface epitopes of the bacteria, but no antibodies to its IgA1 protease. This may occur as a result of prior encounter at a mucosal surface with microorganisms that carry cross-reactive capsules (e.g., *Escherichia coli* K100 in the case of *H. influenzae* type b, and *E. coli* K1 or *Moraxella nonliquefaciens* in the case of *N. meningitidis* group B). At the subsequent colonisation, the actual pathogen encounters preexisting IgA1 antibodies to its surface epitopes but no antibodies that will neutralize its IgA1 protease. This situation enables the pathogen not only to evade the protective functions of IgA1 but also to take advantage of IgA1 antibodies by becoming coated with Fabα fragments with the consequences discussed above. As discussed elsewhere, this hypothetical model is in agreement with several hitherto unexplained observations both in humans and in animal models (Kilian and Reinholdt, 1987).

As mentioned, no accessible animal model allows evaluation of the biological significance of IgA1 proteases. However, two studies have used human organ culture models. Isogenic strains of *N. gonorrhoeae* and *H. influenzae* expressing, or lacking, IgA1 protease activity (as a result of *iga* gene deletion) were used to infect human fallopian tube and nasopharyngeal organ cultures. No difference in the ability to attach to or penetrate the mucosal tissues was observed in these models (Cooper et al., 1984; Farley et al., 1986). Although these studies indicate that other unknown properties of IgA1 proteases may not be important for infection, they do not elucidate any effect related to IgA, as the models lacked specific IgA1 antibodies. The same problem may apply to recent studies of the initial phase of gonococcal infection in human male volunteers, which also failed to detect differences in the infectivity of these isogenic strains (Johannsen et al., 1999).

It is less clear what role IgA1 proteases play in commensal Gram-positive bacteria, and why the property has been conserved through evolution in species like *S. sanguis*, *Streptococcus oralis*, some *S. mitis*, and in *Gemella haemolysans*, but not in several other commensal streptococci including *Streptococcus gordonii*. The fact that 80–90% of the streptococci that initiate the colonization of tooth enamel produce IgA1 protease indirectly suggested that this property plays a role (for review, see Kilian and Russell, 1999). Nevertheless, the colonisation pattern of tooth surfaces in adults shows no difference between normal and selectively IgA-deficient adults who compensate with S-IgM in saliva (Reinholdt et al., 1993), although the selective advantage of IgA1 protease-producing streptococci with regard to IgA antibodies is missing in the latter group.

It has been suggested that IgA1 protease production may be particularly important early in life when the oral microbiota become established under the influence of S-IgA antibodies in mother's milk (Cole et al., 1994), but the exact biological significance of IgA1 proteases in commensal streptococci is still unclear.

5.7 ALTERNATIVE SUBSTRATES AND FUNCTIONS OF IgA PROTEASES

Searches for protein sequences with similarity to the susceptible hinge region of IgA1 resulted in the demonstration that the lysosomal/phagosomal membrane protein Lamp1 is cleaved by the type 2 gonococcal IgA1 protease, thereby promoting intracellular survival of the bacteria (Hauck and Meyer, 1997; Lin et al., 1997). The significance of this activity is further supported by the demonstration that *iga* mutants have a statistically significant and reproducible defect in their ability to traverse monolayers of polarized epithelial cells *in vitro* (Hopper et al., 2000).

The IgA1 proteases range in size from 100 to 200 kDa, which is an unusually large size among proteases. This raises the question as to whether these proteins have functions unrelated to their proteolytic activity. Indeed, recent studies demonstrated significant immuno-modulatory properties of IgA1 protease, some of which are unrelated to the protease activity. Thus, native gonococcal type 2 IgA1 protease, but not denatured protein, induces release of proinflammatory cytokines such as TNF-α, IL-1β, IL-6, and IL-8 from peripheral blood mononuclear cells (Lorenzen et al., 1999). Furthermore, the same protease is capable of inhibiting TNF-α-mediated apoptosis of a human myelomonocytic cell line, an effect conceivably related to the concurrent cleavage by the protease of TNF-α receptor II (Beck and Meyer, 2000). Further evidence for proteases controlling cytokine networks is provided in Chapter 10.

The concept that IgA1 proteases play a role in infections, in addition to cleavage of IgA1, is strongly suggested by the finding that inactivation of IgA1 protease in *S. pneumoniae* leads to reduced virulence in a mouse infection model despite the fact that the protease does not cleave murine IgA (Polissi et al., 1998). Further studies are needed to fully elucidate the spectrum of potential substrates and other properties of IgA proteases. In this context, it will be important to examine each group of IgA protease including the two distinct cleavage specificities produced by strains of some pathogenic species (Fig. 5.1).

5.8 IgA1 PROTEASE FAMILY OF BACTERIAL PROTEINS

The serine type IgA1 proteases are the prototypes of a growing family of IgA1 protease-like extracellular proteins that generally appear to be involved in colonisation and invasion in a diverse group of Gram-negative pathogens (for review see Henderson et al., 1998). Among these proteins the Hap protein of *H. influenzae*, the Sep A protein of *Shigella flexneri*, the Ssp of *Serratia marcescens*, the Tsh protein of avian-pathogenic *E. coli*, the EspC protein of enteropathogenic *E. coli*, and the EspP protein of enterohaemorrhagic *E. coli* have putative serine protease motifs, which in some of the proteins is crucial for autoproteolytic cleavage like in the serine type IgA1 proteases (Pohlner et al., 1987). However, none of them appears to be able to cleave human IgA1.

Like *H. influenzae*, the *S. pneumoniae* genome contains several, but apparently variable numbers of genes with significant homology to its *iga* gene, one of which encodes a zinc metalloprotease that plays a crucial role for the regulated expression of surface proteins with choline-binding properties (Novak et al., 2000).

5.9 OTHER MICROBIAL PROTEASES WITH IMMUNOGLOBULIN-CLEAVING ACTIVITY

Several microorganisms produce broad-spectrum proteases that can degrade immunoglobulins, complement factors, and many other proteins involved in the protection of the human body (see Chapter 4). Broad spectrum proteases capable of cleaving immunoglobulins, including IgA, have been demonstrated in *Porphyromonas gingivalis, Pseudomonas aeruginosa, Proteus mirabilis, Staphylococcus aureus*, the yeast, *Candida albicans*, and the parasites *Entamoeba histolytica, Trichomonas vaginalis*, and *Paragonimus westermani* among many others (for review, see Kilian and Russell, 1999; Prokesová et al., 1992; Shin et al., 2001). However, to what extent these proteases are able to function in the presence of physiological protease inhibitors is not entirely clear, although some are capable of degrading protease inhibitors too. Neither are the consequences of potential cleavage, because the exact cleavage sites often have not been determined. Two interesting exceptions that may resemble IgA proteases by their biological consequence are the cysteine protease, SpeB, of *Streptococcus pyogenes*, which cleaves human IgG in the hinge region (between residues Gly236 and Gly237) and releases intact monomeric Fab fragments (Collin and Olsén, 2001), and the IgE-cleaving protease of *Shistosoma mansoni* (Pleass et al., 2000).

5.10 EFFECTS OF BACTERIAL GLYCOSIDASES

Almost all key molecules involved in the innate and adaptive immune system are glycoproteins. Thus, all five immunoglobulin classes and most of the complement components are glycosylated. These sugar compounds contribute to the stability and correct folding of the proteins and are involved in numerous recognition events that are crucial to the biological functions of the glycoproteins (Rudd et al., 2001). Carbohydrate residues on immunoglobulins play a decisive role for the affinity for their target antigen, complement activating properties, interactions with Fc receptor, and their net charge and hydrophilicity (Chuang and Morrison, 1997; Radaev and Sun, 2001; Rudd et al., 2001). IgA, and in particular S-IgA, are heavily glycosylated with both N-linked and O-linked glycan chains (Mattu et al., 1998) that may be attacked by bacterial glycosidases. *S. pneumoniae* and the oral streptococcal species *S. oralis* and *S. mitis* are capable of stripping IgA molecules of all their carbohydrate moieties *in vitro* (Kilian et al., 1980; Reinholdt et al., 1990). Other microorganisms selectively remove part of the oligosaccharide chains, such as the terminal sialic acid residues (Frandsen, 1994; Reinholdt et al., 1990).

Collin and Olsén (2001) recently demonstrated that a newly identified protein, EndoS, secreted by *S. pyogenes* is capable of hydrolysing the chitobiose core of N-linked carbohydrate side chains on IgG, when the bacteria are grown in the presence of human plasma. This activity may well be relevant also for IgA and other glycoproteins of the immune system (see Chapter 4).

It is conceivable that partial or complete hydrolysis of the glycosylation of IgA molecules and other immune factors by bacterial glycosidases can interfere with normal functions of the immune system during infections. However, the *in vivo* extent of such activities and their significance in host–parasite relationships have not been examined.

5.11 Ig-BINDING PROTEINS

Another potentially important immune escape mechanism is represented by immunoglobulin-binding surface proteins. Such proteins, which bind various immunoglobulin classes and components in a nonimmune way, are widespread among Gram-positive bacteria (Boyle, 1990). Although their biological roles are still incompletely understood, it has been demonstrated that the IgG-binding proteins (A) on *S. aureus* and (H) on *S. pyogenes* inhibit IgG-mediated complement activation on IgG-coated targets *in vitro* (Berge et al., 1997). Functionally analogous IgG-binding proteins (protein G) are present on group C and group G streptococci of human origin. A protein (protein L) expressed by *Peptostreptococcus magnus* binds to the variable region of κ light chains on all immunoglobulin classes (Boyle, 1990). Nonimmune Ig-binding surface proteins have also been demonstrated in several Gram-negative species associated with humans and various animal species, and some of these proteins show specificity for immunoglobulin molecules of their respective hosts (for review see Forsgren et al., 2001).

The IgG binding protein A of *S. aureus* also binds a fraction of IgA but via the Fab portion of molecules belonging to the $V_H III$ subgroup (Sasso et al., 1991). Likewise, several *Streptococcus* species express surface proteins that bind to IgA molecules in an antibody-independent manner. The most extensively studied are those expressed by some of the pyogenic streptococci. Protein Arp and the related protein Sir are surface molecules expressed by strains of *S. pyogenes*. These two proteins are members of the heterogeneous M protein family and bind human IgA of both subclasses including S-IgA (Frithz et al., 1989; Stenberg et al., 1994). A structurally and immunochemically unrelated IgA binding protein, the β protein ("Bac protein"), is expressed by *Streptococcus agalactiae*. Like the proteins in *S. pyogenes*, the β protein binds both IgA subclasses via Fcα but shows week affinity for S-IgA (Jerlström et al., 1991). Recent elegant studies demonstrate that binding of any of the three proteins to IgA blocks its interaction with the Fcα receptor (CD89) on neutrophils, eosinophils, and macrophages, and thereby interferes with an important effector mechanism of IgA (Pleass et al., 2001). Interestingly, both protein Arp and protein Sir bind also the C4b-binding protein (C4BP), which is an inhibitor of the classical pathway C3 convertase C4bC2a (Thern et al., 1995). Whether or not simultaneous IgA-binding plays a role in this context is unknown.

Both *Helicobacter pylori* and *S. pneumoniae* express surface proteins that bind S-IgA through the secretory component (Borén et al., 1993; Hammerschmidt et al., 1997). Studies of the binding specificities of *H. pylori* revealed that this gastric pathogen can interact with fucose residues on S-IgA molecules but not with serum IgA (Borén et al., 1993). The surface protein (SpsA) of *S. pneumoniae* that binds to SC is expressed by two thirds of strains and is surprisingly conserved between different serotypes (Hammerschmidt et al., 1997).

Similar interactions between bacterial lectins and various of the carbohydrate side chains of IgA have been detected in several species of *Enterobacteriaceae* including *E. coli* (Wold et al., 1990, 1994), and in the oral bacteria *S. gordonii* and *Actinomyces naeslundii* (Ruhl et al., 1996). It is not clear if these interactions represent antibody-independent protective activities of IgA or a bacterial strategy to evade IgA functions. Thus, the interaction between the IgA-binding type 1 fimbriae on *E. coli* and IgA results in agglutination of the bacteria and in inhibition of their attachment to colonic epithelial cells *in vitro* (Wold et al., 1990). Likewise, the attachment of *H. pylori* to gastric surface mucosa cells *in vitro* is inhibited by human colostral S-IgA in an α-L-fucosidase-sensitive manner (Falk et al., 1993). Conversely, the observation that selectively IgA-deficient individuals show reduced carriage in the gut of *E. coli* that express type 1 fimbriae (Friman et al., 1996) and that S-IgA in the acquired pellicle forming on tooth surfaces promotes attachment of some oral streptococci (Liljemark et al., 1979; Kilian et al., 1981) suggest that the advantage is on the bacterial side.

5.12 CONCLUSION

During co-evolution, commensal microorganisms and their respective hosts have evolved to maintain a balance that allows continuous mutualism. Many of the bacterial properties described in this chapter should probably be seen in this context. However, some of the properties are strikingly associated with a pathogenic phenotype, although there is little direct evidence for their significance in infections. IgA1 proteases are produced by both overt pathogens and by typical members of the resident microflora. Yet, it is hardly a coincidence that all three major causes of bacterial meningitis possess the same property in spite of the fact that these three bacterial species belong to distinct taxonomic groups. Recent studies indicate that IgA1 proteases are multifunctional proteins, but more detailed studies of IgA1 proteases with different cleavage specificity and from different bacteria are required to fully understand the significance and extent of such functions. The application of transgenic animal models expressing relevant receptors and immunoglobulins combined with appropriate deletion mutants of bacteria are likely to help in understanding the biological significance of IgA proteases and some of the other microbial properties described in this chapter.

REFERENCES

Ahl, T. and Reinholdt, J. (1991). Detection of immunoglobulin A1 protease-induced Fabα fragments on dental plaque bacteria. *Infection and Immunity* **59**, 563–569.

Beck, S.C. and Meyer, T.F. (2000). IgA1 protease from *Neisseria gonorrhoeae* inhibits TNFα-mediated apoptosis of human monocytic cells. *FEBS Letters* **472**, 287–292.

Berge, A., Kihlberg, B.-M., Sjöholm, A.G., and Björck, L. (1997). Streptococcal protein H forms soluble complement-activating complexes with IgG, but inhibits complement activation by IgG-coated targets. *Journal of Biological Chemistry* **272**, 20,774–20,781.

Biesbrock, A.R., Reddy, M.S., and Levine, M.J. (1991). Interaction of a salivary mucinsecretory immunoglobulin A complex with mucosal pathogens. *Infection and Immunity* **59**, 3492–3497.

Borén, T., Falk, P., Roth, K.A., Larson, G., and Normark, S. (1993). Attachment of *Helicobacter pylori* to human gastric epithelium mediated by blood group antigens. *Science* **262**, 1892–1895.

Boyle, M. D. P. (1990). ed. *Bacterial Immunoglobulin-Binding Proteins: Microbiology, Chemistry, Biology.* San Diego: Academic Press.

Bronson, R.A., Cooper, G.W., Rosenfeld, D.L., Gilbert, J.V., and Plaut, A.G. (1987). The effect of an IgA1 protease on immunoglobulins bound to the sperm surface and sperm cervical mucus penetrating ability. *Fertility and Sterility* **47**, 985–991.

Brooks, G.F., Lammel, C.J., Blake, M.S., Kusecek, B., and Achtman, M. (1992). Antibodies against IgA1 protease are stimulated both by clinical disease and asymptomatic carriage of serogroup A *Neisseria meningitidis. Journal of Infectious Diseases* **166**, 1316–1321.

Chuang, P.D. and Morrison, S.L. (1997). Elimination of N-linked glycosylation sites from the human IgA1 constant region: effects on structure and function. *Journal of Immunology* **158**, 724–732.

Cole, M.F., Evans, M., Fitzsimmons, S., Johnson, J., Pearce, C., Sheridan, M.J., Wientzen, R., and Bowden, G. (1994). Pioneer oral streptococci produce immunoglobulin A1 protease. *Infection and Immunity* **62**, 2165–2168.

Collin, M. and Olsén, A. (2001). EndoS, a novel secreted protein from *Streptococcus pyogenes* with endoglycosidase activity on human IgG. *The European Molecular Biology Organisation Journal* **20**, 3046–3055.

Cooper, M.D., McGhee, Z.A., Mulks, M.H., Koomey, J.M., and Hindman, T.L. (1984). Attachment to and invasion of human fallopian tube mucosa by an IgA1 protease-deficient mutant of *Nesseria gonorrhoeae* and its wild-type parent. *Journal of Infectious Diseases* **150**, 737–744.

Falk, P., Roth, K.A., Boren, T., Westblom, T.U., Gordon, J.I., and Normark, S. (1993). An in vitro adherence assay reveals that *Helicobacter pylori* exhibits cell lineage-specific tropism in the human gastric epithelium. *Proceedings of the National Academy of Sciences USA* **90**, 2035–2039.

Farley, M.M., Stephens, D.S., Mulks, M.H., Cooper, M.D., Bricker, J.V., Mirra, S.S., and Wright, A. (1986). Pathogenesis of IgA1 protease-producing and -nonproducing *Haemophilus influenzae* in human nasopharyngeal organ cultures. *Journal of Infectious Diseases* **154**, 752–759.

Forsgren, A., Brant, M., Möllenkvist, A., Muyombwe, A., Janson, H., Woin, N., and Riesbeck, K. (2001). Isolation and characterization of a novel IgD-binding protein from *Moraxella catarrhalis*. *Journal of Immunology* **167**, 2112–2120.

Frandsen, E.V.G. (1994). Carbohydrate depletion of immunoglobulin A1 by oral species of Gram-positive rods. *Oral Microbiology and Immunology* **9**, 352–358.

Frandsen, E.V.G., Reinholdt, J., and Kilian, M. (1995). In vivo cleavage of immunoglobulin A1 by immunoglobulin A1 proteases from *Prevotella* and *Capnocytophaga* species. *Oral Microbiology and Immunology* **10**, 291–296.

Friman, V., Adlerberth, I., Connell, H., Svanborg, C., Hanson, L.-A., and Wold, A.E. (1996). Decreased expression of mannose-specific adhesins by *Escherichia coli* in the colonic microflora of immunoglobulin A-deficient individuals. *Infection and Immunity* **64**, 2794–2798.

Frithz, E., Hedén, L.-O., and Lindahl, G. (1989). Extensive sequence homology between IgA receptor and M proteins in *Streptococcus pyogenes*. *Molecular Microbiology* **3**, 1111–1119.

Fujiyama, Y., Iwaki, M., Hodohara, K., Hosoda, S., and Kobayashi, K. (1986). The site of cleavage in human alpha chains of IgA1 and IgA2:A2m(1) allotype paraproteins by the clostridial IgA protease. *Molecular Immunology* **23**, 147–150.

Gilbert, J.V., Plaut, A.G., Longmaid, B., and Lamm, M.E. (1983). Inhibition of microbial IgA proteases by human secretory IgA and serum. *Molecular Immunology* **20**, 1039–1049.

Hajishengallis, G., Nikolova, E., and Russell, M.W. (1992). Inhibition of *Streptococcus mutans* adherence to saliva-coated hydroxyapatite by human secretory immunoglobulin A (S-IgA) antibodies to cell surface protein antigen I/II: reversal by IgA1 protease cleavage. *Infection and Immunity* 60, 5057–5064.

Halter, R., Pohlner, J., and Meyer, T.F. (1989). Mosaic-like organization of IgA protease genes in *Neisseria gonorrhoeae* generated by horizontal genetic exchange *in vivo*. *The European Molecular Biology Organisation Journal* 8, 2737–2744.

Hammerschmidt, S., Talay, S.R., Brandtzaeg, P., and Chatwal, G.S. (1997). SpsA, a novel pneumococcal surface protein with specific binding to secretory immunoglobulin A and secretory component. *Molecular Microbiology* 25, 1113–1124.

Hauck, C.R. and Meyer, T.F. (1997). The lysosomal/phagosomal membrane protein h-lamp-1 is a target of the IgA1 protease of *Neisseria gonorrhoeae*. *FEBS Letters* 405, 86–90.

Hedges, S.R., Mayo, M.S., Kallman, L., Mestecky, J., Hook III, E.W., and Russell, M.W. (1998). Evaluation of immunoglobulin A1 (IgA1) protease and IgA1 protease-inhibitory activity in human female genital infection with *Neisseria gonorrhoeae*. *Infection and Immunity* 66, 5826–5832.

Henderson, I.R., Navarro-Garcia, F., and Nataro, J.P. (1998). The great escape: structure and function of the autotransporter proteins. *Trends in Microbiology* 6, 370–378.

Hohwy, J., Reinholdt, J., and Kilian, M. (2001). Population kinetics of *Streptococcus mitis* in its natural habitat. *Infection and Immunity* 69, 6055–6063.

Hopper, S., Vasquez, B., Merz, A., Clary, S., Wilbur, J.S., and So, M. (2000). Effects of the IgA1 protease on *Neisseria gonorrhoeae* trafficking across polarized T84 epithelial monolayers. *Infection and Immunity* 68, 906–911.

Jarvis, G.A. and Griffiss, J.M. (1991). Human IgA1 blockade of IgG-initiated lysis of *Neisseria meningitidis* is a function of antigen-binding fragment binding to the polysaccharide capsule. *Journal of Immunology* 147, 1962–1967.

Jerlström, P.G., Chhatwal, G.S., and Timmis, K.N. (1991). The IgA-binding beta antigen of the c protein complex of group B streptococci: sequence determination of its gene and detection of two binding regions. *Molecular Microbiology* 5, 843–849.

Johannsen, D.B., Johnston, D.M., Koymen, H.O., Cohen, M.S., and Cannon, J.G. (1999). A *Neisseria gonorrhoeae* immunoglobulin A1 protease mutant is infectious in the human challenge model of urethral infection. *Infection and Immunity* 67, 3009–3013.

Kapatais-Zoumbos, K., Chandler, D.K., and Barile, M.F. (1985). Survey of immunoglobulin A protease activity among selected species of *Ureaplasma* and

Mycoplasma specificity for host immunoglobulin A. *Infection and Immunity* **47**, 704–709.

Kett, K., Brandtzaeg, P., Radl, J., and Haaijman, J.T. (1986). Different subclass distribution of IgA-producing cells in human lymphoid organs and various secretory tissues. *Journal of Immunology* **136**, 3631–3635.

Kilian, M., Husby, S., Høst, A., and Halken, S. (1995). Increased proportions of bacteria capable of cleaving IgA1 in the pharynx of infants with atopic disease. *Pediatric Research* **38**, 182–186.

Kilian, M., Mestecky, J., Kulhavy, R., Tomana, M., and Butler, W.T. (1980). IgA1 proteases from *Haemophilus influenzae*, *Streptococcus pneumoniae*, *Neisseria meningitidis*, and *Streptococcus sanguis*: comparative immunochemical studies. *Journal of Immunology* **124**, 2596–2600.

Kilian, M., Mestecky, J., and Schrohenloher, R.E. (1979). Pathogenic species of *Haemophilus* and *Streptococcus pneumoniae* produce immunoglobulin A1 protease. *Infection and Immunity* **26**, 143–149.

Kilian, M. and Reinholdt, J. (1987). A hypothetical model for the development of invasive infection due to IgA1 protease-producing bacteria. *Advances in Experimental Medicine and Biology* **216B**, 1261–1269.

Kilian, M., Roland, K., and Mestecky, J. (1981). Interference of secretory immunoglobulin A with sorption of oral bacteria to hydroxyapatite. *Infection and Immunity* **31**, 942–951.

Kilian, M. and Russell, M.W. (1999). Microbial evasion of IgA functions. In *Handbook of Mucosal Immunology*, 2nd edn, ed. P.L. Ogra, J. Mestecky, M.E. Lamm, W. Strober, J.B. Bienenstock and J.R. McGhee, pp. 241–251. San Diego: Academic Press.

Kilian, M. and Thomsen, B. (1983). Antigenic heterogeneity of immunoglobulin A1 proteases from encapsulated and non-encapsulated *Haemophilus influenzae*. *Infection and Immunity* **42**, 126–132.

Kirkeby, L., Rasmussen, T.T., Reinholdt, J., and Kilian, M. (2000). Immunoglobulins in nasal secretions of healthy humans: Structural integrity of secretory immunoglobulin A1 (IgA1) and occurrence of neutralizing antibodies to IgA1 proteases of nasal bacteria. *Clinical and Diagostic Laboratory Immunology* **7**, 31–39.

Kobayashi, K., Fujiyama, Y., Hagiwara, K., and Kondoh, H. (1987). Resistance of normal serum IgA and secretory IgA to bacterial IgA proteases: evidence for the presence of enzyme-neutralizing antibodies in both serum and secretory IgA, and also in serum IgG. *Microbiology and Immunology* **31**, 1097–1106.

Liljemark, W.F., Bloomquist, C.G., and Ofstehage, J.C. (1979). Aggregation and adherence of *Streptococcus sanguis*: role of human salivary immunglobulin A. *Infection and Immunity* **26**, 1104–1110.

Lin, L., Ayala, P., Larson, J., Mulks, M., Fukuda, M., Carlsson, S.R., Enns, C., and So, M. (1997). The *Neisseria* type 2 IgA1 protease cleaves LAMP1 and promotes survival of bacteria within epithelial cells. *Molecular Microbiology* **24**, 1083–1094.

Lomholt, H. (1995). Evidence of recombination and an antigenically diverse immunoglobulin A1 protease among strains of *Streptococcus pneumoniae*. *Infection and Immunity* **63**, 4238–4243.

Lomholt, H., Poulsen, K., and Kilian, M. (1995). Comparative characterization of the *iga* gene encoding IgA1 protease in *Neisseria meningitidis*, *Neisseria gonorrhoeae* and *Haemophilus influenzae*. *Molecular Microbiology* **15**, 495–506.

Lomholt, H., van-Alphen, L., and Kilian, M. (1993). Antigenic variation of immunoglobulin A1 proteases among sequential isolates of *Haemophilus influenzae* from healthy children and patients with chronic obstructive pulmonary disease. *Infection and Immunity* **61**, 4575–4581.

Lorenzen, D.R., Düx, F., Wölk, U., Tsirpouchtsidis, A., Hass, G., and Meyer, T.F. (1999). Immunogloubulin A1 protease, an exoenzyme of pathogenic *Neisseriae*, is a potent inducer of proinflammatory cytokines. *Journal of Experimental Medicine* **190**, 1049–1058.

Male, C. (1979). Immunoglobulin A1 protease production by *Haemophilus influenzae* and *Streptococcus pneumoniae*. *Infection and Immunity* **26**, 254–261.

Mattu, T.S., Pleass, R.J., Willis, A.C., Kilian, M., Wormald, M.R., Lellouch, A.C., Rudd, P.M., Woof, J.M., and Dwek, R.A. (1998). The glycosylation and structure of human serum IgA1, Fab and Fc regions and the role of N-glycosylation on FcαR interactions. *Journal Biological Chemistry* **273**, 2260–2272.

Mestecky, J. and Russell, M.W. (1986). IgA subclasses. *Monographs in Allergy* **19**, 277–301.

Meyer, T.F., Pohlner, J., and Putten, J.P. (1994). Biology of the pathogenic Neisseriae. *Current Topics in Microbiology and Immunology* **192**, 283–317.

Morelli, G., del Valle, J., Lammel, C.J., Pohlner, J., Muller, K., Blake, M., Brooks, G.F., Meyer, T.F., Koumare, B., Brieske, N., et al. (1994). Immunogenicity and evolutionary variability of epitopes within IgA1 protease from serogroup A *Neisseria meningitidis*. *Molecular Microbiology* **11**, 175–187.

Mulks, M.H., Plaut, A.G., and Lamm, M. (1980). Gonococcal IgA protease reduces inhibition of bacterial adherence by human secretory IgA. In *Genetics and Immunobiology of Pathogenic Neisseria*, ed. S. Normark and D. Danielsson, pp. 217–220. University of Umeå: Umeå, Sweden.

Novak, R., Charpentier, E., Braun, J.S., Park, E., Murti, S., Tuomanen, E., and Masure, R. (2000). Extracellular targeting of choline-binding proteins in *Streptococcus pneumoniae* by a zinc metalloprotease. *Molecular Microbiology* **36**, 366–376.

Ogra, P.L., Mestecky, J., Lamm, M.E., Strober, W., Bienenstock, J., and McGhee, J.R. (1999). *Mucosal Immunology*, 2nd edn. San Diego: Academic Press.

Plaut, A.G., Genco, R.J., and Tomasi, Jr. T.B. (1974). Isolation of an enzyme from *Streptococcus sanguis* which specifically cleaves IgA. *Journal of Immunology* **113**, 289–291.

Plaut, A.G., Gilbert, J.V., Artenstein, M.S., and Capra, J.D. (1975). *Neisseria gonorrhoeae* and *Neisseria meningitidis*: extracellular enzyme cleaves human immunoglobulin A. *Science* **193**, 1103–1105.

Plaut, A.G., Qiu, J., Grundy, F., and Wright, A. (1992). Growth of *Haemophilus influenzae* in human milk: synthesis, distribution and activity of IgA protease as determined by study of *iga*+ and mutant *iga*-cells. *Journal of Infectious Diseases* **166**, 43–52.

Pleass, R.J., Areschoug, T., Lindahl, G., and Woof, J.M. (2001). Streptococcal IgA-binding proteins bind in the C2-C3 interdomain region and inhibit binding of IgA to human CD89. *Journal of Biological Chemistry* **276**, 8197–8204.

Pleass, R.J., Kusel, J.R., and Woof, J.M. (2000). Cleavage of human IgE mediated by *Schistosoma mansoni*. *International Archives of Allergy and Immunology* **121**, 194–204.

Polissi, A., Pontiggia, A., Feger, G., Altieri, M., Mottl, H., Ferrari, L., and Simon, D. (1998). Large-scale identification of virulence genes from *Streptococcus pneumoniae*. *Infection and Immunity* **66**, 5620–5629.

Pohlner, J., Halter, R., Beyreuther, K., and Meyer, T.F. (1987). Gene structure and extracellular secretion of *Neisseria gonorrhoeae* IgA protease. *Nature* **325**, 458–462.

Poulsen, K., Reinholdt, J., and Kilian, M. (1992). A comparative genetic study of serologically distinct *Haemophilus influenzae* type 1 immunoglobulin A1 proteases. *Journal of Bacteriology* **174**, 2913–2921.

Poulsen, K., Reinholdt, J., Jespersgaard, C., Boye, K., Brown, T.A., Hauge, M., and Kilian, M. (1997). A comprehensive genetic study of streptococcal IgA1 protease: Evidence for recombination within and between species. *Infection and Immunity* **66**, 181–190.

Prokesová, L., Potuzniková, B., Potempa, J., Zikán, J., Radl, J., Hachova, L., Baran, K., Porwit-Bobr, Z., and John, C. (1992). Cleavage of human immunoglobulins by serine proteinase from *Staphylococcus aureus*. *Immunology Letters* **31**, 259–266.

Qiu, J., Brackee, G.P., and Plaut, A.G. (1996). Analysis of the specificity of bacterial immunoglobulin A (IgA) protease by a comparative study of ape serum IgAs as substrates. *Infection and Immunity* **64**, 933–937.

Qiu, J., Hendrixson, D.R., Baker, E.N., Murphy, T.F., St. Geme III, J.W., and Plaut, A.G. (1998). Human milk lactoferrin inactivates two putative colonization factors expressed by *Haemophilus influenzae*. *Proceedings of the National Academy of Sciences USA* **95**, 12,641–12,646.

Radaev, S. and Sun, P.D. (2001). Recognition of IgG by Fcγ receptor. *Journal of Immunology* **276**, 16,478–16,483.

Reinholdt, J. and Kilian, M. (1987). Interference of IgA protease with the effect of secretory IgA on adherence of oral streptococci to saliva-coated hydroxyapatite. *Journal of Dental Research* **66**, 492–497.

Reinholdt, J. and Kilian, M. (1997). Comparative analysis of immunoglobulin A1 protease activity among bacteria representing different genera, species, and strains. *Infection and Immunity* **65**, 4452–4459.

Reinholdt, J., Friman, V., and Kilian, M. (1993). Similar proportions of immunoglobulin A1 (IgA1) protease-producing streptococci in initial dental plaque of selectively IgA-deficient and normal individuals. *Infection and Immunity* **61**, 3998–4000.

Reinholdt, J., Tomana, M., Mortensen, S.B., and Kilian, M. (1990). Molecular aspects of immunoglobulin A1 degradation by oral streptococci. *Infection and Immunity* **58**, 1186–1194.

Rudd, P.M., Elliott, T., Cresswell, P., Wilson, I.A., and Dwek, R.A. (2001). Glycosylation and the immune system. *Science* **291**, 2370–2376.

Ruhl, S., Sandberg, A.L., Cole, M.F., and Cisar, J.O. (1996). Recognition of immunoglobulin A1 by oral actinomyces and streptococcal lectins. *Infection and Immunity* **64**, 5421–5424.

Russell, M.W., Kilian, M., and Lamm, M.E. (1999). Biological activities of IgA. In *Mucosal Immunology*, 2nd edn., ed. P.L. Ogra, J. Mestecky, M.E. Lamm, W. Strober, J.B. Bienenstock, and J.R. McGhee, pp. 225–240. San Diego: Academic Press.

Russell, M.W., Reinholdt, J., and Kilian, M. (1989). Anti-inflammatory activity of human IgA antibodies and their Faba fragments: inhibition of IgG-mediated complement activation. *European Journal of Immunology* **19**, 2243–2249.

Saltzman, W.M., Radomsky, M.L., Whaley, K.J., and Cone, R.A. (1994). Antibody diffusion in human cervical mucus. *Biophysical Journal* **66**, 508–515.

Sasso, E.H., Silverman, G.J., and Mannik, M. (1991). Human IgA and IgG F(ab′)$_2$ that bind to staphylococcal protein A belong to the V_HIII subgroup. *Journal of Immunology* **147**, 1877–1883.

Senda, S., Fujiyama, Y., Ushijima, T., Hodohara, K., Bamba, T., Hosoda, S., Kobayashi, K. (1985). *Clostridium ramosum*, an IgA protease-producing species and its ecology in the human intestinal tract. *Microbiology and Immunology* **29**, 203–207.

Shin, M.H., Kita, H., Park, H.Y., and Seoh, J.Y. (2001). Cysteine protease secreted by *Paragonimus westermani* attenuates effector functions of human eosinophils stimulated with immunoglobulin G. *Infection and Immunity* **69**, 1599–1604.

Sørensen, C.H. and Kilian, M. (1984). Bacterium-induced cleavage of IgA in nasopharyngeal secretions from atopic children. *Acta Pathologica, Microbiologica, and Immunologica Scandinavica, Section C* **92**, 85–87.

Stenberg, L., O'Toole, P.W., Mestecky, J., and Lindahl, G. (1994). Molecular characterization of protein Sir, a streptococcal cell surface protein that binds both IgA and IgG. *Journal of Biological Chemistry* **269**, 13,458–13,464.

Thern, A., Stenberg, L., Dahlbäck, B., and Lindahl, G. (1995). Ig-binding surface proteins of *Streptococcus pyogenes* also bind human C4b-binding protein (C4BP), a regulatory component of the complement system. *Journal of Immunology* **154**, 375–386.

van Egmond, M., Damen, C.A., van Spriel, A.B., Vidarsson, G., van Garderen, E., and van de Winkel, J.G.J. (2001). IgA and the IgA Fc receptor. *Trends in Immunology* **22**, 205–211.

Vitovski, S., Read, R.C., and Sayers, J.R. (1999). Invasive isolates of *Nesseria meningitidis* possess enhanced immunoglobulin A1 protease activity compared to colonizing strains. *FASEB Journal* **13**, 331–337.

Wold, A.E., Motas, C., Svanborg, C., and Mestecky, J. (1994). Lectin receptors on IgA isotypes. *Scandinavian Journal of Immunology* **39**, 195–201.

Wold, A., Mestecky, J., Tomana, M., Kobata, A., Ohbayashi, H., Endo, T., and Svanborg Edén, C. (1990). Secretory immunoglobulin A carries oligosaccharide receptors for *Escherichia coli* type 1 fimbrial lectin. *Infection and Immunity* **58**, 3073–3077.

CHAPTER 6

Evasion of antibody responses: Bacterial phase variation

Nigel J. Saunders

6.1 INTRODUCTION

The generation of diversity within bacterial populations is important in the evolution and development of bacterial species and also in the adaptability of bacteria to their changing environments. Diversity is generated by a combination of programmed and random events that occur at different rates and confer different types of variability on the population. At one extreme there are random point mutations that occur throughout the coding and intergenic sequences that alter the expression, structure, and function of bacterial components. At the other extreme, there are regulated responses that allow bacteria to control the expression of genes whenever the appropriate environmental conditions are encountered. Between these there is a variety of processes that adds to the capacity of a population to diversify, including the presence and movement of insertion sequences that affect expression, mobile genetic elements that can move within and between populations, and the horizontal transfer of DNA between individual bacteria. One process that lies between the mutations that occur randomly throughout the genome and the programmed regulation of environmentally responsive genes is phase variation. This process involves alterations in the cell at the level of DNA but in a way that generates predictable and predetermined adaptability for the bacterial population.

6.2 SECTION 1: PHASE VARIATION, ITS CHARACTERISTICS AND HOW IT WORKS

Phase variation has been recognised as a process associated with diversification since the early days of medical bacteriology (Andrewes, 1922). There are several examples, described in this chapter, that have been extensively investigated, including the variation in flagella in *Salmonella* spp., the major antigens of *Borrelia recurrentis* in the relapsing fevers, and some surface components of *Haemophilus influenzae* and *Neisseria* spp. However, the range of bacterial species and the number of genes that display phase variation are much broader than this, and phase variation is recognised to be a common mechanism capable of generating enormous diversity within clonal populations of many bacterial species. While different species make use of genetic switches that are mechanistically unrelated, the phenotypic consequences are the same. For example, coliform bacteria tend to use sequence inversions in promoter regions while Gram-negative, naturally transformable, species tend to use instability within simple sequence repeats. Species recognised to use phase variation include *Bordetella* spp., *Borrelia* spp., *Campylobacter jejuni*, *Citrobacter* spp., *Escherichia coli*, *Haemophilus* spp., *Helicobacter pylori*, *Klebsiella pneumoniae*, *Moraxella* spp., *Mycoplasma* spp., *Neisseria* spp., *Pasteurella haemolytica*, *Proteus mirabilis*, *Pseudomonas atlantica*, *Salmonella* spp., *Serratia marcescens*, *Vibrio cholerae*, and *Yersinia pestis*.

Phase variation describes a process of reversible switching between phenotypes, mediated by a genetic reorganisation, mutation, or modification, which is not associated with a loss of coding potential, and which occurs at a comparatively high frequency. This results in the continuous generation of alternative phenotypes within a clonal population that facilitate adaptation to changing environmental conditions. Genes that undergo phase variation have been called "contingency genes" (Moxon et al., 1994), which emphasises the evolutionary and functional implications of the variability of this subset of hypermutable genes.

Contingency genes are quite difficult to define but they share a number of features:

    i. Their expression is reversibly variable.

    ii. The mechanism of switching involves a genetic mutation, rearrangement, or modification resulting in a structural change in the DNA.

    iii. The switching process occurs stochastically.

    iv. The switching process occurs at a rate higher than the point mutation rate.

    v. Their variation generates sub-populations within a clonal population, usually providing adaptations to different microenvironmental conditions, upon which selection acts to determine the subsequent population structure with respect to each characteristic.

vi. They permit a clonal bacterial population to explore a variety of phenotypic solutions to the various microenvironmental conditions that they encounter, while leaving the majority of the population, which is already established, unchanged.

vii. They often encode determinants of surface structures that interface with the environment and are frequently under immunological pressures.

The switching processes are an unusual mixture of programmed and random events. The elements that undergo mutation or reorganisation do so randomly in the sense that the generation at which the event will occur cannot be predicted. However, the phenomenon is predictable because it is an inevitable consequence of the presence and behaviour of genetically unstable structures. It is even possible for the rate and direction of the event to be influenced, as is the case for the inversion event of a section of the pilin gene of *Salmonella* spp. and *Escherichia coli* in which environmental conditions affect the behaviour of the switch while the timing of the event remains essentially stochastic.

The use of the term "stochastic" in this context (while technically synonymous with random) attempts to convey the concept that whereas the actual mutations that mediate the switching occur in most situations at random, the genes that are switched in this way are predetermined. So, although the genes that will switch can be predicted, which genes in which cells and in what order cannot be. In this sense it can be considered as a "programmed random" behaviour.

In the context of immune evasion this has several aspects. First, the gene that is switched may be directly involved in some form of immune evasion mechanism. Second, a varied gene may be a prominent immunological target and the switching of the gene can mediate immune evasion by allowing the organism to evade learned antibody responses. Third, the varied genes can affect the behaviour of the organisms such that it can gain access to immunologically privileged sites, such as the intracellular compartment. Fourth, genes that are varied may allow a colonizing population to explore a number of phenotypes that include epitopes that will be seen as "self" by the host and evade immune recognition in this fashion. Finally, it can allow colonizing bacteria to evade preexisting immune responses raised against related or unrelated bacteria with similar antigenic properties.

It is actually striking that the role of phase variation in immune evasion is not as well established experimentally as its importance would suggest. This is probably a result of a combination of factors. It is comparatively straightforward to study the consequences of a stable phenotype, and in this context the different properties of cells that are expressing variable components have been assessed. It is more problematic, and there has been an insufficient theoretical framework, to investigate unstable traits. Also, when the switching of a gene that alters a bacterial behaviour, such as a binding specificity or other phenotypically important features, is also involved in immune evasion, it is not trivial to be able to ascribe the fitness advantages that are conferred by each of the selection pressures. It is therefore necessary to put together a picture of the role of phase variation in immune evasion by pulling together a body of information from diverse sources of bacterial systems that illustrate its various aspects.

6.2.1 The capacity to generate diversity

In order to appreciate the potential contribution of phase variation to bacterial population behaviour, it is necessary to recognise the scale of its capacity to generate diversity. The first genome scale analysis to identify the complete complement of phase variable genes that allows this to be addressed was done using the *Haemophilus influenzae* strain Rd genome sequence (Hood et al., 1996). In this species the contribution of repeat tract length variation, particularly in simple sequence repeats composed of tetramers, was well established. The investigators used a method that looked for each potential tri-, tetra-, and pentameric repeat sequence and to perform effectively an *in silicio* Southern blot analysis of the complete genome sequence. Using this approach, twelve genes were identified with characteristic features of phase variability. Each switch in each gene is entirely independent of other similar switches, and theoretically these genes could be expressed in any combination. This generates a potential phenotypic diversity of 2^{12} (around 4,000) phenotypes. There is also evidence of a small number of additional phase variable genes in other strains. A more comprehensive search methodology was subsequently developed and used to assess the complete genome sequences of *H. pylori* strain 2669 (Saunders et al., 1998), and *N. meningitidis* strain MC58 (Saunders et al., 2000; Tettelin et al., 2000). These searches revealed twenty-six and forty-four genes that are very likely to be phase variable. These could theoretically generate around 67 million, and 17 trillion combinations of expressed genes, respectively. These figures are certainly overestimates of the number of phenotypes that are actually expressed. A number of the phase-variable genes are frequently associated with restriction modification systems that would not be expected to affect host–bacterium interactions. A number of genes are part of related biosynthetic pathways such that the phenotype associated with one gene is dependent upon the expression of others. However, it is in no way unreasonable to consider that some of these species are capable of expressing many thousands and perhaps millions of phenotypes that directly influence host interactions and in antigenic variation that mediates immune evasion.

6.2.2 Stability in the presence of instability

In order to understand the role of phase variation, it is necessary to consider the rates at which it occurs. It has to be sufficiently high to generate the phenotypes that are likely to confer adaptive advantages within the colonizing bacterial population. It also has to be sufficiently high that genes that are switched OFF are not subject to other mutations that disrupt the function of the gene before they are switched ON again. On the other hand, it has to be sufficiently slow not to continuously disrupt the phenotype of a well-adapted stable colonizing bacterial population, a point that becomes particularly relevant in bacterial species that have many phase-variable genes.

It has been proposed that there is an optimal rate of phase variation that occurs as the reciprocal of the frequency of transition between the alternate environments for which the varied phenotype is adaptive (Moxon et al., 1994). It can also be argued that during certain critical steps of transmission and colonisation, for example in a small inoculum that has to establish infection, diversification at a high rate that permits colonisation may be adaptive. The principal requirement for success for a microbial parasite is to survive as a propagating population in the host (Finlay and Falkow, 1989). The energy expended in this process or even the proportion of the population that is lost is irrelevant, as long as the residual population can expand to survive and be transmitted to new hosts (Wise, 1993).

There is some confusion in the literature about what is actually measured and described as a rate of phase variation, which has led to some overestimation of the actual rates (Saunders et al., 2001a). Rates as high as 10^{-2} are probably not typical of phase variation. When the data can be addressed directly and analysed with methods specifically to address this type of rapidly switching, reversible system rates are typically in the range of 10^{-3} to 10^{-5} per generation per cell. It is possible to extend these models to address the influences of phase-variation rate and selective advantages on population composition (Saunders et al., 2001b). These reveal that these lower rates of variation are consistent with a biologically interesting balance between flexibility and stability for a bacterial population. Conclusions about the influence of phase variation using these models can be summarized as follows:

1. Over time, in the absence of selection, phase-variable populations will tend toward an equilibrium state that is proportionate to the switching rates for each phenotype.

2. In the absence of selection, the time taken until changes in population phenotype composition occur is determined by the variation rate.
3. Phase variation, at the rates observed *in vitro*, will result in relatively stable phenotypic population composition in the absence of selection over biologically relevant time periods.
4. The rate of change in the population composition is largely determined by the relative fitness of the alternate phenotypes.
5. When fitness differences are present, the proportion of the "residual" phenotype increases with the phase-variation rate.

The third and fourth points are of central importance. A population will be stable over biologically relevant numbers of generations, in the absence of selection. When a change in environmental conditions occurs, the presence of relatively small numbers of variants will be sufficient to allow the population to respond.

Once these points are appreciated one can see how this type of system can work, albeit in a slightly counter intuitive fashion. For example, if one considers any one of a number of recognised systems in which a large number of antigenic surface structures are phase varied, in either a bacterial or a parasitic context, it might seem as if an organism is progressively displaying its surface repertoire to the host. However, once the rate of variation and proportion of the actual alternate populations is considered, this is not actually a problem for the pathogen. There are limitations of epitope restriction in the immune response such that it cannot focus on a thousand things at once. Until the predominant clonal family in an infection has a selective pressure mounted against it by the immune system – which of course takes time – then the minor populations will remain very minor and will not become the focus of the learned immune responses. They will however be immediately available once an immune response occurs (see the section on *Borrelia* below).

6.3 SECTION 2: EXAMPLES OF THE ROLE OF PHASE VARIATION IN IMMUNE EVASION

6.3.1 Phase variation in *Borrelia*

One of the earliest and most dramatic consequences of phase variation during infection is illustrated in the *Borrelia* spp. Antigenic variations in *Borrelia* spp. involve the reversible selection of an expressed phenotype of its major surface proteins called variable major proteins (VMPs) from a repertoire of possible genes within each strain. The consequences of this variation are most dramatically evident in the relapsing fevers. In relapsing fevers, the infected individual experiences periods of fever that are interspaced by intervals of well-being. When the fever occurs, large numbers of spirochetes can be found in blood smears and the borreliae disappear as the patient responds to them with specific antibodies. The waxing and waning of the borreliae populations is associated with the antigenic variation of the VMPs in the population within the host (Meleney, 1928). A studied strain of B. *hermsii*, an agent of relapsing fever, was found to express twenty-six antigenic variants in a mouse model with a rate of switching estimated to be 10^{-3} to 10^{-4} (Stoenner et al., 1982; Barbour et al., 1982). VMPs differ in their molecular weights, peptide maps, and reactives with serotype-specific antibodies (Barbour et al., 1982, 1983). Importantly, a serotype eliminated by neutralising antibodies from a first host may reappear in the populations in a nonimmune second host (Meleney, 1928; Coffey and Eveland, 1967), demonstrating the reversibility of the variation process.

Interestingly, this is an example of phase variation that is mediated by recombination, as are similar types of variation in other *Borrelia* spp. As emphasised previously, phase variation is a process that has to be reversible. When allelic variation occurs through recombination, this is frequently not the case. For example, the recombination between the silent pilus gene cassettes and the expressed pilus gene in *Neisseria gonorrhoeae* and *Neisseria meningitidis* cannot be considered to be a form of phase variation. In this case, the recombination, while continuously generating variants does so in such a way that it also generates continuous change in the pilus coding sequence. It is possible that the nonreciprocal recombination may recreate previous phenotypes but the process is not such that this can be reasonably predicted. In the *Borrelia* spp., there is a substantial region of identity in the 5′ region of the genes within which recombination occurs – and then the whole of the remainder of the gene is exchanged. In this way, the phenotypic variation can recapitulate the previous phenotypes, as is thought to occur each time a new host is infected, and therefore fulfil the reversibility criterion.

6.3.2 The association between phase variation and antigenic variation

Probably as a reflection of the types of gene that are phase varied, i.e., those that interact with the host directly and are the targets of immune responses, there is a close link between genes that are antigenically variable and those that are phase varied. The combined effect of these two processes leads to an exponential increase in the diversity of cell surface structures that can be generated by a population derived from a single infecting clone.

Phase variability and antigenic variability are frequently combined to greatly increase the potential for surface diversification in a bacterial population. *Mycoplasma* spp. have no cell wall and no LPS containing outer membrane. Their single membrane presents a unique surface for interaction with their environment and host. As a group, the mycoplasmas share a spectrum of phase and antigenically variable structures likely to play a role in both bacterially directed interactions with the host and immune evasion. The various *Mycoplasma* spp. have functionally similar, but mechanistically varied, systems of this type.

The first study of phase variation in *Mycoplasma* reported high frequency, reversible, changes in colony morphology, opacity and expression of a lipoprotein in *Myc. hyorrhinis* (Rosengarten and Wise, 1990). *Myc. hyorrhinis* is a species found in the respiratory tract of pigs and can cause rhinitis and arthritis. *Myc. hyorrhinis* surface proteins include a group of three lipid modified proteins, VlpA, VlpB, and VlpC, which undergo phase variation and size variation resulting from duplication, deletion, and recombination within a highly repetitive C-terminal region (Rosengarten and Wise, 1990, 1991; Yogev et al., 1991). Vlps have been shown to be targets for immune damage to mycoplasmas, to be involved in interactions with host cells (Rosengarten and Wise, 1991) and both mechanisms of variation generate escape variants from growth-inhibiting antibodies (Citti et al., 1997). The combinatorial effect of size and phase variation has been estimated to be able to generate over 10^4 structural permutations of Vlps. Further, that estimate was made prior to the demonstration that individual strains could have six or seven *vlp* genes (Yogev et al., 1995) and does not consider the possibility of recombination between them.

Mycoplasma bovis (Myc. bovis) is a species that usually exists as a harmless commensal in the respiratory tract and can also be isolated from the milk of healthy cattle. It also causes bovine mastitis, arthritis, pneumonia, subcutaneous abscesses, meningitis, and infertility. *Myc. bovis* has the capacity to express a repertoire of variable surface expressed lipoproteins called Vsps. These are different from the Vlps described above (Behrens et al., 1994). This variability includes both size and antigenic variation, resulting from alterations in the number of repetitive structural units of different sizes located predominantly at the exposed C-terminal end of the proteins, as well as phase variation of expression (Rosengarten et al., 1994). Infection studies demonstrated that there is substantial variation in the surface antigens expressed in serial isolates from single animals, and *Myc. bovis*-infected cattle develop strong preferential antibody reactions to these proteins (Rosengarten et al., 1994) demonstrating that the variability that can be detected *in vitro* occurs during natural infection and that this is potentially a means of immune evasion. *In vitro* the phase variation of each Vsp was independent. In addition to the capacity to vary the expression of Vsps, *Myc. bovis* colony immuno-staining exhibits additional phase variation that is not due to size variation or to altered expression of the proteins as detected by Western blotting, suggesting the presence of a phase-variable process that can mask Vsp epitopes (Behrens et al., 1994). The Vsps of *Myc. bovis* and the Vlps of *Myc. hyorhinis* share several features: (1) they are both anchored to the surface as lipoproteins, (2) there are some shared epitopes, (3) they both have a surface exposed C-terminal region with an extensive and size variable structure, (4) they exhibit independent phase variation with the capacity for combinatorial expression, and (5) they are the major antigens on the membrane surface. However, there are also differences: (1) carboxypeptidase digestion reveals a digest pattern that shows a regular pattern in Vlps (suggesting tandem repeats of very similar subunits) and an irregular pattern in Vsps (suggesting sets of multiple sets of nonsimilar repeats), (2) Vsps are much more resistant to trypsin digestion, and (3) Vsps are able to alter their size by variations in sites other than the C-terminal end. It is not possible to determine, on the basis of the available evidence, whether this represents great divergence in a common ancestral system or whether it represents an example of convergent evolution within two *Mycoplasma* spp. Either alternative suggests that phase variation is important in the generation of diversity in *Mycoplasma* spp. and in host interactions.

6.3.3 *Haemophilus* LPS variation *in vivo*

Lipopolysaccharide (LPS) is the predominant constituent of the outer membrane of Gram-negative bacteria (Nikaido, 1996) and is therefore one of the bacterial structures that is available to mediate interactions between bacterium and host. *H. influenzae* has an LPS that lacks the long O-side chain that comprises the O-antigen of most enterobacteriaceae. Therefore, it does not have the capacity to mask its core LPS structures by expressing a variety of O-side chains, as seen for example in *Salmonella* spp. and *E. coli*. Instead, it is the sugars and substitutions of the LPS core that are available to interact with the host. There is variability in these structures between strains and also within a single strain and there is both coordinate and independent phase variation of multiple LPS epitopes as indicated by binding to anti-LPS monoclonal antibodies (Kimura and Hansen, 1986; Weiser et al., 1989). The study of the function of individual phenotypes is complicated by the extent of the LPS variability. However, there is some evidence that particular phenotypes affect survival *in vivo* (Kimura and Hansen, 1986) and susceptibility to serum killing (Gilsdorf and Ferrieri, 1986), and that isolates from the nasopharynx and systemic sites often differ in LPS phenotype (Mertsola et al., 1991).

Demonstration of the presence and biological role of phase variation in the more complex multigene systems, such as LPS biosynthetic genes, is problematic. This is not surprising when one considers that no two colonies on a plate will have an identical phenotypic composition, and therefore the starting material in no two experiments will ever be the same. However, there is strong evidence that these systems are variable and important *in vivo* from studies of similar variable genes in animals.

Variation in the LPS of *H. somnus* is known to involve repeat associated genes in a fashion similar to that present in *H. influenzae* (Inzana et al., 1997). Phase variation of these LPS phenotypes was observed to occur in a calf infection model (Inzana et al., 1992). Lung-infected cattle tended to clear infections over a 10-week period during which marked fluctuations in the bacterial colonization were observed. Apparently random changes in the LPS profiles were observed to occur rapidly (detected by weekly sampling) and were common. This contrasted with relative stability during repeated *in vitro* subculture. Sequential changes in antibody reactivity were observed and although each calf reacted differently, the generation of specific immune responses followed by the counter-selection of the corresponding LPS phenotypes was observed. The particular importance of this experiment is that it demonstrates phase variation in a bacterium during long-term colonisation of that bacterium's natural host. It also demonstrates that the LPS phenotypes that are present and expressed are immunogenic in the natural host and that there is a selective advantage conferred upon the subpopulations that exhibit phase variation, resulting in prolonged colonisation.

In human pathogens, such as *H. influenzae*, similar experiments are not possible. In animal model systems using human pathogens, variation in population composition has been demonstrated but the extent to which the variation is present in the initial inoculum or occurs during colonisation is difficult to determine (e.g., Weiser et al., 1998). Evidence that variation in the number of repeats in these genes occurs *in vivo* comes from studies of outbreaks of epidemiologically associated *H. influenzae* infection in which strains that appear to be clonally related when assessed using traditional methods have different lengths of repeat. In this situation, variation in repeat numbers can be observed in isolates from different individuals (van Belkum et al., 1997a, 1997b). This does not identify when the variation occurs but it does demonstrate that these repeats are unstable and alter the expression phenotype *in vivo*.

lic-1 is another phase variable gene in *H. influenzae* which is thought to be more directly involved in immune evasion. *lic*-1 has homology with eukaryotic choline kinases and *H. influenzae* has the capacity to link choline acquired from the environment to its LPS in a phase variable fashion (Weiser et al., 1997; Risburg et al., 1997; Schweda et al., 1997). This gene is thought to act in the substitution of LPS with choline. Strains expressing the phosphorylcholine (ChoP) epitope are more sensitive to serum killing involving C-reactive protein (CRP) (Weiser et al., 1997; Weiser et al., 1998). For more details of the bacterial cell surface and evasion of complement activation refer to Chapter 4. There is a bias toward expression of ChoP in human isolates and there is evidence that suggests that the presence of ChoP contributes to persistence in the respiratory tract in animal models (Weiser et al., 1998). It has subsequently been proposed that variation between LPS, which carries the structure determined by *lic2* (which confers some resistance to antibody mediated serum bactericidal activity) and the ChoP substituted LPS mediated by *lic*-1 (with resistance to CRP), adapts *H. influenzae* to different host microenvironments (Weiser and Pan, 1998). The contribution of ChoP to survival in the host may be more complex than solely its effects on CRP dependent killing. For example ChoP is known, in other contexts, to affect persistence and invasiveness and to influence lymphocyte responsiveness and cytokine production (reviewed in Harnett and Harnett, 1999). In this context it is noteworthy that this gene is associated with one of the most variable repeat loci in studies of related strains (van Belkum et al., 1997a, 1997b), indicating that this switching process is one of the most frequent during transmission and colonization events.

6.3.4 *Neisseria meningitidis* and *Neisseria gonorrhoeae*

There are many parallels between phase variation in *Neisseria* spp., and *Haemophilus* spp., especially with regard to variability of the LPS. The most dramatic aspect of phase variability in *Neisseria* spp. is the sheer number and range of the genes that are phase varied. In this regard each species probably has in the region of fifty phase-variable genes (Saunders et al., 2000; Tettelin et al., 2000). The potential for diversification through different combinations, as described in Section 1, is theoretically vast. This is an area in which the role of gene variation in immune evasion is not as well defined as it might be. This is perhaps a reflection, in the case of meningococci, of their usual presence as harmless commensals with the host rather than as pathogens. This is less frequently the case with gonococci, but in women there can also be prolonged asymptomatic carriage and an absence of features of acute infection. The immune response and its role in the normal interactions of bacterium and host in these contexts is poorly defined. However, it should be noted that patients with gonococci who are treated are not subsequently immune to reinfection, even from the same strain, which is why treatment of all sexual partners is of such importance.

Bacterial cell surface sialic acids interfere with the immune system through the alternative complement activation pathway (Fearon, 1978; Jarvis, 1995), which is required to respond to *N. meningitidis* (Nicholson and Lepow, 1979) and the genes in the capsule locus confer serum resistance upon meningococci (Hammerschmidt et al., 1994). The presence of capsule also reduces adherence and uptake into macrophages and delays or prevents the killing of phagocytosed bacteria (McNeil et al., 1994; Read et al., 1996). This is discussed in more detail in Chapter 4. In this way the capsule is thought to provide an immune evasion mechanism in invasive disease isolates. The bacterial capsule of *N. meningitidis* is perhaps the major determinant of bacterial survival once organisms are disseminated in the blood stream (DeVoe, 1982) and is the basis of the division of the species into twelve serogroups (Jennings et al., 1977). Only a few of these serogroups are associated with disease, and these are the ones associated with sialic acid. Serogroup A causes the majority of epidemic-associated cases in sub-Saharan Africa while the majority of disease in the northern hemisphere is caused by serogroups B and C. Serogroup B and C capsules are composed of $\alpha2$–8 and $\alpha2$–9 linkages respectively of polysialic acid. The importance of these structures in association with invasive disease and meningitis is emphasised by the observation that other bacteria that cause meningitis, *E. coli*, and group B streptococci, have a similar capsule to serogroup B meningococci (Kasper et al., 1973, 1983). Several of the capsular structures are immunogenic and form the basis of meningococcal vaccines. The structure of the group B capsule is present on N-CAM, which is present within the human brain (Finne et al., 1983) and in other sites (Finne et al., 1987). Thus, it is not recognised as a foreign epitope, is not immunogenic in man, and is not a useful vaccine candidate. Therefore, each of the capsules associated with the virulent serogroups confer serum resistance, and a mechanism of immune evasion, the serogroup B capsule, has the superadded characteristic of not being immunogenic.

Some of the phase-varied LPS components are also sialylated. The LPS of *N. gonorrhoeae* can be externally modified by the sialylation of the terminal sugar residues (reviewed in Smith, 1991), an event that occurs *in vivo* (Apicella et al., 1987; Parsons et al., 1990) and which confers resistance to killing and opsonisation by normal human serum (Parsons et al., 1989; Gill et al., 1996). It also impedes adherence and hence uptake and killing by neutrophils (Kim et al., 1992; Rest and Frangipane, 1992). The mechanism by which serum sensitive and resistant phenotypes varied was found to be due to phase variation of the LPS side chain that was substituted with the sialic acid (van Putten, 1993), which was subsequently shown to be due to variation of expression of *lsi-2* (also called *lgtA*) (Danaher et al., 1995). This study showed that the phase variation affected the behaviour of the cells in cell culture. LPS variants with little sialic acid invaded efficiently but were susceptible to complement mediated killing (as discussed in Chapter 4). Phase variation resulted in highly sialylated, equally adherent, but entry-deficient bacteria, which were resistant to killing by antibodies and complement because of altered complement activation and also masking of some LPS and protein epitopes (Poolman et al., 1988; Judd and Shafer, 1989; van Putten, 1993; de la Paz et al., 1995). The situation in *N. meningitidis* is similar but not identical.

Despite the lack of a good overall picture of the relative roles of each variable gene in immune evasion, it is possible to put together a picture that indicates the central role of phase variation in this process. The surface that *N. meningitidis* presents to its environment and to the hosts' immune responses is a highly fluid and dynamic structure. Phase-variable structures in *Neisseria* include capsule, which confers serum resistance and affects cell interactions (DeVoe, 1982; Virji et al., 1992, 1993a, 1993b; Stephens et al., 1993; Hammerschmidt et al., 1994, 1996), pili and pilus modifications, which affect adhesion (Stephens and McGee 1981; Virji et al., 1991, 1993b; Nassif et al., 1994; Rudel et al., 1992, 1995; Weiser and Pan, 1998; Jennings et al., 1998; Waldbeser et al., 1994; Kupsch et al., 1993; McNeil et al., 1994), several surface proteins including Opas, Opc, PorA and iron binding proteins which have roles in adhesion, formation of surface pores, and nutrient acquisition (Virji et al., 1992, 1993a; Sparling et al., 1986; Stern et al., 1984, 1986; Achtman et al., 1988; Bhat et al., 1991; Sarkari et al., 1994; Tommassen et al., 1990; Poolman et al., 1980; Hopman et al., 1994; van der Ende et al., 1995; Schryvers and Stojiljkovic, 1999; Chen et al., 1996, 1998; Lewis et al., 1999), and lipopolysaccharide (LPS) (Apicella et al., 1987; Schneider et al., 1988, 1991; Weel et al., 1989; van Putten and Robertson, 1995; Jennings et al., 1999). It is noteworthy, in this context, that one of the few stable surface structures, Class 4 proteins, elicits blocking antibodies that may serve to protect the organism from the immune response (Munkley et al., 1991).

The full repertoire of phase-variable genes is now recognised to be far more extensive than this list of studied examples (Saunders et al., 2000). As the roles and functions of these new genes are determined, new aspects of immune evasion can be expected to be revealed and the central nature of these processes of switching in controlling the host–pathogen interaction can be determined.

6.3.5 Evasion of preexisting immune responses

To establish colonization or infection an organism not only has to evade responses that occur against it during the current or previous exposures, it also has to evade responses to common antigens on unrelated organisms. For example, *Salmonella enterica* serogroup *typhimurium* cannot evade immune responses to recent exposures to itself, but does evade cross-immunity to serotypes with different O-antigens. However, despite such differences, the flagellar components are frequently antigenically conserved between otherwise dissimilar strains (Norris and Baumler, 1999). Immunization of mice with flagellate organisms results in selection against phase-ON organisms during subsequent challenges, but does not protect against colonization or mortality. This evasion does not occur if phase variation is prevented (Nicholson and Baumler, 2001), clearly indicating that phase variation of fimbriae is a mechanism to evade cross-immunity between otherwise different bacteria. The extent to which this type of immune evasion is important in other species has yet to be determined but there are many similar systems present in other bacterial species and it is unlikely to be limited to the enteric bacteria.

6.4 CONCLUSION

Phase variation is a common strategy used by bacteria to adapt to changing environmental conditions. In pathogenic bacteria, these transitions include the movement from one host to another and movement between microenvironmental niches within the host. The immune response represents an additional process of this type, in which the environment changes while the location of the organism remains the same. The range of genes that are varied in this way has been intensively studied in some species, and these examples reveal that there are many ways in which this type of stochastic process can mediate immune evasion within a colonizing population. The recent definition and investigation of the complete repertoires will almost certainly reveal further novel strategies of this type.

ACKNOWLEDGMENT

NJS is supported by a Wellcome Advanced Research Fellowship.

REFERENCES

Achtman, M., Neibert, M., Crowe, B.A., Strittmatter, W., Kusecek, B., Weyse, E., Walsh, M.J., Slawig, B., Morelli, G., Moll, A., and Blake, M. (1988). Purification and characterization of eight class 5 outer membrane protein variants from a clone of *Neisseria meningitidis* serogroup A. *Journal of Experimental Medicine* **168**, 507–525.

Andrewes, F.W. (1922). Studies in group-agglutination. I. The *Salmonella* group and its antigenic structure. *Journal of Pathology and Bacteriology* **25**, 505–521.

Apicella, M.A., Shero, M., Jarvis, G.A., Griffiss, J.M., Mandrell, R.E., and Schneider, H. (1987). Phenotypic variation in epitope expression of the *Neisseria gonorrhoeae* lipooligosaccharide. *Infection and Immunity* **55**, 1755–1761.

Barbour, A.G., Tessier, S.L., and Stoenner, H.G. (1982). Variable major proteins of *Borrelia hermsii*. *Journal of Experimental Medicine* **156**, 1312–1324.

Barbour, A.G., Barrera, O., and Judd, R.C. (1983). Structural analysis of the variable major proteins of *Borrelia hermsii*. *Journal of Experimental Medicine* **158**, 2127–2140.

Behrens, A., Heller, M., Kirchhoff, H., Yogev, D., and Rosengarten, R. (1994). A family of phase- and size-variant membrane surface lipoprotein antigens (Vsps) of *Mycoplasma bovis*. *Infection and Immunity* **62**, 5075–5084.

Bhat, K.S., Gibbs, C.P., Barrera, O., Morrison, S.G., Jahnig, F., Stern, A., Kupsch, E.-M., Meyer, T.F., and Swanson, J. (1991). The opacity proteins of *Neisseria gonorrhoeae* strain MS11 are encoded by a family of 11 complete genes. *Molecular Microbiology* **5**, 1889–1901. Also see erratum: *Molecular Microbiology* **6**, 1073–1076.

Chen, C.-J., Sparling, P.F., Lewis, L.A., Dyer, D.W., and Elkins, C. (1996). Identification and purification of a hemoglobin-binding outer membrane protein from *Neisseria gonorrhoeae*. *Infection and Immunity* **64**, 5008–5014.

Chen, C.J., Elkins, C., and Sparling, P.F. (1998). Phase variation of hemoglobin utilization in *Neisseria gonorrhoeae*. *Infection and Immunity* **66**, 987–993.

Citti, C., Kim, M.F., and Wise, K.S. (1997). Elongated versions of Vlp surface lipoproteins protect *Mycoplasma hyrhinis* escape variants from growth-inhibiting host antibodies. *Infection and Immunity* **65**, 1773–1785.

Coffey, E.M. and Eveland, W.C. (1967). Experimental relapsing fever initiated by *Borrelia hermsii*. II. Sequential appearance of major serotypes in the rat. *Journal of Infectious Diseases* **177**, 29–34.

Danaher, R.J., Levin, J.C., Arking, D., Burch, C.L., Sandlin, R., and Stein, D.C. (1995). Genetic basis of *Neisseria gonorrhoeae* lipooligosaccharide antigenic variation. *Journal of Bacteriology* **177**, 7275–7279.

de la Paz, H., Cooke, S.J., and Heckels, J.E. (1995). Effect of sialylation of lipopolysaccharide of *Neisseria gonorrhoeae* on recognition and complement-mediated killing by monoclonal antibodies directed against different outer-membrane antigens. *Microbiology* **141**, 913–920.

DeVoe, I.W. (1982). The meningococcus and mechanisms of pathogenicity. *Microbiology Reviews* **46**, 162–190.

Fearon, D.T. (1978). Regulation by membrane sialic acid of ß1H-dependent decay-dissociation of amplification C3 convertase of the alternative complement pathway. *Proceedings of the National Academy of Sciences USA* **75**, 1971–1975.

Finlay, B.B. and Falkow, S. (1989). Common themes in microbial pathogenicity. *Microbiology Reviews* **53**, 210–230.

Finne, J., Leinonen, M., and Makela, P.H. (1983). Antigenic similarities between brain components and bacteria causing meningitis. Implications for vaccine development and pathogenesis. *Lancet* **2**, 355–357.

Finne, J., Bitter-Suermann, D., Goridis, C., and Finne, U. (1987). An IgG mono-clonal antibody to group B meningococci cross-reacts with developmentally regulated polysialic acid units of glycoproteins in neural and extraneural tissues. *Journal of Immunology* **138**, 4402–4407.

Gill, M.J., McQuillen, D.P., van Putten, J.P.M., Wetzler, L.M., Bramley, J., Crooke, H., Parsons, N.J., Cole, J.A., and Smith, H. (1996). Functional characteriza-tion of a sialyltranferase-deficient mutant of *Neisseria gonorrhoeae*. *Infection and Immunity* **64**, 3374–3378.

Gilsdorf, J.R. and Ferrieri, P. (1986). Susceptibility of phenotypic variants of *Haemophilus influenzae* type b to serum bactericidal activity: relationship to surface lipopolysaccharide. *Journal of Infectious Diseases* **153**, 223–231.

Hammerschmidt, S., Birkholtz, C., Zahringer, U., Robertson, B.D., van Putten, J., Ebeling, O., and Frosch, M. (1994). Contribution of genes from the cap-sule gene cluster complex (*cps*) to lipopolysaccharide biosynthesis and serum resistance in *Neisseria meningitidis*. *Molecular Microbiology* **11**, 885–896.

Hammerschmidt, S., Hilse, R., van Putten, JPM., Gerardy-Schahn, R., Unkmeir, A., and Frosch, M. (1996). Modulation of cell surface sialic acid expression in *Neisseria meningitidis* via a transposable genetic element. *European Molec-ular Biology Organisation Journal* **15**, 192–198.

Harnett, W. and Harnett, M.M. (1999). Phosphorylcholine: friend or foe of the immune system? *Immmunology Today* **20**, 125–129.

Hood D.W., Deadman, M.E., Jennings, M.P., Biscercic, M., Fleischmann, R.D., Venter, J.C., and Moxon, E.R. (1996). DNA repeats identify novel virulence genes in *Haemophilus influenzae*. *Proceedings of the National Academy of Sciences USA* **93**, 11,121–11,125.

Hopman, C.T.P., Dankert, J., and van Putten, J.P.M. (1994). Variable expression of the class 1 protein of *Neisseria meningitidis*, In, ed. C.J. Conde-Glez, S. Morse, P. Rice, F. Sparling, and E. Calderon. pp. 513–517. *Pathology and immunobiology of Neisseriaceae*, Instituto Nacional de Salud Publica Cuernavaca, Mexico.

Inzana, T.J., Gogolewski, R.P., and Corbeil, L.B. (1992). Phenotypic phase variation in *Haemophilus somnus* lipopolysaccharide during bovine pneumonia and after in vitro passage. *Infection and Immunity* **60**, 2943–2951.

Inzana, T.J., Hensley, J., McQuiston, J., Lesse, A.J., Campagnari, A.A., Boyle, S.M., and Apicella, M.A. (1997). Phase variation and conservation of lipopolysaccharide epitopes in *Haemophilus somnus*. *Infection and Immunity* **65**, 4675–4681.

Jarvis, G.A. (1995). Recognition and control of neisserial infection by antibody and complement. *Trends in Microbiology* **3**, 198–201.

Jennings, H.J., Bhattacharjee, A.K., Bundle, D.R., Kenny, C.P., Martin, A., and Smith, I.C. (1977). Structures of the capsular polysaccharides of *Neisseria meningitidis* as determined by ^{13}C-nuclear magnetic resonance spectroscopy. *Journal of Infectious Diseases* **136**, S78–S83.

Jennings, M.P., Virji, M., Evans, D., Foster, V., Srikhanta, Y.N., Steeghs, L., van der Ley, P., and Moxon, E.R. (1998). Identification of a novel gene involved in pilin glycosylation in *Neisseria meningitidis*. *Molecular Microbiology* **29**, 975–984.

Jennings, M.P., Srithanta, Y.N., Moxon, E.R., Kramer, M., Poolman, J.T., Kuipers, B., and van der Ley, P. (1999). The genetic basis of the phase variation repertoire of lipopolysaccharide immunotypes in *Neisseria meningitidis*. *Microbiology* **145**, 3013–3021.

Judd, R.S. and Shafer, W.M. (1989). Topographical alterations in proteins I of *Neisseria gonorrhoeae* correlated with lipooligosaccharide variation. *Molecular Microbiology* **3**, 637–642.

Kasper, D.L., Winkelhake, J.L., Zollinger, W.D., Brandt, B.L., and Artenstein, M.S. (1973). Immunological similarity between polysaccharide antigens of *Escherichia coli* O7:K1 (L): NM and group B *Neisseria meningitidis*. *Journal of Immunology* **110**, 262–268.

Kasper, D.L., Baker, C.J., Galdes, B., Katzenellenbogen, E., and Jennings, H.J. (1983). Immunological analysis and immunogenicity of the type II group B streptococccal capsular polysaccharide. *Journal of Clinical Investigation* **72**, 260–269.

Kim, J.J., Zhau, D., Mandrell, R.E., and Griffiss, J.M. (1992). Effects of endogenous sialylation of the lipooligosaccharide of *Neisseria gonorrhoeae* on opsonophagocytosis. *Infection and Immunity* **60**, 4439–4442.

Kimura, A. and Hansen, E.J. (1986). Antigenic and phenotypic variations of *Haemophilus influenzae* type b lipopolysaccharide and their relationship to virulence. *Infection and Immunity* **51**, 69–79.

Kupsch, E.-M., Knepper, B., Kuroki, T., Heuer, I., and Meyer, T.F. (1993). Variable opacity (Opa) outer membrane proteins account for the cell tropisms displayed by *Neisseria gonorrhoeae* for human leukocytes and epithelial cells. *European Molecular Biology Organisation Journal* **12**, 641–650.

Lewis, L.A., Gipson, M., Hartman, K., Ownbey, T., Vaughn, J., and Dyer, D.W. (1999). Phase variation of HpuAB and HmbR, two distinct haemoglobin receptors of *Neisseria meningitidis* DNM2. *Molecular Microbiology* **32**, 977–989.

McNeil, G., Virji, M., and Moxon, E.R. (1994). Interactions of *Neisseria meningitidis* with human monocytes. *Microbial Pathogenesis* **16**, 153–163.

Meleney, H.E. (1928). Relapse phenomena of *Spironema recurrentis. Journal of Experimental Medicine* **48**, 65–82.

Mertsola, J., Cope, L.D., Saez-Lorenz, X., Ramilo, O., Kennedy, W., McCraken, G.H., Jr., and Hansen, E.J. (1991). *In vivo* and *in vitro* expression of *Haemophilus influenzae* type b lipooligosaccharide epitopes. *Journal of Infectious Diseases* **164**, 555–563.

Moxon, E.R., Rainey, P.B., Nowak, M.A., and Lenski, R.E. (1994). Adaptive evolution of highly mutable loci in pathogenic bacteria. *Current Biology* **4**, 24–33.

Munkley, A., Tinsley, R., Virji, M., and Heckels, J.E. (1991). Blocking of bactericidal killing of *Neisseria meningitidis* by antibodies directed against class 4 outer membrane protein. *Microbial Pathogenesis* **11**, 447–452.

Nassif, X., Beretti, J.-L., Lowy, J., Stenberg, P., O'Gaora, P., Pfeifer, J., Normark, S., and So, M. (1994). Roles of pilin and PilC in adhesion of *Neisseria meningitidis* to human epithelial and endothelial cells. *Proceedings of the National Academy of Sciences USA* **91**, 3769–3773.

Nicholson, A. and Lepow, I.H. (1979). Host defence against *Neisseria meningitidis* requires a complement-dependent bactericidal activity. *Science* **205**, 298–299.

Nicholson, T.L. and Baumler, A.J. (2001). *Salmonella enterica* serotype typhimurium elicits cross-immunity against a *Salmonella enterica* serotype enteritidis strain expressing LP fimbriae from the *lac* promoter. *Infection and Immunity* **69**, 204–212.

Nikaido, H. (1996). Outer membrane. In *Escherichia coli and Salmonella*, 2nd ed., ed. F.C. Neidhardt. pp. 29–47. Washington, DC: ASM Press.

Norris, T.L. and Baumler, A.J. (1999). Phase variation of the *lpf* operon is a mechanism to evade cross-immunity between *Salmonella* serotypes. *Proceedings of the National Academy of Sciences USA* **96**, 13,393–13,398.

Parsons, N.J., Andrade, J.R.C., Patel, P.V., Cole, J.A., and Smith, H. (1989). Sialylation of lipopolysaccharide and loss of absorption of bactericidal antibody during conversion of gonococci to serum resistance by cytidine 5'-monophospho-N-acetyl neuraminic acid. *Microbial Pathogenesis* 7, 63–72.

Parsons, N.J., Cole, J.A., and Smith, H. (1990). Resistance to human serum of gonococci in urethral exudates is reduced by neuraminidase. *Proceedings of Royal Society of London B* 241, 3–5.

Poolman, J.T., de Marie, S., and Zanen, H.C. (1980). Variability of low-molecular-weight, heat-modifiable outer membrane proteins of *Neisseria meningitidis*. *Infection and Immunity* 30, 642–648.

Poolman, J.T., Timmermans, H.A.M., Hopman, C.T.P., Teerlink, T., van Vught, P.A.M., Witvliet, M.H., and Beuvery, E.C. (1988). In *Gonococci and Meningococci*, ed. J.T. Poolman, H.C. Znen, T.F. Meyer, J.E. Heckels, P.R.H. Makela, H. Smith and E.C. Bauvery. pp. 159–166. Dortrecht: Kluwer Academic.

Read, R.C., Zimmerli, S., Broaddus, V.C., Sanan, D.A., Stephens, D.S., and Ernst, J.D. (1996). The ($\alpha2 \rightarrow$ 8)-linked polysialic acid capsule of group B *Neisseria meningitidis* modifies multiple steps during interaction with human macrophages. *Infection and Immunity* 64, 3210–3217.

Rest, R.F. and Frangipane, J.V. (1992). Growth of *Neisseria gonorrhoeae* in CMP-N-acetylneuraminic acid inhibits nonopsonic (opacity-associated outer membrane protein-mediated) interactions with human neutrophils. *Infection and Immunity* 60, 989–997.

Risberg, A., Schweda, E.K.H., and Jansson, P.-E. (1997). Stuctural studies of the cell-envelop oligosaccharide from lipopolysaccharide of *Haemophilus influenzae* strain RM 118–28. *European Journal of Biochemistry* 243, 701–707.

Rosengarten, R. and Wise, K.S. (1990). Phenotypic switching in Mycoplasmas: Phase variation of diverse surface lipoproteins. *Science* 247, 315–318.

Rosengarten, R. and Wise, K.S. (1991). The Vlp system of *Mycoplasma hyorhinis*: combinatorial expression of distinct size variant lipoproteins generating high-frequency surface antigenic variation. *Journal of Bacteriology* 173, 4782–4793.

Rosengarten, R., Behrens, A., Stetefeld, A., Heller, M., Ahrens, M., Sachse, K., Yogev, D., and Kirchhoff, H. (1994). Antigenic heterogeneity among isolates of *Mycoplasma bovis* is generated by high-frequency variation of diverse membrane surface proteins. *Infection and Immunity* 62, 5066–5074.

Rudel, T., van Putten, J.P.M., Gibbs, C.P., Haas, R., and Meyer, T.F. (1992). Interaction of two variable proteins (PilE and PilC) required for pilus mediated adherence of *Neisseria gonorrhoeae* to human epithelial cells. *Molecular Microbiology* 6, 3439–3450.

Rudel, T., Boxberger, H.-J., and Meyer, T.F. (1995). Pilus biogenesis and epithelial adherence of *Neisseria gonorrhoeae pilC* double knock-out mutants. *Molecular Microbiology* 17, 1057–1071.

Sarkari, J., Pandit, N., Moxon, E.R., and Achtman, M. (1994). Variable expression of the Opc outer membrane protein in *Neisseria meningitidis* is caused by size variation of a promoter containing poly-cytidine. *Molecular Microbiology* 13, 207–217.

Saunders, N.J., Peden, J.F., Hood, D.W., and Moxon, E.R. (1998). Simple sequence repeats in the *Helicobacter pylori* genome. *Molecular Microbiology* 27, 1091–1098.

Saunders, N.J., Jeffries, A.C., Peden, J.F., Hood, D.W., Tettelin, H., Rappouli, R., and Moxon, E.R. (2000). Repeat associated phase variable genes in the complete genome sequence of *Neisseria meningitidis*. *Molecular Microbiology* 37, 207–215.

Saunders, N.J., Moxon, E.R., and Gravenor, M. (2002a). The determination of mutation rates associated with phase variation. – submitted.

Saunders, N.J., Moxon, E.R., and Gravenor, M. (2002b). The influence of phase variation and selection on population structure. – submitted.

Schneider, H., Hammack, C.A., Apicella, M.A., and Griffiss, J.M. (1988). Instability of expression of lipooligosaccharides and their epitopes in *Neisseria gonorrhoeae*. *Infection and Immunity* 56, 942–946.

Schneider, H., Griffiss, J.M., Boslego, J.W., Hitchcock, P.J., Zahos, K.M., and Apicella, M.A. (1991). Expression of paragloboside-like lipooligosaccharides may be a necessary component of gonococcal pathogenesis in men. *Journal of Experimental of Medicine* 174, 1601–1605.

Schryvers, A.B. and Stojiljkovic, I. (1999). Iron acquisition systems in the pathogenic *Neisseria*. *Molecular Microbiology* 32, 1117–1123.

Schweda, E.K.H., Masoud, H., Martin, A., Risberg, A., Hood, D.W., Moxon, E.R., Weiser, J.N., and Richards, J.C. (1997). Phase variable expression and characterisation of phosphorylcholine oligosaccharide epitopes in *Haemophilus influenzae* lipopolysaccharides. *Glycoconjugate Journal* 14 (Suppl), S23.

Smith, H. (1991). The Leeuwenhoek Lecture, 1991. The influence of the host on microbes that cause disease. *Proceedings of Royal Society of London B* 246, 97–105.

Sparling, P., Cannon, J., and So, M. (1986). Phase and antigenic variation of pili and outer membrane protein II of *Neisseria gonorrhoeae*. *Journal of Infectious Diseases* 153, 196–201.

Stephens, D.S. and McGee, Z.A. (1981). Attachment of *Neisseria meningitidis* to human mucosal surfaces: influence of pili and type of receptor cell. *Journal of Infectious Diseases* 143, 525–532.

Stephens, D.S., Spellman, P.A., and Swartley, J.S. (1993). Effect of the ($\alpha 2 \rightarrow 8$) -linked polysialic acid capsule on adherence of *Neisseria meningitidis* to human mucosal cells. *Journal of Infectious Diseases* **167**, 475–479.

Stern, A., Nickel, P., Meyer, T., and So, M. (1984). Opacity determinants of *Neisseria gonorrhoeae*: gene expression and chromosomal linkage to the gonoccocal pilus gene. *Cell* **37**, 447–456.

Stern, A., Brown, M., Nickel, P., and Meyer, T. (1986). Opacity genes in *Neisseria gonorrhoeae*: control of phase and antigenic variation. *Cell* **47**, 61–71.

Stoenner, H.G., Dodd, T., and Larsen, C. (1982). Antigenic variation in *B. hermsii*. *Journal of Experimental Medicine* **156**, 1297–1311.

Tettelin, H., Saunders, N.J., Heidelberg, J., Jeffries, A.C., Nelson, K.E., and 37 other authors. (2000). The complete genome sequence of *Neisseria meningitidis* serogroup B strain MC58. *Science* **287**, 1809–1815.

Tommassen, J., Vermeij, P., Struyve, M., Benz, R., and Poolman, J.T. (1990). Isolation of *Neisseria meningitidis* mutants deficient in class 1 (PorA) and class 3 (PorB) outer membrane proteins. *Infection and Immunity* **58**, 1355–1359.

van Belkum, A., Scherer, S., van Leeuwen, W., Willemse, D., van Alphen, L., and Verbrugh, H. (1997a). Variable number of tandem repeats in clinical strains of *Haemophilus influenzae*. *Infection and Immunity* **65**, 5017–5027.

van Belkum, A., Melchers, W.J.G., Ijsseldijk, C., Nohlmans, L., Verbuch, H., and Meis, J.F.G.M. (1997b). Outbreak of amoxycillin-resistant *Haemophilus influenzae* type b: variable number of tandem repeats as novel molecular markers. *Journal of Clinical Microbiology* **35**, 1517–1520.

van der Ende, A., Hopman, C.T.P., Zaat, S., Oude Essink, B.B., Berkhout, B., and Dankert, J. (1995). Variable expression of class 1 outer membrane protein in *Neisseria meningitidis* is caused by variation in the spacing between the −10 and −35 regions of the promoter. *Journal of Bacteriology* **177**, 2475–2480.

van Putten, J.P.M. (1993). Phase variation of lipopolysaccharide directs interconversion of invasive and immuno-resistant phenotypes of *Neisseria gonorrhoeae*. *European Molecular Biology Organisation Journal* **12**, 4043–4051.

van Putten, J.P.M. and Robertson, B.D. (1995). Molecular mechanisms and implication for infection of lipopolysaccharide variation in *Neisseria*. *Molecular Microbiology* **16**, 847–853.

Virji, M., Kayhty, H., Fergusson, D.J.P., Alexandrescu, C., Heckles, J.E., and Moxon, E.R. (1991). The role of pilin in the interactions of pathogenic *Neisseria* with cultured human endothelial cells. *Molecular Microbiology* **5**, 1831–1841.

Virji, M., Alexandrescu, C., Fergusson, D.J.P., Saunders, J.R., and Moxon, E.R. (1992). Variations in the expression of pili: the effect on adherence of *Neisseria meningitidis* to human epithelial and endothelial cells. *Molecular Microbiology* **6**, 1271–1279.

Virji, M., Makepeace, K., Ferguson, D.J., Achtman, M., and Moxon, E.R. (1993a). Menigococcal Opa and Opc proteins: their role in colonisation and invasion of human epithelial and endothelial cells. *Molecular Microbiology* 10, 499–510.

Virji, M., Saunders, J.R., Sims, G., Makepeace, K., Maskell, D., and Ferguson, D.J.P. (1993b). Pilus-facilitated adherence of *Neisseria meningitidis* to human epithelial and endothelial cells: modulation of adherence phenotype occurs concurrently with changes in the primary amino acid sequence and the glycosylation status of pilin. *Molecular Microbiology* 10, 1013–1028.

Waldbeser, L.S., Ajioka, R.S., Merz, A.J., Puaoi, D., Lin, L., Thomas, M., and So, M. (1994). The Opah locus of *Neisseria gonorrhoeae* MS11A is involved in epithelial cell invasion. *Molecular Microbiology* 13, 919–928.

Weel, J.F.L., Hopman, C.T.P., and van Putten, J.P.M. (1989). Stable expression of lipooligosaccharide antigens during attachment, internalization, and intracellular processing of *Neisseria gonorrhoeae* in infected epithelial cells. *Infection and Immunity* 57, 3395–3402.

Weiser J.N., Lindberg, A.A., Manning, E.J., Hansen, E.J., and Moxon, E.R. (1989). Identification of a chromosomal locus for expression of lipopolysaccharide epitopes in *Haemophilus influenzae*. *Infection and Immunity* 57, 3945–3052.

Weiser, J.N., Shchepetov, M., and Chong, S.T.H. (1997). Decoration of lipopolysaccharide with phosphorylcholine: a phase-variable characteristic of *Haemophilus influenzae*. *Infection and Immunity* 65, 943–950.

Weiser, J.N. and Pan, N. (1998). Adaptation of *Haemophilus influenzae* to acquired and innate humoral immunity based on phase variation of lipopolysaccharide. *Molecular Microbiology* 30, 767–775.

Weiser, J.N., Pan, N., McGowan, K.L., Musher, D., Martin, A., and Richards, J. (1998a). Phosphorylcholine on the lipopolysaccharide of *Haemophilus influenzae* contributes to persistence in the respiratory tract and sensitivity to serum killing mediated by C-reactive protein. *Journal of Experimental Medicine* 187, 631–640.

Wise, K.S. (1993). Adaptive surface variation in mycoplasmas. *Trends in Microbiology* 1, 59–63.

Yogev, D., Rosengarten, R., Watson-McKown, R., and Wise, K.S. (1991). Molecular basis of *Mycoplasma* surface antigenic variation: a novel set of divergent genes undergo spontaneous mutation of periodic coding regions and 5' regulatory sequences. *European Molecular Biology Organisation Journal* 10, 4069–4079.

Yogev, D., Watson-McKown, R., Rosenbgarten, R., Im, J., and Wise, K.S. (1995). Increased structural and combinatorial diversity in an extended family of genes encoding Vlp surface proteins of *Mycoplasma hyorhinis*. *Journal of Bacteriology* 177, 5636–5643.

PART III   Evasion of cellular immunity

CHAPTER 7

Type III secretion and resistance to phagocytosis

Åke Forsberg, Roland Rosqvist, and Maria Fällman

7.1 INTRODUCTION

Phagocytosis is an essential first line of defence, which normally efficiently clears and destroys microorganisms. This process is mostly attributed to professional phagocytes: macrophages, monocytes, and neutrophils, which express specialised receptors that promote phagocytosis (Rabinovitch, 1995; Aderem and Underhill, 1999; see Chapter 1). These receptors recognise opsonins such as IgG and products of complement that bind to bacterial surfaces (see Chapter 4). Following phagocytic uptake, bacteria are normally killed and destroyed inside phagosomes, especially after maturation to phagolysosomes. Maturation is caused by fusion of the phagosome with endocytic vesicles causing an increasingly acidic environment, and finally fusion with lysosomes that contain digestive enzymes, mainly acid hydrolases (Tjelle et al., 2000). The professional phagocytes can also produce reactive oxygen and nitrogen (in macrophages) intermediates that contribute to killing (Hampton et al., 1998; Vazquez-Torres et al., 2000a). When activated, these cells secrete pro-inflammatory cytokines, which in turn stimulate other immune cells. Macrophages also serve as antigen presenting cells enabling generation of specific cellular and humoral defences (Morrisette et al., 1999; see Chapter 2). Therefore, it is not surprising that many microorganisms have developed strategies to circumvent phagocyte activity. The pathogens discussed in this chapter, *Yersinia* and *Pseudomonas aeruginosa*, directly block the engulfment process and remain extracellular, while other pathogens invade phagocytes and remodel the vesicle fusion events to promote persistence and replication. This latter mechanism, employed for example by *Mycobacterium tuberculosis*, will not be discussed, and this chapter will detail the use of a highly specialised mechanism for delivery of antihost factors into host cells, namely type III secretion systems.

7.2 TYPE III SECRETION SYSTEMS

Several Gram-negative bacterial pathogens share an important mechanism that serves to target virulence effectors into host cells. This mechanism is type III secretion/translocation, which mediates secretion and delivery of antihost factors into eukaryotic cells via a contact-dependent mechanism. Type III secretion/translocation systems have been identified in a number of animal and plant pathogens and several components of these systems also show homology to components of the secretion/assembly apparatus of the flagellar system (reviewed by Hueck, 1998). Type III secretion of effectors involves about twenty proteins that are believed to assemble into a structure that spans the two bacterial membranes. Eleven core components appear to be conserved in the secretion systems of Gram-negative plant and animal pathogens as well as in the flagellar export apparatus. Most of these proteins are associated with the cytoplasmic membrane and are likely to be involved in recognition and targeting of secretion substrates to the secretion apparatus (Hueck, 1998; Plano et al., 2001). Motility and chemotaxis are believed to be ancient properties essential for free-living organisms and this suggests that type III secretion systems have evolved from the flagellar export/assembly system. Structurally, the type III apparatus also resembles the basal body of flagella, with a "needle-like" structure at the tip replacing the hook and flagellar filament. Such a "micro-injection syringe" has been isolated for the system encoded by pathogenicity island one (SPI-1) of *Salmonella typhimurium* (Kubori et al., 1998, 2000) and a similar structure has also been visualised in *Shigella flexneri* (Blocker et al., 1999). The structure isolated and visualised by electron microscopy appears to include the components required for secretion of proteins in a continuous process across the two bacterial membranes without processing of the secreted proteins. The needle-like structure, which extends from the bacterial surface, has been suggested to be a hollow structure and is required for the secretion system to be functional.

The type III secreted proteins can be divided into two functionally distinct groups. One is the virulence effectors with their respective targets inside host cells. The other group of proteins are either directly involved in translocation of effector proteins across the target cell membrane or in controlling the process. These proteins appear to localise at the site of contact between the pathogen and the target cell (Håkansson et al., 1996a and b; Holmström et al., 1997; Knutton et al., 1998; Pettersson et al., 1999). There is evidence from several systems that the overall mechanism of secretion and translocation is functionally conserved. This is supported by studies showing that type III targeted effector proteins from different pathogens can be secreted and translocated by heterologous type III secretion/translocation systems (Rosqvist et al., 1995; Frithz-Lindsten et al., 1997; Rossier et al., 1999; Anderson et al., 1999).

Even if the type III secretion systems of different bacteria are functionally conserved and show a high degree of similarity, the biological activity of the targeted effector proteins varies between different pathogens. Intracellular delivery of key virulence effectors, such as YopE, YopH, and YopJ (for _Yersinia_ outer proteins) of pathogenic _Yersinia_ species into eukaryotic cells, results in a general reduction of phagocytic capability as well as suppression of induction of inflammatory cytokines in response to infection (Fällman et al., 1995; Rosqvist et al., 1988, 1990; Palmer et al., 1998; Schesser et al., 1998). The action of the Yops in inhibiting NF-κB is discussed in detail in Chapter 12. In contrast, some of the effector proteins delivered by the type III system (SPI-1) of _S. typhimurium_ and _S. flexneri_ trigger uptake and a massive inflammatory response (Parsot and Sansonetti, 1996; Galán and Zhou, 2000). Interestingly, in _S. typhimurium_ there is a second type III secretion system, encoded by pathogenicity island two (SPI-2), which delivers effector proteins that promote intracellular survival of this pathogen (Hensel et al., 1995, 1998; Ochman et al., 1996; Cirillo et al., 1998). This suggests that the effect on virulence and the role of the type III secretion system in disease is mainly determined by the activity of the delivered effector proteins. An interesting observation is that type III secretion systems that deliver effector proteins with similar activity also tend to show a higher degree of overall similarity extending beyond the common core components. This is the case for the systems of _S. typhimurium_ (SPI-1) and _Shigella_ where some of the secreted proteins induce uptake of the respective pathogen (Nhieu and Sansonetti, 1999; Galán and Zhou, 2000). Therefore, it appears that there is also a need to control the secretion/delivery of the effector proteins to optimise the effect of the respective antihost factors in order to establish infection.

7.3 SECRETION AND DELIVERY OF EFFECTOR PROTEINS BY _YERSINIA_ AND _P. AERUGINOSA_

For pathogenic _Yersinia_ species and _P. aeruginosa_ essentially all the proteins involved in the type III-delivery show similarity (Yahr et al., 1996, 1997; Frank, 1997). This suggests that the mechanism of secretion and delivery of effector proteins into the host cell is similar for these two pathogens. The secretion and translocation of Yop effectors have been studied in some detail in *Yersinia*. The N-terminal 15-17 amino acids of the secreted proteins are sufficient to direct export of reporter proteins (Sory et al., 1995; Schesser et al., 1996; Woestyn et al., 1996). The fact that several frameshift mutations in the N-terminal region that completely altered the amino acid sequence did not prevent secretion led to the suggestion that the secretion signal resided in predicted stem-loop structures in the 5′ end of mRNA rather than in the amino acid sequence (Anderson and Schneewind, 1997). However, recent data from Lloyd et al. (2001) on YopE show that the secretion signal probably resides in the amino acid sequence, as mutations that extensively altered the mRNA sequence without affecting the amino acid sequence were still secreted. Instead, it was suggested that the common motif recognised by type III secretion systems is an N-terminal amphipathic region, and in support of this theory a synthetic amphipathic sequence of alternating serine and isoleucine residues was functional in secretion (Lloyd et al., 2001).

Some of the secreted Yop effectors, such as YopE and YopH, have a second domain downstream of the N-terminal secretion signal. Specific chaperones bind to this region, and strains lacking the chaperone secrete/translocate less of the respective Yop effector (Wattiau and Cornelis, 1993; Frithz-Lindsten et al., 1995; Schesser et al., 1996; Wattiau et al., 1996; Cheng et al., 1997). However, the chaperone interaction mechanism is not essential for targeting of effectors into host cells, as YopE mutants lacking the chaperone binding domain can be translocated into eukaryotic cells by strains lacking other Yop effectors (Boyd et al., 2000). In addition, other Yop effectors like YopM and YpkA, appear to lack specific chaperones but are still translocated into host cells. Therefore, it is possible that the chaperones provide a mechanism by which Yop effectors are sequentially delivered. Lloyd et al. (2001) have also shown that the YopE chaperone, YerA (SycE), is required for rapid release of YopE when secretion is induced *in vitro*. Therefore it is possible that effector proteins, which have a cognate chaperone, are delivered by a different mechanism that allows rapid delivery.

Three of the secreted proteins, YopB, YopD, and LcrV are required for delivery of effector proteins into the eukaryotic cell (Rosqvist et al., 1991, 1994; Håkansson et al., 1996b; Boland et al., 1996; Pettersson et al., 1999). These proteins mediate pore formation on infected cells and it has been suggested that translocation occurs via a pore in the host cell membrane (Håkansson et al., 1996b; Holmström et al., 1997; Neyt and Cornelis, 1999). In support of this model is the finding that purified LcrV forms channels in artificial lipid bilayer membranes (Holmström et al., 2001). Exoenzyme S, a type III secreted effector protein of *P. aeruginosa*, can be secreted and translocated by the heterologous type III system of *Yersinia pseudotuberculosis*

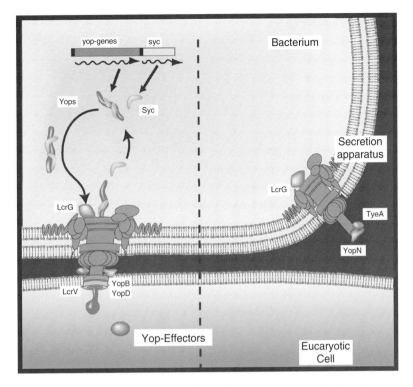

Figure 7.1. Secretion and translocation of Yop effector proteins by *Yersinia*. Secretion/ translocation only occurs where the bacterium is in contact with the eukaryotic cell. LcrG, YopN, and TyeA are all involved in controlling this so-called polarised delivery of Yops into the target cell. LcrV, YopB, and YopD are all required for the delivery of effector proteins across the host cell membrane, possibly via formation of a pore. The type III system of *P. aeruginosa* is very similar to that of *Yersinia* and this pathogen expresses proteins that are homologous to LcrG, YopN, TyeA, LcrV, YopB, and YopD. Therefore, the mechanism of type III mediated delivery of effector proteins is likely to occur by the same mechanism as in the model depicted in this figure.

(Frithz-Lindsten et al., 1997, 1998; Pettersson et al., 1999). Similar to Yop effectors, delivery of ExoS into HeLa cells depends on the translocated proteins YopB, YopD, and LcrV. This argues that the overall mechanism of secretion as well as targeting of proteins into to host cell is conserved between these two pathogens (Fig. 7.1).

The YopB, YopD, and LcrV homologues in *P. aeruginosa* PopB, PopD, and PcrV show an overall homology of around 40% identity at the amino acid level. Mutation in either of the genes encoding these proteins renders *P. aeruginosa* unable to target effector exoenzymes into infected cells (Yahr et al., 1997; Sawa et al., 1999). Moreover, the genes encoding *popB, popD*, and *pcrV* complement *yopB, yopD,* and *lcrV* mutants of *Y. pseudotuberculosis* with respect to delivery of Yop effectors into infected HeLa cells (Frithz-Lindsten et al., 1997; Pettersson et al., 1999).

Studies of effector protein translocation by *Yersinia* have revealed that the process is tightly regulated. The delivery occurs by a polarised mechanism and effector proteins are secreted and translocated only into the interacting host cell and not into the extracellular milieu (Rosqvist et al., 1994; Persson et al., 1995). Several virulence plasmid-encoded components are involved in this targeted delivery; Inactivation of *yopN* or *lcrG* results in loss of polarised delivery; that is, effector proteins are secreted into the extracellular environment (Forsberg et al., 1991; Rosqvist et al., 1994; Persson et al., 1995; Boland et al., 1996, Nilles et al., 1997). TyeA appears to have similar role to YopN and LcrG in that loss of this protein results in secretion of Yop effectors to the extracellular environment. However, there is also a unique feature of *tyeA* mutants in that they are unable to deliver YopE and YopH into host cells (Iriarte et al., 1998). YopE and YopH share the fact that they contain a secretion/translocation domain that bind to specific chaperones. Binding of the chaperones to these Yop effectors provides a second mechanism for secretion (Woestyn et al., 1996; Cheng et al., 1997; Lloyd et al., 2001). It is possible that TyeA is part of a specific delivery mechanism for a subset of Yop effectors that are secreted via a chaperone-dependent mechanism and that this mechanism has evolved to deliver effectors that block uptake by phagocytic cells. The role of the cognate chaperones and TyeA might be to ensure that these effectors are delivered rapidly after cell contact (see also below for YopH).

The delivery process has not been studied as extensively in *P. aeruginosa* as in *Yersinia*, but it is clear that PopB, PopD, and PcrV are all essential for delivery of the effector exoenzymes into host cells (Yahr et al., 1997; Frithz-Lindsten et al., 1998; Sawa et al., 1999). ExoS secretion is more efficient if the upstream encoded chaperone-like gene (homologous to *year/sycE*) is present (Frithz-Lindsten et al., 1997), and it is likely that this chaperone serves a similar function as the Yop effector chaperones. Importantly, there are also homologues in *P. aeruginosa* for YopN, LcrG, and TyeA (Frank, 1997), which makes it likely that the translocation of effector proteins occurs by similar mechanisms (Fig. 7.1). It is possible that a common mechanism has evolved that allows these two pathogens to rapidly deliver effectors that block uptake.

7.4 ROLE OF TYPE III SECRETION SYSTEMS IN BACTERIAL INFECTIONS

For a number of pathogens, type III secretion systems have a major impact on the host–parasite interaction and the nature of the infection. For instance, the large virulence plasmids of pathogenic *Yersinia* species and *Shigella* that were initially recognised to be essential for virulence were subsequently shown to encode type III secretion systems. The large gene cluster in *S. typhimurium* originally identified for its role in promoting invasion into epithelial cells was later shown to encode a type III secretion system. The two type III secretion systems of *S. typhimurium* were both subsequently shown to be present on pathogenicity islands, and the type III secretion system of enteropathogenic *E. coli* (EPEC) has also been shown to be encoded by a pathogenicity island denoted LEE (locus of enterocyte effacement). Thus, the type III secretion systems are encoded by large gene clusters either present on pathogenicity islands or virulence plasmids.

In *S. typhimurium* (SPI-1) and *Shigella*, delivery of effector proteins via the type III secretion system promotes uptake of the pathogen via a mechanism involving extensive rearrangement of the actin cytoskeleton and membrane ruffling (Nhieu and Sansonetti, 1999; Galán and Zhou, 2000). In *S. typhimurium*, this is achieved by translocation of effector proteins which stimulate the Rho GTPases, Cdc42 and Rac1 (Galán and Zhou, 2000), two host proteins with a central role in signalling pathways that control actin rearrangement in host cells. Thus, in this case the secretion system is crucial for internalisation of the pathogens and promotes their intracellular lifestyle.

The second type III secretion system of *S. typhimurium* (Hensel et al., 1995) clearly has a different role in the infection compared to the system encoded by SPI-1. Inactivation of genes essential for this second secretion system results in strains that are highly attenuated in the systemic mouse infection model. The major role of the system encoded by SPI-2 is to enable intracellular survival (Hensel et al., 1995, 1998; Ochtman et al., 1996; Cirillo et al., 1998; Vasquez-Torres et al., 2000b). In the latter case, it is likely that secretion and translocation occurs from an intracellular location and that effector proteins are targeted from the phagosome into the cytosol. This is also likely to be the case for the recently discovered type III secretion system of the intracellular pathogen *Chlamydia* (Hsia et al., 1997; Subtil et al., 2000). This pathogen enters host cells as a small infectious particle, the elementary body, which after internalisation differentiates into a larger replicating form, the reticulate body, that continues to proliferate inside vacuoles denoted inclusions (Moulder, 1991; Hackstadt, 2000). The inclusion vesicle does not fuse with endosomes or lysosomes but instead fuses with exocytic vesicles in the peri-Golgi (Hackstadt et al., 1995, 1996). The inclusion membrane has been shown to contain structures similar to those associated with type III secretion organelles in *Salmonella* and *Shigella* (Bavoil and Hsia, 1998). These structures are seen early after uptake of the elementary bodies and suggests that type III secretion of components into or across the inclusion membrane is involved in modulating the host cell to allow differentiation and replication of *Chlamydia*. Interestingly, Inc proteins that localise to the inclusion membrane have been shown to be secreted via a type III secretion mechanism in the heterologous host *Shigella flexneri* (Subtil et al., 2001). One of the Inc proteins expressed by *Chlamydia psittaci,* IncG, interacts with the host cell protein 14-3-3β via a C-terminal domain that extends into the host cell cytosol (Scidmore and Hackstadt, 2001). This argues for a mechanism where type III secreted effector proteins are targeted to or across the inclusion membrane where it interacts with host cell proteins. These protein interactions in turn could interfere with host cell signalling and trafficking of inclusion vesicles to allow replication of the pathogen.

The LEE pathogenicity island of EPEC encodes proteins involved in forming the characteristic attaching and effacing lesions at the intestinal surface (reviewed in Vallance and Finlay, 2000). The intimate attachment of EPEC to enterocytes requires type III-mediated translocation of proteins into the interacting cell. One of the proteins, Tir, is translocated into and inserted in the host cell membrane where it functions as the receptor to which the bacteria intimately bind (Kenny et al., 1997). The EPEC translocated proteins also interfere with the host cell actin cytoskeleton leading to formation of pedestal-like structures beneath the adherent bacteria. Hence, in the case of EPEC, the type III system appears to be needed for tethering of the bacteria to the host intestinal surface. Recent studies have shown that translocated effectors required for lesion formation are crucial for EPEC pathogenesis in humans (Tacket et al., 2000) and for similar symptoms in rabbits caused by a rabbit EPEC strain (Marchés et al., 2000).

The type III secretion systems of pathogenic *Yersinia* species and *P. aeruginosa* show a high degree of similarity extending essentially to all proteins involved in secretion across the bacterial membranes (see also above). Interestingly, some of the secreted effector proteins also show similarity and share biological function in that they prevent uptake by host cells, thereby contributing to the extracellular lifestyle of these pathogens (see below).

7.4.1 *P. aeruginosa* infections

The natural habitats of *P. aeruginosa* are soil and water, but it can also be isolated from plants and animals. In humans, *P. aeruginosa* is an opportunistic pathogen that causes infections only when epithelial barriers are damaged or the host immune system is impaired. Susceptible individuals include those with cystic fibrosis, burn wounds, or leukaemia (Mandell et al., 1995). In some cases the infections can develop into serious conditions, such as severe tissue damage of lungs (cystic fibrosis) or eyes and systemic infections after burn wounds. A large number of proteins produced by *P. aeruginosa* have been suggested to contribute to virulence, for example exotoxin A, elastase, phospholipase C, alginate, and type IV pili (Nicas and Iglewski, 1985; Hahn, 1997). The type III secretion system of *P. aeruginosa* mediates secretion and translocation of the exoenzymes, ExoS, ExoT, ExoU, and ExoY, into host cells (Yahr et al., 1996, 1998; Finck-Barbancon et al., 1997; Frithz-Lindsten et al., 1997). Assessment of virulence and pathogenic mechanisms for *P. aeruginosa* infections is difficult as the animal infection models used in most cases do not fully reflect the infections caused in humans. However, most studies support that *P. aeruginosa* is mainly extracellular during infection and there is also evidence suggesting that the type III secretion system is important for virulence (Dacheux et al., 1999, 2000; Sawa et al., 1999; Garrity-Ryan et al., 2000). In cell infection models, some strains of *P. aeruginosa* are described as invasive while other strains remain extracellular and have a cytotoxic effect on the infected cells (Fleiszig et al., 1997). It is however clear that for both types of strains, the type III effector proteins of *P. aeruginosa* act to prevent uptake of the pathogen by epithelial and phagocytic cells (Frithz-Lindsten et al., 1997; Cowell et al., 2000; Garrity-Ryan et al., 2000). It is therefore likely that this mechanism is important early during establishment of *P. aeruginosa* infections in humans including patients with cystic fibrosis. However, during the later chronic stages of *P. aeruginosa* infections in these patients, the type III secretion system is likely to be dispensable as some *P. aeruginosa* strains isolated from patients with chronic infections lack a functional type III secretion system as a result of mutations in the transcriptional activator ExsA (Dacheux et al., 2001). Expression of a complex system, such as the type III secretion/translocation system, is an energetically expensive process. There-fore, once the system is not required for maintaining the infection, there is a selective advantage to shut down the system. The role of quorum sensing in the control of *P. aeruginosa* infections is detailed in Chapter 9.

7.4.2 *Yersinia* infections

Yersinia pestis, the causative agent of bubonic plague, is usually transmit-ted to humans by an infected rodent flea and thereafter it invades lymphatic tissue and proliferates in lymph nodes. The other pathogenic *Yersinia* species, *Yersinia enterocolitica* and *Y. pseudotuberculosis,* both cause enteric infections that are usually self-limiting in humans (Kornhoof et al., 1999). Despite a different route of infection compared to *Y. pestis,* these orally transmitted pathogens also exhibit tropism for lymphoid tissue. Their primary site of infection is the lymphoid follicles of the small intestine, which they enter through the M cells overlying the surface of the Peyer's patches (Grutzkau et al., 1990). Enteropathogenic yersiniae encode a protein, invasin, a ligand for β_1-integrin receptors, and entry into this niche is facilitated by binding of this adhesin to β_1-integrins on the lumenal surface of M cells (Isberg and Leong, 1990; Autenrieth and Firsching, 1996; Marra and Isberg, 1997; Clark et al., 1998). At this location *Yersinia* proliferate in the extracellular fluid during infection (Hanski et al., 1989; Simonet et al., 1990). This extracellular life style, which depends on the ability to survive and proliferate in lymphoid tissues, is an essential virulence property shared by all three pathogenic *Yersinia* species. The virulence proteins of importance are all encoded on the common ~70 kb virulence plasmid; the components of the type III secretion, the effector proteins, and proteins involved in regulating translocation. The Yop effectors serve different functions during infection. Upon contact of phagocytic cells with *Yersinia*, neither engulfment nor release of inflammatory mediators occur (see Chapter 12), and there is no activation of the oxidative burst (Lian and Pai, 1985; Hartland et al., 1994; Ruckdeschel et al., 1996).

Pathogenic *Yersinia* cause lethal systemic infections in mice and therefore provide a suitable infection model to determine the importance of potential virulence effectors. Bacterial mutants defective in Yop expression are usually avirulent and rapidly cleared by the primary immune defence. YopH and YopE are effectors which, by distinct mechanisms, mediate blockage of phagocytosis (discussed below). Another translocated effector protein, YopJ, is involved in shutting down stress response signalling in host cells, involving blockage of MAP kinases, NF-κB and induction of TNF-α and IL-8 (Ruckdeschel et al., 1997; Schesser et al., 1998; Orth et al., 2000), as discussed in detail in Chapter 12. Other Yop effectors that are essential for virulence are YopM (Leung et al., 1990) and YpkA (Galyov et al., 1993), but their specific functions in the infection process are not fully established. YopM, which is an acidic protein containing leucine rich repeats, has been shown to be targeted to the host cell nucleus by a vesicle-associated mechanism (Skrzypek et al., 1998). YpkA is a serine/threonine kinase that localises to the host cell plasma membrane and interferes with the actin cytoskeleton (discussed below).

7.5 INHIBITION OF PHAGOCYTOSIS

Although phagocytosis can be mediated by different receptors that activate signalling pathways within cells, a common feature is regulation by tyrosine phosphorylation and the Rho-family of GTPases (Ernst, 2000; Chimini and Chavrier, 2000). Tyrosine kinases/phosphatases are important players in many processes involving receptor-mediated changes of the cytoskeleton as well as in regulating cellular adhesion events, which are closely related to changes in cell shape and migration. Phosphorylations/dephosphorylations of tyrosine residues in proteins generate or abolish interaction sites for SH2-containing, and other phosphotyrosine binding proteins alternatively activate/deactivate proteins involved in cell signalling. Like many other processes involving the cytoskeleton, β_1 integrin-mediated, Fc receptor-mediated and many other types of phagocytosis are impaired in the presence of tyrosine kinase inhibitors (Rosenshine et al., 1992; Greenberg et al., 1993; Magae et al., 1994; Andersson et al., 1996).

Rho family GTPases, which are a subgroup of the Ras superfamily of GTPases, regulate a variety of cellular functions, including actin reorganisation (reviewed in Hall, 1998). Like all members of the Ras superfamily, Rho proteins cycle between an inactive conformation bound to GDP and an active conformation bound to GTP. This cycling is regulated by guanine nucleotide exchange factors (GEFs) and GTPase activating proteins (GAPs). GEFs have an activating effect by promoting GDP dissociation and GTP binding, whereas GAPs counteract this by stimulating the intrinsic GTPase activity of these proteins. Rho induces assembly of contractile F-actin bundles (stress fibres) that are anchored to focal adhesion structures. Other members of the family, Rac and Cdc42, are involved in regulation of protrusive actin structures, such as formation of lamellipodia and filopodia, respectively. In phagocytosis, different Rho GTPases operate in different types of receptor-mediated phagocytosis. Phagocytosis via complement receptors requires Rho but not Rac and Cdc42, whereas uptake via Fcγ receptors requires Rac and Cdc42 but not Rho (Caron and Hall, 1998).

In the mid 1950s, Burrows and Bacon (1956) conducted studies demonstrating that virulent strains of *Y. pestis* resisted engulfment by professional phagocytes. Rosqvist and co-workers (1988, 1990) confirmed and extended these studies using *Y. pseudotuberculosis*. It was clearly demonstrated that the ability to block phagocytosis was linked to the expression of the virulence plasmid and that the effectors were the plasmid encoded determinants YopH and YopE. In contrast, strains not expressing YopH or YopE were avirulent in mice (Straley and Bowmer, 1986; Bölin and Wolf-Watz, 1988; Forsberg and Wolf-Watz, 1988), underscoring the coupling between virulence and the ability to prevent phagocytosis. Today we also know that antiphagocytosis by *Yersinia* involves blocking of the uptake of the bacteria via various phagocytic receptors (e.g., Fc receptors and integrin receptors) by both macrophages and granulocytes (Fällman et al., 1995; Visser et al., 1995; Ruckdeschel et al., 1996). Hence, the pathogen can overcome opsonisation by both complement and IgG, two important host defence molecules that are abundant in lymphoid tissue. Moreover, pathogenic *Yersinia* circumvent phagocytosis by interfering with both tyrosine kinase and GTPase signalling of host cells through the actions of YopH and YopE.

Phagocytic uptake by professional phagocytes is a rapid process that is activated immediately as the bacterium interacts with receptors on the phagocyte surface. From this, it follows that the mechanisms used by a microbe to block phagocytosis must be utilised very rapidly. It should be kept in mind that the effectors have to be translocated from the extracellularly located bacterium into the interior of the host cell to perform their tasks. In line with this, studies done on the role of the *Yersinia* effectors implies that at least YopH exerts a nearly instantaneous effect on immediate early signalling in phagocytes (Andersson et al., 1999). YopH has been found to impede β_1-integrin-mediated elevations in the intracellular concentration of free Ca^{2+} in human neutrophils (Andersson et al., 1999). That Ca^{2+}-signal is triggered at almost the same moment that *Yersinia* binds to the surface of the cell. Thus, the ability to block such a rapid signal requires a strategy that ensures that the site of action is reached instantaneously.

Although *Yersinia* is generally assumed as the paradigm concerning antiphagocytosis, other pathogens have been suggested to exhibit the ability to block phagocytosis by interfering with intracellular signalling of phagocytes. These are *P. aeruginosa*, EPEC, and *Helicobacter pylori*. The latter does not, however, use the type III secretion machinery, instead it appears to involve type IV secretion components (Ramaro et al., 2000). Although phagocytic blocking by EPEC appears to require a functional type III secretion machinery, the specific mechanism behind it is different from that of *Yersinia* in that the onset of the blocking effect is less rapid (Andersson et al., 1999; Goosney et al., 1999). As mentioned above, *P. aeruginosa* appears to use a mechanism that is to some extent analogous to *Yersinia* comprising a homologous type III apparatus mediating injection of effectors, ExoS and ExoT. These exoenzymes display similarities with YopE, and like YopE they prevent uptake by different types of cells including phagocytes (Yahr et al., 1997; Frithz-Lindsten 1998; Sawa et al., 1999; Cowell et al., 2000; Garrity-Ryan et al., 2000).

7.6 SUBVERSION OF HOST CELL TYROSINE KINASE SIGNALLING

7.6.1 The *Yersinia* effector YopH

In 1990, Guan and Dixon (1990) performed a database search for proteins sharing sequence identity with the human protein tyrosine phosphatase

(PTPase) PTP1 and identified YopH as one homologue. Further characterisation and structural analysis of YopH revealed that this bacterial virulence effector shared considerable homology with eukaryotic PTPases and was by far the most active PTPase identified (Guan and Dixon, 1990; Zhang et al., 1992; Denu et al., 1996; Tonks and Neel, 1996). Accordingly, because tyrosine kinases/phosphatases were known as important players in receptor-mediated changes of the cytoskeleton, it was generally assumed that *Yersinia*, when blocking its own phagocytosis via YopH, subverted tyrosine kinase signalling of importance for the process. Furthermore, because phagocytosis is initiated almost immediately upon binding of a bacterium to the cell surface, YopH was expected to have a very rapid effect. Experiments designed to verify these assumptions showed that YopH indeed interrupted a very early, infection-induced, phosphotyrosine signal and that a phosphotyrosine protein of approximately 120–125 kD was the primary target of YopH in macrophages (Andersson et al., 1996). The YopH-mediated resistance of *Yersinia* uptake is however not restricted to phagocytes. In HeLa cells, the invasin-promoted uptake is blocked by YopH, which also in this case acts by interrupting phosphotyrosine signalling induced by bacterial binding to the β_1-integrin receptor (Persson et al., 1997). In studies investigating molecular targets of YopH, a so-called "substrate trapping" method was used, in which YopH-interacting proteins were co-precipitated with a point-mutated inactive variant of YopH from lysates of infected cells (Bliska et al., 1992). A highly tyrosine-phosphorylated form of p130 Crk-associated substrate (Cas) was found as a common substrate of YopH in both macrophages and HeLa cells (Black and Bliska, 1997; Persson et al., 1997; Hamid et al., 1999). However, Cas did not constitute the only or major substrate for YopH in macrophages; another protein, namely Fyn-binding protein (FYB) appeared as the primary substrate in these cells (da Silva et al., 1997a, 1997b; Hamid et al., 1999). Both Cas and FYB are known to participate in signal transduction from the β1-integrin receptor to the cytoskeleton (Hunter et al., 2000; O'Neill et al., 2000), but whether this signalling concerns phagocytic uptake remains to be clarified. It is however clear that the effects exerted by YopH on these proteins are very rapid: the invasin-stimulated phosphotyrosine signal is blocked within one minute of infection (Andersson et al., 1996), and the association of YopH with its substrates can be detected after only two minutes (Persson et al., 1997). Thus, by interfering with these host cell molecules the signal transduction from the integrin receptor is effectively down-regulated.

Other host cell proteins, such as paxillin and focal adhesion kinase (FAK – only in HeLa cells), have been implicated as YopH substrates (Andersson et al., 1996; Persson et al., 1997; Black et al., 1998). Paxillin is rapidly dephosphorylated in cells infected with strains expressing YopH and in HeLa cells, FAK was identified together with Cas in the YopH substrate trap assay. However, in contrast to Cas and FYB, a direct interaction of paxillin with YopH has only been seen *in vitro*, and FAK does not interact directly with YopH (Black et al., 1998; De Leuil and Fällman unpublished). FAK can bind to the SH3 domain of Cas via its proline rich domain (Polte and Hanks, 1995), and this could explain the detection of FAK in YopH immuno-precipitates. Like Cas, both FAK and paxillin are found in cell adhesion structures, such as focal adhesions. These cellular structures are rapidly destroyed by YopH upon infection of cultured cells (Black and Bliska, 1997; Persson et al., 1997). This is seen as a rounding up and subsequent detachment of infected cells, denoted YopH-mediated cytotoxicity, which is distinct from the well-characterised YopE-mediated cytotoxicity (discussed below and Fig. 7.2).

7.6.2 Role of the YopH targets in normal cell function

Cas and FAK share many features indicative of function in a common signalling pathway: (i) they interact with each other, (ii) are tyrosine phosphorylated upon clustering of β1-integrins, and (iii) are found within focal complex structures, where they function as docking proteins in recruitment of signalling molecules. There are many indications that Cas and FAK are involved in receptor-mediated regulation of the actin cytoskeleton. Both proteins promote migration when overexpressed in cells (Cary et al., 1996, 1998; Klemke et al., 1998), and results from gene-disruption studies in mice indicated that they are essential for survival and are involved in actin-associated cell adhesion events (Ilic et al., 1995, 1996; Honda et al., 1998, 1999). FAK-/- cells exhibited reduced motility because of impaired turnover of focal adhesions. Cas-/- cells exhibited impaired stress fibre formation as a result of disturbed cytoskeletal organisation at sites of actin anchoring. Hence, the involvement in these cytoskeletal and adhesion-associated adhesion events makes both FAK and Cas potential regulators of bacterial uptake and thereby suitable targets for an antiphagocytic bacterial virulence effector. Support for this came from studies in which cells were infected with *Yersinia* strains lacking YopH. In this case, the bacterial infection stimulated invasin-dependent tyrosine phosphorylation of both Cas and FAK, and these proteins were also recruited to focal complex structures lining the edges of the cell (Persson et al., 1997). These events take part close to where the bacteria initially bind. Therefore, it likely that in the absence of YopH, infection-induced phosphorylated forms of Cas and FAK associate to focal complex structures and participate in the β1-integrin-mediated uptake of the interacting bacterium. However,

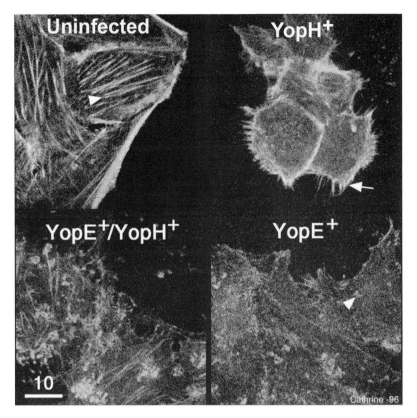

Figure 7.2. YopH and YopE have distinct effects on host cell cytoskeleton. Confocal images of HeLa cells, not infected, or infected with a *Yersinia* multiple *yop* mutant strain expressing either YopH or YopE, or infected with *Yersinia* wt. YopE fragments the F-actin cytoskeleton whereas YopH affects the integrity of focal adhesions and associated stress fibers. The combined effect is seen with the *Yersinia* wt strain (NB. this strain translocates less amount of the effectors compared to the multiple mutant strain). Cellular F-actin was visualized by staining with fluorescein-conjugated phalloidin (green); vinculin-containing focal adhesions were visualized by indirect immunofluorenscence (red). The yellow color represents colocalization of microfilaments and vinculin. Vinculin-containing focal adhesions (arrow heads) and vinculin-containing retraction fibers (arrows) are shown. All sections were scanned under identical conditions and show the basolateral side of the cells. Scale bar: 10 μm.

when the roles of FAK and Cas in uptake subsequently were addressed by utilising the knock-out cells, somewhat divergent results were obtained. FAK-/- fibroblasts were found to be severely impaired in their ability to mediate bacterial uptake, and similar results were obtained when a dominant negative variant of FAK was overexpressed (Altrutz and Isberg, 1998; McGee and Fallman, unpublished). On the other hand, Cas-/- fibroblasts were un-affected in bacterial uptake, but a reduction was seen when a dominant-negative variant of Cas was overexpressed (Weidow et al., 2000, McGee and Fallman, unpublished). This suggests a role for FAK in the uptake process of these cells, whereas the importance of Cas remains obscure. It is important to bear in mind that all experiments regarding FAK and Cas were done in nonprofessional phagocytes. FAK is expressed at very low levels in macrophages and has not been found as a substrate of YopH in these cells (Choi et al., 1993; Lin et al., 1994; Hamid et al., 1999). It is possible that professional phagocytes utilise a somewhat different mechanism for uptake than do fibroblasts. Importantly, these cells display a different cytoskeletal organisation with fine actin cables rather than stress fibres, and the focal contact structures are smaller and more dynamic (Gumbiner, 1996). Interestingly, FYB, which was found as the target substrate of YopH in macrophages, is specifically expressed in cells of hematopoetic origin (Jücker et al., 1997; da Silva et al., 1997a). The exact role of FYB in these cells is not known, but its cell type-specific expression suggests an immunological function. FYB, together with Cas, constitute specific substrates for the Fes tyrosine kinase in macrophages (Jücker et al., 1997), and FYB has been implicated as participating in β1 integrin signalling, although not as a focal adhesion protein. Interestingly, it was recently shown that FYB interacts with the actin-binding/regulating protein VASP (Krause et al., 2000), a finding that links this YopH substrate to regulation of the actin cytoskeleton in these cells.

7.6.3 Immediate early targeting of focal complex structures by YopH

The amino acid sequences of PTPases situated outside the catalytic domain are thought to be involved in localisation and/or regulation of these enzymes and thereby contribute to their *in vivo* substrate specificities (Mauro and Dixon, 1994). The amino acid sequence in the N-terminal noncatalytic region of YopH does not have any obvious homology with known proteins. However, it is known that the seventy most N-terminal amino acids are required for secretion and translocation of YopH (Sory et al., 1995). Furthermore, it has been demonstrated that the 130 most N-terminal amino acids contain the binding site for Cas and FYB, although the biological importance of this interaction still has to be clarified (Black et al., 1998; Montagna et al., 2001; DeLeuil and Fallman unpublished). In addition, a more central region is of importance for YopH-mediated effects on host cells. *Yersinia* expressing YopH lacking four amino acids within this region fail to block phagocytosis by macrophages and HeLa cells, and are no longer virulent in mice (Persson et al., 1999). This region is distinct from the Cas binding site and no interaction partner has been identified so far. This mutation also rendered YopH unable to localise to peripheral focal complexes in infected cells (Fig. 7.3), suggesting that these structures are important for the phagocytic process. It was also established that this region of YopH is required for early effects on host cells. A *Yersinia* strain expressing this mutated variant of YopH failed to block the immediate early Ca^{2+} response in neutrophils (Persson et al., 1999). Thus, fully PTPase active YopH that cannot localise to peripheral focal complex structures fails to block immediate early signalling in the phagocyte and cannot promote phagocytosis and infection. It is likely that this region allows YopH entering via the type III translocation apparatus to "hook up to" active integrin-associated signalling complexes situated beneath the extracellularly located bacteria. By doing so, YopH disarms the signalling complexes involved in mediating uptake of the very bacterium that targeted the PTPase into the host cell.

7.7 SptP OF SALMONELLA

When the *S. typhimurium* chromosomal region SPI-1 was characterised, an open reading frame encoding a protein sharing homology with both *Yersinia* and *P. aeruginosa* effectors was identified (Kaniga et al., 1996). The protein was called SptP and consisted of two functional domains. The C-terminal domain exhibited significant homology with the catalytic domain of YopH, and the N-terminal part shared similarities with YopE and exoenzymes S and T of *P. aeruginosa* (illustrated in Fig. 7.4). SptP is translocated into host cells, but the biological role of its PTPase domain is not clear (Fu and Galán, 1988). It exhibits PTPase activity, but does not dephosphorylate Cas or disrupt focal adhesion structures of cells, and neither does it mediate an antiphagocytic effect (Kaniga et al., 1996; Fu and Galán, 1998). As discussed above, the substrate specificity of PTPases are likely to be influenced by sequences outside their catalytic domains, which mediate their interactions with other proteins and subcellular localisations. Accordingly, SptP with its N-terminal GAP-domain (discussed below) and YopH with its focal adhesion localisation have different subcellular distributions (Black and Bliska, 1997; Persson et al., 1997; Fu and Galán, 1998). Structural analyses of SptP showed that the PTPase region was different from YopH with respect to shape and charge distribution. The homology with YopH spanned the active site and the hydrophobic core, whereas amino acids responsible for the surface properties were different (Stebbins and Galán, 2000). Therefore, it is not surprising that

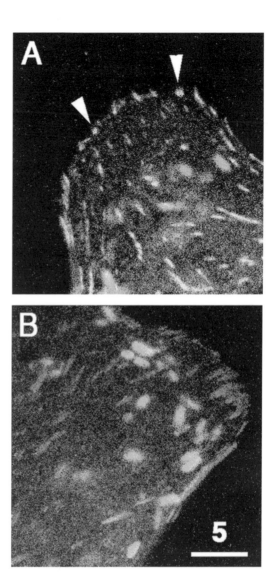

Figure 7.3. YopH contains an inherent sequence that mediates localisation to host cell peripheral focal complex structures. Confocal images of HeLa cells infected with a *Yersinia* multiple *yop* mutant strain expressing the PTPase inactive YopHC403A (A) or an in frame deletion mutant thereof, YopHC403AΔ223–226 (B) (region of importance for localisation to focal complex structures). Arrowheads indicate representative focal complexes where YopH and vinculin colocalise (yellow). NB: some bacteria are stained by the YopH antiserum. All sections were scanned under identical conditions and shows the basolateral side of cells. Scale bar: 5 μm.

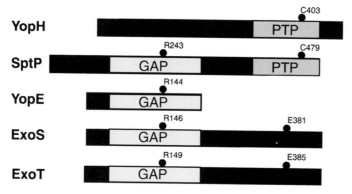

Figure 7.4. Bacterial effectors with PTPase and GAP activities. Schematic illustrations of YopH (*Yersinia*), SptP (*Salmonella*), YopE (*Yersinia*), ExoS, and ExoT (*P. aeruginosa*), which represent homologous type III translocated effector proteins exhibiting PTPase and/or GAP activities. The dark grey boxes indicate homologous regions encoding PTPase domains, whereas light grey boxes indicate homologous regions encoding GAP domains. The position of residues essential for activity and which include cysteine in the PTPase domains, arginine in the GAP domains and others important for ADP ribosylation are indicated.

the PTPases of *Salmonella* and *Yersinia* have different substrate specificities and thereby distinct effects on host cells.

7.8 INTERFERENCE WITH HOST CELL GTPases

The small GTP-binding proteins of the Rho subfamily, Rho, Rac, and Cdc42, play important roles in various cellular events including actin cytoskeleton rearrangements, membrane trafficking, transcriptional regulation, and cell growth control (Hall, 1998; Van Aelst and D'Souza-Schorey, 1997). A number of bacterial pathogens have evolved toxins that specifically targets and covalently modify the Rho GTPases resulting in either activation or inactivation (Aktories, 1997; Aktories et al., 2000). There are also several bacterial effectors that instead are potent regulators of the GTPases, acting as GAP or GEF proteins (see also above).

7.8.1 *Yersinia* effectors interfering with Rho GTPases

The type III secretion system of *Yersinia* is known to translocate six Yop effector proteins into host target cells. Of these, YpkA, YopT, and YopE target host proteins of the Rho-GTPase family. Thus, potentially all these effectors could be involved in preventing phagocytosis, but this has been clearly established only for YopE (see below).

YpkA is a serine/threonine kinase with sequence similarity to eukaryotic kinases (Galyov et al., 1993). Strains mutated for *ypkA* are avirulent in the mouse infection model (Galyov et al., 1994). YpkA is targeted into infected cells via the type III system and localises to the host cell plasma membrane (Håkansson et al., 1996a). Interestingly, YpkA was recently shown to bind RhoA and Rac1 and cause disruption of the host cell cytoskeleton (Håkansson et al., 1996a; Barz et al., 2000; Dukuzumuremyi et al., 2000; Juris et al., 2000). However, the effect of YpkA occurs relatively late and despite the effect on the host cell cytoskeleton, no anti-phagocytic effect has been reported.

YopT is targeted into host cells by *Y. enterocolitica* and its effect is similar to that of YopE, i.e., disruption of actin microfilaments (see below; Iriarte and Cornelis, 1998). YopT modifies RhoA causing its redistribution, which is associated with disruption of stress fibres (Zumbihl et al., 1999). No studies have yet been undertaken to elucidate if YopT, similar to YopE, is involved in blocking uptake. In a mouse infection model, a *yopT* mutant of *Y. enterocolitica* was unaffected in the ability to colonise the Peyer's patches of orally infected mice (Iriarte and Cornelis, 1998). Not all strains of *Y. pseudotuberculosis* express YopT and the strains lacking YopT are still fully virulent in the systemic mouse infection model (Forsberg and Cherepanov, unpublished results). Therefore, the role of YopT in *Yersinia* infections remains to be established.

7.8.2 The *Yersinia* effector YopE

YopE was one of the first secreted virulence plasmid-encoded proteins of *Yersinia* that was proven essential for virulence in the systemic mouse infection model (Straley and Bowmer, 1986; Forsberg and Wolf-Watz, 1988). The previously recognised virulence plasmid encoded cytotoxic effect of *Yersinia* on eukaryotic cells (Rosqvist and Wolf-Watz, 1986; Goguen et al., 1986) was later shown to be mediated by YopE (Rosqvist et al., 1990). Work by Rosqvist et al. (1991) established that delivery of purified YopE into HeLa cells via microinjection could reproduce the effect of bacterial infection of cultured cells. This made it clear that the target for YopE was intracellular. However, the challenge at the time was to elucidate how YopE was targeted into the cell. Addition of purified secreted proteins had no effect on cell morphology, which made it unlikely that YopE was delivered via a similar mechanism as other bacterial toxins. One important observation was that a strain mutated for one of the secreted proteins, YopD, did not induce a cytotoxic response on infected HeLa cells even though YopE secretion was unaffected (Rosqvist et al., 1991). In subsequent work, YopE delivery into infected cells was shown to be mediated by adhering extracellular bacteria involving at least three of the secreted proteins, YopB, YopD, and LcrV (Rosqvist et al., 1994; Sory and Cornelis, 1994; Boland et al., 1996; Håkansson et al., 1996b; Pettersson et al., 1999).

As YopE was responsible for the effect on cell morphology (cell rounding) it seemed plausible that YopE would have a major effect on phagocytosis. However, when *yopE* mutants were tested in a macrophage phagocytosis assay, only a minor effect was observed (Rosqvist et al., 1990). The effect on phagocytosis was much more dramatic when *yopH* mutants were tested. Interestingly, when a double *yopE/yopH* mutant was analysed, a synergistic effect on uptake was obvious, indicating that YopE and YopH acted in concert to block phagocytosis. As discussed earlier in this chapter, phagocytosis is a rapid process that is initiated immediately upon contact between pathogen and phagocyte. Therefore, to block uptake, the action of YopE/YopH must be both rapid and specific. This suggests that a pool of YopE/YopH is present before contact with the phagocytic cell and ready to be rapidly delivered into the host cell upon binding. *In vitro*, removal of Ca^{2+} mimics cell contact and YopE is rapidly secreted immediately after shifting the bacteria from secretion-inhibiting ($+Ca^{2+}$) to secretion-permissive conditions (no Ca^{2+}). This rapid secretion of YopE was dependent on the cognate chaperone YerA (SycE), but independent of *de novo* protein synthesis (Lloyd et al., 2001). The amount of translocated YopE in the absence of *de novo* protein synthesis was also sufficient to induce a cytotoxic effect on target cells (Lloyd et al., 2001). Thus, there appears to be a pool of Yop proteins in noninduced bacteria that is ready for translocation immediately upon contact with the phagocytic cell.

When the effect on cell morphology was studied in more detail, YopE was found to mediate disruption of the host cell actin microfilaments (Rosqvist et al., 1991). However, there was no direct effect on actin polymerisation *in vitro* and YopE did not have any effect on Ca^{2+} or cyclic AMP levels in the cell (Rosqvist et al., 1991). Rather, this suggested that YopE targeted factors involved in regulating the actin cytoskeleton. This was later confirmed when it was shown that YopE acts as a GAP for Rho GTPases (von Pawel-Rammingen et al., 2000; Black and Bliska, 2000). The GAP activity of YopE was shown to be essential for its effect on: (i) the actin cytoskeleton, (ii) uptake of *Y. pseudotuberculosis* in infected HeLa cells, and (iii) virulence in mice (von Pawel-Rammingen et al., 2000; Black and Bliska, 2000). Experiments using HeLa cells transfected with constitutively active RhoA or Rac1 indicated that down-regulation of RhoA by YopE is important for disruption of actin filaments. On the other hand, YopE-mediated down-regulation of Rac1, but not RhoA, is important for blockage of bacterial uptake into HeLa cells (Black and Bliska, 2000). Therefore, the immediate effect of the YopE GAP-activity (in concert with YopH) results in blockage of phagocytosis while the changed morphology (disruption of actin microfilaments) occurs only after prolonged exposure of the cell to YopE.

7.8.3 The *Pseudomonas aeruginosa* effectors Exoenzyme S and T

ExoS was one of the first recognised secreted exoenzymes of *P. aeruginosa* with ADP-ribosylating activity *in vitro*. It was shown to modify several eukaryotic proteins, in particular small GTP binding proteins (Coburn et al., 1989; Coburn and Gill, 1991) and required an intracellular eukaryotic host protein for ADP-ribosyltranferase activity *in vitro*, identified as factor activating exoenzyme S (FAS) (Coburn et al., 1991). FAS was subsequently shown to be a member of the 14-3-3 family of proteins (Fu et al., 1993), which are known to activate eukaryotic signalling molecules (Fu et al., 2000). It was also clear that ExoS did not have the typical features of the two-subunit or two-domain toxins such as exotoxin A. Exotoxin A is another ADP-ribosyltransferase secreted by *P. aeruginosa*, which is targeted into host cell via a receptor-mediated mechanism (Kounnas et al., 1992). The intracellular targeting of ExoS was subsequently shown to be mediated by a contact-dependent type III secretion mechanism (Frithz-Lindsten et al., 1997; Frank, 1997; Sawa et al., 1999). Direct proof for intracellular delivery of ExoS via a type III secretion mechanism came from work where ExoS was expressed in the heterologous host *Y. pseudotuberculosis*. In a similar way to the Yop effectors, ExoS was targeted into infected cells by the type III secretion system of *Yersinia*. Interestingly, ExoS elicited a cytotoxic response involving disruption of host cell actin microfilaments. This effect was similar to that induced by YopE but it was more drastic and had a significant effect on host cell viability (Frithz-Lindsten et al., 1997). The observed similarity in the biological effect of ExoS and YopE was not unexpected as the N-terminal section of ExoS is homologous to YopE, whereas the C-terminal domain that contains the active site for the ADP-ribosylating activity is unique for ExoS (illustrated in Fig. 7.4; Yahr et al., 1995). Intracellular delivery of ExoS was also found to prevent uptake of *Yersinia* by macrophages at a level comparable to YopE (Frithz-Lindsten et al., 1997). The glutamic acid residue in position 381 is essential for the ADP-ribosyltransferase activity of ExoS and replacement of this amino acid with alanine resulted in a more than 2000-fold reduction of enzymatic activity. However, delivery of ExoS-E381 still elicited a cytotoxic response involving disruption of actin microfilaments (Frithz-Lindsten et al., 1997). The enzymatically inactive form of ExoS also promoted a certain level of phagocytosis blockage, albeit at a somewhat lower level compared to wild-type ExoS. This suggested that ExoS is a bifunctional protein with an N-terminal domain with an activity similar to that of YopE and a C-terminal domain with ADP-ribosylating activity, which also affects cell viability.

The bifunctional property of ExoS was confirmed in studies where the N-terminal and C-terminal domains of ExoS were expressed in CHO cells. Expression of the C-terminal 222 amino acids of ExoS in CHO cells induced a cytotoxic response (cell rounding) and down-regulation of protein synthesis and cell death (Pederson and Barbieri, 1998). This effect was only seen for ExoS derivatives with ADP-ribosyltransferase activity. The effect on cell viability of ExoS on host cells is likely the result of inactivation of Ras via ADP-ribosylation (McGuffie et al., 1998; Vincent et al., 1999; Henriksson et al., 2000). ExoS has been shown to interfere with Ras-mediated signal transduction pathways, which in turn has the potential to affect multiple signalling pathways involved in survival and proliferation as well as the cytoskeleton. The fact that targeting of a derivative of ExoS with essentially no ADP-ribosyl transferase activity still induced disruption of actin microfilaments strongly argued that ExoS had two distinct activities (Frithz-Lindsten et al., 1997). Indeed, expression of the N-terminal 234 amino acids of ExoS in CHO cells, or bacterial delivery of the ExoS N-terminal, caused disruption of actin microfilaments (Pederson et al., 1999; Henriksson et al., 2000). In a similar manner to the homologous proteins, YopE and SptP, the ExoS N-terminal domain was shown to be a GAP for Rho GTPases (Goerhing et al., 1999). Support for the idea that Rho GTPases are targets of the ExoS GAP activity came from work showing that disruption of actin microfilaments could be blocked by the *E. coli* cytotoxic necrotizing factor 1 (CNF-1) (Pederson et al., 1999). The CNF-1 toxin specifically activates Rho GTPases via a deamidation of a glutamine residue (Schmidt et al., 1997) and it is possible that the GAP activity of ExoS is unable to inactivate the CNF-1 activated form of Rho.

ExoT was originally described as an enzymatically less active form of ExoS, as the proteins were antigenically highly related. However, it was subsequently shown that ExoT was encoded by a separate gene and the two proteins showed an overall homology of 76% identity at the amino acid level (Yahr et al., 1996). Despite the high level of homology, ExoT only displays about 1% ADP-ribosyltransferase activity on Ras *in vitro* compared to ExoS (Lui et al., 1997). ExoT appears to be present in most clinical strains (Fleiszig

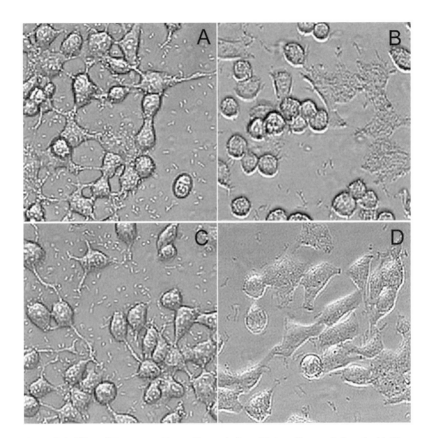

Figure 7.5. Effect of Exoenzyme T on cell morphology. HeLa cells were infected with *Y. pseudotuberculosis* expressing different forms of ExoT of *P. aeruginosa*. (A) wt ExoT. (B) GAP negative ExoT (R146K). (C) ADP-ribosyltransferase negative ExoT (E385A). (D) GAP/ADP-ribosyltransferase negative ExoT (R146K/E385A). The enzymatically inactive form of ExoT (D) has no effect on cell morphology, whereas the wt (A) and the ADP-ribosyltransferase negative form (C; has only GAP activity) has a similar effect on cell morphology as YopE, i.e., the cells round up but do not detach from the surface. The GAP-negative form of ExoT (B) shows a more profound effect on cell morphology, which suggests that ExoT targets a yet unidentified protein that is involved in regulating actin microfilaments in host cells.

et al., 1997; Sundin et al., 2001), which suggests that this exoenzyme has an important role in human infections. Similar to ExoS, intracellular targeting of ExoT results in a rounded cell morphology (Fig. 7.5), albeit less pronounced than for ExoS (Vallis et al., 1999; Garrity-Ryan et al., 2000; Sundin et al., 2001). In addition, similarly to the related proteins ExoS and YopE, expression of ExoT has been noted to inhibit uptake by corneal cells (Cowell et al., 2000) as well as epithelial cells and phagocytes (Garrity-Ryan et al., 2000). The N-terminal region of ExoT is homologous to ExoS, YopE, and SptP (Fig. 7.4), and it was recently shown that ExoT also exhibits GAP activity on Rho proteins (Krall et al., 2000; Garrity-Ryan et al., 2000; Sundin et al., 2001). The ExoT GAP activity is important for disruption of actin microfilaments as well as for the antiphagocytic effect (Garrity-Ryan et al., 2000). In contrast to ExoS, ExoT has no major effect on cell viability and does not modify Ras *in vivo* (Sundin et al., 2001). From this one could expect that the main biological activity of ExoT is the GAP activity on Rho GTPases and that the low level of ADP-ribosyltransferase activity does not have any significant biological role *in vivo*. However, it was recently shown that a GAP-negative variant of ExoT still induced a cytotoxic response on cells (Fig. 7.5; Sundin et al., 2001). This argues that *in vivo* there is an as yet unrecognised target for the ADP-ribosylating activity of ExoT. It is, however, not known if this activity is involved in blocking uptake. Some strains of *P. aeruginosa* are invasive in cell infection models whereas other strains are noninvasive but cytotoxic to cells. For both invasive and noninvasive strains, the biological effect of ExoT prevents uptake (Cowell et al., 2000; Garrity-Ryan et al., 2000; Sundin et al., 2001). A recent study by Kazmierczak et al. (2001) demonstrated that uptake of *P. aeruginosa* by epithelial cells correlates with an increase in RhoGTP levels, and expression of a constitutively active form of RhoA increased the internalisation of the bacteria. In contrast, no increased uptake was seen when constitutively active forms of Rac1 and Cdc42 were expressed. This suggests that RhoA is important for uptake of *P. aeruginosa* by epithelial cells and that blocking of uptake is caused by the GAP activity of ExoT.

7.8.4 SptP of Salmonella

SptP, the bifunctional effector of *S. typhimurium*, is also targeted into host cells via a type III secretion mechanism. The N-terminal region of SptP shows homology to YopE and the amino terminal domains of ExoS and ExoT (Fig. 7.4). SptP was the first of these proteins that was shown to inactivate Rho GTPases by acting as a GAP protein (Fu and Galán, 1999).[*] Even though *S. typhimurium* translocates SptP with its GAP activity into host cells, the overall biological effect of the type III secretion system encoded by SPI-1 is to promote uptake and invasion rather than blocking uptake. *Salmonella* triggers the activation of Cdc42 and Rac by injecting SopE and SopB through the SPI-1 encoded type III secretion system. SopE acts as a potent exchange

---

[*] The three-dimensional structures of ExoS, SptP, and YopE are shown in Fig. 7.6.

Structure and specificity of the bacterial GAP proteins

A   B

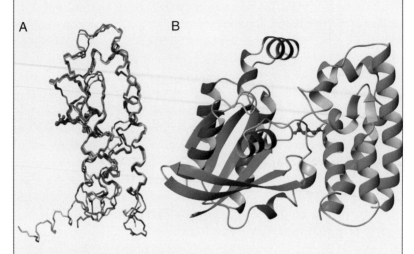

Figure 7.6. Structure of bacterial GAP proteins. (A) The three-dimensional structures of ExoS (yellow), SptP (red), and the calculated structure of YopE (blue), superimposed on top of each other. The N-terminal part of the respective protein is located to the left in the image. The arginine-finger extends out to the left in the middle of the protein structure. (B) Ribbon diagram of modeled YopE in complex with Rac. Rac1 is shown in blue-green and YopE in red-yellow. The main building block of the GAP proteins is a right-handed anti-parallel four-helix bundle with a characteristic up-down-up-down topology. Two characteristic bulges (I and II) and the neighbouring helices encompass the two most conserved regions of the bacterial GAPs (Stebbins and Galán, 2000; Wurtele et al., 2001).

The crystal structure of a complex between Rac and the GAP domain of SptP and ExoS, respectively, was recently published (Stebbins and Galán, 2000; Würtele et al., 2001). This work established that the amino acid sequence and tertiary fold of ExoS and SptP and their interactions with Rho proteins are different compared to that of eukaryotic RhoGAPs (Scheffzek et al., 1998). This implies that bacterial and eukaryotic RhoGAPs have evolved through convergent evolution. Since YopE shows high degree of similarity to the GAP domains of SptP (33% identity, 44% similarity) and ExoS (27% identity, 53% similarity), the three-dimensional structure of YopE can be modelled. By the use of the SWISS-model program (Peitsch, 1995; Guex and Peitsch, 1997; Guex et al., 1999) and the x-ray co-ordinate files of SptP and ExoS a three dimensional structure of YopE was created and superimposed on the SptP and ExoS structures. The overall structures of the three GAP proteins are very similar. The main building block of the GAP proteins is a right-handed antiparallel four-helix bundle with a characteristic up-down-up-down topology. Two characteristic bulges (loops) and the neighbouring helices encompass the two most conserved regions of the bacterial GAPs. The three bacterial GAPs, YopE, ExoS, and SptP, have different substrate specificities *in vitro*. YopE and ExoS show similar activities for Rho, Rac, and Cdc42, while SptP is equally activity on Rac and Cdc42 but have a 1000-fold lower activity on Rho (Fu and Galán, 1999; Goerhing et al., 1999; Black and Bliska, 2000; von Pawel-Rammingen et al., 2000). This difference in substrate specificity may be explained by that the three GAPs have diverged in several elements involved in binding to the RhoGTPases.

factor (Hardt et al., 1998) and SopB is an inositol polyphosphate phosphatase (Norris et al., 1998; Zhou et al., 2001). The delivery of SopE and SopB into epithelial cells induces membrane ruffling leading to internalisation of the bacteria. The cellular responses stimulated by *Salmonella* are short-lived, and immediately after bacterial entry, the cells regain normal architecture. It has been suggested that the SptP GAP activity is involved in turning off the early activation of the Rho GTPases and thereby normalising the cytoskeleton of the infected cell (Fu and Galán, 1999). The activities of the effector proteins of *S. typhimurium* are probably related to how the effector proteins are delivered by the SPI-1 encoded type III secretion system. It is likely that the effectors are delivered in a specific order where the proteins that activate the Rho GTPases are delivered before SptP.

7.9 CONCLUSIONS

Pathogenic *Yersinia* species and *P. aeruginosa* target specific virulence effectors into host cells to prevent uptake. This mechanism is part of an essential virulence mechanism for these pathogens, which strongly contributes to their resistance to phagocytosis and extracellular replication during infection. In order to prevent uptake by host cells, these pathogens have evolved a strategy where key molecules, which control uptake of bacteria, are targeted by virulence effectors. YopH targets host cell signalling at the level of tyrosine phosphorylation, a key signal used by eukaryotic cells to activate/deactivate proteins in signal transduction. YopE and ExoS/T instead target host cell GTPases that have important roles in regulating the host cell cytoskeleton. These effector proteins modulate the activity of the GTPases by modifications or by acting as GAP proteins to inactivate the GTPases.

The examples from type III systems of different pathogens also demonstrate that not only the biological activity of the delivered proteins is important; the timing of the delivery of effector proteins is also essential. Apparently, the type III secretion systems of *Yersinia* and *P. aeruginosa* have evolved to optimise rapid delivery of effector proteins. Phagocytosis is a rapid process and this requires that the delivery of blocking effectors occurs instantly after bacterial-cell contact. This is most likely achieved by a mechanism involving accessory proteins such as chaperones and TyeA, which are specifically required for delivery of the effector proteins involved in preventing uptake of these pathogens.

ACKNOWLEDGMENTS

We thank our technical associates, students, fellows, and colleagues for support, effort, and direct contributions to work cited. We are grateful to R. Nordfelth, C. Sundin, and M. Aili for help with figures. This work was supported by the Swedish Medical Research Council, the Swedish Foundation for Strategic Research (SSF), the SSF-based Infection & Vaccinology Program, the Royal Academy of Sciences, the King Gustaf Vth 80 year Foundation, and the Medical Faculty Research Foundation at Umeå university.

REFERENCES

Aderem, A. and Underhill, D.M. (1999). Mechanisms of phagocytosis in macrophages. *Annual Review of Immunology* **17**, 593–623.

Aktories, K. (1997). Bacterial toxins that target Rho proteins. *Journal of Clinical Investigation* **99**, 827–829.

Aktories, K., Schmidt, G., and Just, I. (2000). Rho GTPases as targets of bacterial toxins. *Biological Chemistry* **381**, 421–426.

Altrutz, M. and Isberg, R. (1998). Involvement of focal adhesion kinase in invasin-mediated uptake. *Proceedings of the National Academy of Sciences USA* **95**, 13,658–13,663.

Anderson, D.M. and Schneewind, O. (1997). A mRNA signal for the type III secretion of Yop proteins by *Yersinia enterocolitica*. *Science* **278**, 1140–1143.

Anderson, D.M., Fouts, D.E., Collmer, A., and Schneewind, O. (1999). Reciprocal secretion of proteins by the bacterial type III machines of plant and animal pathogens suggests universal recognition of mRNA targeting signals. *Proceedings of the National Academy of Sciences USA* **96**, 12,839–12,843.

Andersson, K., Carballeira, N., Magnusson, K., Persson, C., Stendahl, O., Wolf-Watz, H., and Fällman, M. (1996). YopH of *Yersinia pseudotuberculosis*

Å. FORSBERG, R. ROSQVIST, AND M. FÄLLMAN

interrupts early phosphotyrosine signalling associated with phagocytosis. *Molecular Microbiology* **20**, 1057–1069.

Andersson, K., Magnusson, K.-E., Majeed, M., Stendahl, O., and Fallman, M. (1999). *Yersinia pseudotuberculosis*-induced calcium signaling in neutrophils is blocked by the virulence effector YopH. *Infection and Immunity* **67**, 2567–2574.

Autenrieth, I. and Firsching, R. (1996). Penetration of M cells and destruction of Peyer's patches by *Yersinia enterocolitica*: an ultrastructural and histological study. *Journal of Medical Microbiology* **44**, 285–294.

Barz, C., Abahji, T.N., Trulzsch, K., and Heesemann, J. (2000). The Yersinia Ser/Thr protein kinase YpkA/YopO directly interacts with the small GTPases RhoA and Rac-1. *FEBS Letters* **482**, 139–143.

Bavoil, P.M. and Hsia, R.C. (1998). Type III secretion in *Chlamydia*: a case of deja vu? *Molecular Microbiology* **28**, 860–862.

Black, D. and Bliska, J. (1997). Identification of p130Cas as a substrate of *Yersinia* YopH (Yop51), a bacterial protein tyrosine phosphatase that translocates into mammalian cells and targets focal adhesions. *European Molecular Biology Organisation Journal* **16**, 2730–2744.

Black, D.S., Montagna, L.G., Zitsmann, S., and Bliska, J.B. (1998). Identification of an amino-terminal substrate-binding domain in the *Yersinia* tyrosine phosphatase that is required for efficient recognition of focal adhesion targets. *Molecular Microbiology* **29**, 1263–1274.

Black, D.S. and Bliska, J.B. (2000). The RhoGAP activity of the *Yersinia pseudotuberculosis* cytotoxin YopE is required for antiphagocytic function and virulence. *Molecular Microbiology* **37**, 515–527.

Bliska, J., Clemens, J., Dixon, J., and Falkow, S. (1992). The *Yersinia* tyrosine phosphatase: specificity of a bacterial virulence determinant for phosphoproteins in the J774A.1 macrophage. *Journal of Experimental Medicine* **176**, 1625–1630.

Blocker, A., Gounon, P., Larquet, E., Niebuhr, K., Cabiaux, V., Parsot, C., and Sansonetti, P. (1999). The tripartite type III secreton of *Shigella flexneri* inserts IpaB and IpaC into host membranes. *Journal of Cell Biology* **147**, 683–693.

Boland, A., Sory, M.-P., Iriarte, M., Kerbourch, C., Wattiau, P., and Cornelis, G.R. (1996). Status of YopM and YopN in the *Yersinia* Yop virulon: YopM of *Y. enterocolitica* is internalized inside the cytosol of PU5-1.8 macrophages by the YopB, D, N delivery apparatus. *European Molecular Biology Organisation Journal* **15**, 5191–5201.

Bölin, I. and Wolf-Watz, H. (1988). The plasmid-encoded Yop2b protein of *Yersinia pseudotuberculosis* is a virulence determinant regulated by calcium and temperature at the level of transcription. *Molecular Microbiology* **2**, 237–245.

Boyd, A.P., Lambermont, I., and Cornelis, G.R. (2000). Competition between the Yops of *Yersinia enterocolitica* for delivery into eukaryotic cells: role of the SycE chaperone binding domain of YopE. *Journal of Bacteriology* **182**, 4811–4821.

Burrows, T. and Bacon, G. (1956). The basis of virulence in *Pasteurella pestis*: an antigen determining virulence. *British Journal of Experimental Pathology* **37**, 481–493.

Caron, E. and Hall, A. (1998). Identification of two distinct mechanisms of phagocytosis controlled by different Rho GTPases. *Science* **282**, 1717–1721.

Cary, L., Chang, J., and Guan, J. (1996). Stimulation of cell migration by overexpression of focal adhesion kinase and its association with Src and Fyn. *Journal of Cell Science* **109**, 1787–1794.

Cary, L., Han, D., Polte, T., Hanks, S., and Guan, J. (1998). Identification of p130Cas as a mediator of focal adhesion kinase-promoted cell migration. *Journal of Cell Biology* **140**, 211–221.

Cheng, L.W., Anderson, D.M., and Schneewind, O. (1997). Two independent type III secretion mechanisms for YopE in *Yersinia enterocolitica*. *Molecular Microbiology* **24**, 757–765.

Chimini, G. and Chavrier, P. (2000). Function of Rho family proteins in actin dynamics during phagocytosis and engulfment. *Nature Cell Biology* **2**, E191–E196.

Choi, K., Kennedy, M., and Keller, G. (1993). Expression of a gene encoding a unique protein-tyrosine kinase within specific fetal- and adult-derived hematopoietic lineages. *Proceedings of the National Academy of Sciences USA* **90**, 5747–5751.

Cirillo, D.M., Valdivia, R.H., Monack, D.M., and Falkow, S. (1998). Macrophage-dependent induction of the *Salmonella* pathogenicity island 2 type III secretion system and its role in intracellular survival. *Molecular Microbiology* **30**, 175–188.

Clark, M., Hirst, B., and Jepson, M. (1998). M-cell surface beta1 integrin expression and invasin-mediated targeting of *Yersinia* pseudotuberculosis to mouse Peyer's patch M cells. *Infection and Immunity* **66**, 1237–1243.

Coburn, J., Wyatt, R.T., Iglewski, B.H., and Gill, D.M. (1989). Several GTP-binding proteins including p21[ras] are preferred substrates of *Pseudomonas aeruginosa* exoenzyme S. *Journal of Biological Chemistry* **264**, 9004–9008.

Coburn, J. and Gill, D.M. (1991). ADP-ribosylation of p21[ras] and related proteins by *Pseudomonas aeruginosa* exoenzyme S. *Infection and Immunity* **59**, 4259–4262.

Coburn, J., Kane, A.V., Feig, L., and Gill, D.M. (1991). *Pseudomonas aeruginosa* exoenzyme S requires a eukaryotic protein for ADP-ribosyltransferase activity. *Journal of Biological Chemistry* **266**, 6438–6446.

Cowell, B.A., Chen, D.Y., Frank, D.W., Vallis, A.J., and Fleiszig, S.M. (2000). ExoT of cytotoxic *Pseudomonas aeruginosa* prevents uptake by corneal epithelial cells. *Infection and Immunity* **68**, 403–406.

Cunningham, M.W. (2000). Pathogenesis of group A streptococcal infections. *Clinical Microbiology Review* **13**, 470–511.

da Silva, A., Rosenfield, J., Mueller, I., Bouton, A., Hirai, H., and Rudd, C. (1997a). Biochemical analysis of p120/130: a protein-tyrosine kinase substrate restricted to T and myeloid cells. *Journal of Immunology* **158**, 2007–2016.

da Silva, A., Li, Z., de Vera, C., Canto, E., Findell, P., and Rudd, C. (1997b). Cloning of a novel T-cell protein FYB that binds FYN and SH2-domain-containing leukocyte protein 76 and modulates interleukin 2 production. *Proceedings of the National Academy of Sciences USA* **94**, 7493–7498.

Dacheux, D., Attree, I., Schneider, C., and Toussaint, B. (1999). Cell death of human polymorphonuclear neutrophils induced by a *Pseudomonas aeruginosa* cystic fibrosis isolate requires a functional type III secretion system. *Infection and Immunity* **67**, 6164–6167.

Dacheux, D., Toussaint, B., Richard, M., Brochier, G., Croize, J., and Attree, I. (2000). *Pseudomonas aeruginosa* cystic fibrosis isolates induce rapid, type III secretion-dependent, but ExoU-independent, oncosis of macrophages and polymorphonuclear neutrophils. *Infection and Immunity* **68**, 2916–2924.

Dacheux, D., Attree, I., and Toussaint, B. (2001). Expression of ExsA in trans confers type III secretion system-dependent cytotoxicity on noncytotoxic *Pseudomonas aeruginosa* cystic fibrosis isolates. *Infection and Immunity* **69**, 538–542.

Davis, K.J., Fritz, D.L., Pitt, M.L., Welkos, S.L., Worsham, P.L., and Friedlander, A.M. (1996). Pathology of experimental pneumonic plague produced by fraction 1-positive and fraction 1-negative *Yersinia pestis* in African green monkeys (*Cercopithecus aethiops*). *Archives of Pathology and Laboratory Medicine* **120**, 156–163.

Denu, J., Stuckey, J., Saper, M., and Dixon, J. (1996). Form and function in protein dephosphorylation. *Cell* **87**, 361–364.

Drozdov, I.G., Anisimov, A.P., Samoilova, S.V., Yezhov, I.N., Yeremin, S.A., Karlyshev, A.V., Krasilnikova, V.M., and Kravchenko, V.I. (1995). Virulent non-capsulate *Yersinia pestis* variants constructed by insertion mutagenesis. *Journal of Medical Microbiology* **42**, 264–268.

Dukuzumuremyi, J.M., Rosqvist, R., Hallberg, B., Akerstrom, B., Wolf-Watz, H., and Schesser, K. (2000). The *Yersinia* protein kinase A is a host factor inducible RhoA/Rac-binding virulence factor. *Journal of Biological Chemistry* **275**, 35,281–35,290.

Ernst, J. (2000). Bacterial inhibition of phagocytosis. *Cellular Microbiology* 2, 379–386.

Fällman, M., Andersson, K., Håkansson, S., Magnusson, K., Stendahl, O., and Wolf-Watz, H. (1995). *Yersinia pseudotuberculosis* inhibits Fc receptor-mediated phagocytosis in J774 cells. *Infection and Immunity* 63, 3117–3124.

Finck-Barbancon, V., Goranson, J., Zhu, L., Sawa, T., Wiener-Kronish, J.P., and Fleiszig, S.M. (1997). ExoU expression by *Pseudomonas aeruginosa* correlates with acute cytotoxicity and epithelial injury. *Molecular Microbiology* 25, 547–557.

Fleiszig, S.M., Wiener-Kronish, J.P., Miyazaki, H., Vallas, V., Mostov, K.E., Kanada, D., Sawa, T., Yen, T.S., and Frank, D.W. (1997). *Pseudomonas aeruginosa*-mediated cytotoxicity and invasion correlate with distinct genotypes at the loci encoding exoenzyme S. *Infection and Immunity* 65, 579–586.

Forsberg, Å. and Wolf-Watz, H. (1988). The virulence protein Yop5 of *Yersinia pseudotuberculosis* is regulated at transcriptional level by plasmid pIB1 encoded transacting elements controlled by temperature and calcium . *Molecular Microbiology* 2, 121–133.

Forsberg, Å., Viitanen, A.M., Skurnik, M., and Wolf-Watz, H. (1991). The surface-located YopN protein is involved in calcium signal transduction in *Yersinia pseudotuberculosis*. *Molecular Microbiology* 5, 977–986.

Frank, D.W. (1997). The exoenzyme S regulon of *Pseudomonas aeruginosa*. *Molecular Microbiology* 26, 621–629.

Frithz-Lindsten, E., Du, Y., Rosqvist, R., and Forsberg, A. (1997). Intracellular targeting of exoenzyme S of *Pseudomonas aeruginosa* via type III-dependent translocation induces phagocytosis resistance, cytotoxicity and disruption of actin microfilaments. *Molecular Microbiology* 25, 1125–1139.

Frithz-Lindsten, E., Holmström, A., Jacobsson, L., Soltani, M., Olsson, J., Rosqvist, R., and Forsberg, Å. (1998). Functional complementation of the effector protein translocators PopB/YopB and PopD/YopD of *Pseudomonas aeruginosa* and *Yersinia pseudotuberculosis*. *Molecular Microbiology* 29, 1155–1165.

Frithz-Lindsten, E., Rosqvist, R., Johansson, L., and Forsberg, A. (1995). The chaperone-like protein YevA of *Yersinia pseudotuberculosis* stabilises YopE in the cytoplasm but is dispensible for targeting to secretion loci. *Molecular Microbiology* 16, 635–647.

Fu, H., Coburn, J., and Collier, R.J. (1993). The eukaryotic host factor that activates exoenzyme S of *Pseudomonas aeruginosa* is a member of the 14-3-3 protein family. *Proceedings of the National Academy of Sciences USA* 90, 2320–2324.

Fu, Y. and Galán, J.E. (1998). The *Salmonella typhimurium* tyrosine phosphatase SptP is translocated into host cells and disrupts the actin cytoskeleton. *Molecular Microbiology* **27**, 359–368.

Fu, Y. and Galán, J.E. (1999). A *Salmonella* protein antagonizes Rac-1 and Cdc42 to mediate host-cell recovery after bacterial invasion. *Nature* **401**, 293–297.

Fu, H., Subramanian, R.R., and Masters, S.C. (2000). 14-3-3 proteins: Structure, function and regulation. *Annual Review of Pharmacology and Toxicology* **40**, 617–647.

Galán, J. and Zhou, D. (2000). Striking a balance: Modulation of the actin cytoskeleton by *Salmonella*. *Proceedings of the National Academy of Sciences USA* **97**, 8754–8761.

Galyov, E., Håkansson, S., Forsberg, Å., and Wolf-Watz, H. (1993). A secreted protein kinase of *Yersinia pseudotuberculosis* is an indispensable virulence determinant. *Nature* **361**, 730–732.

Galyov, E.E., Håkansson, S., and Wolf-Watz, H. (1994). Characterization of the operon encoding the YpkA Ser/Thr protein kinase and the YopJ protein of *Yersinia pseudotuberculosis*. *Journal of Bacteriology* **176**, 4543–4548.

Garrity-Ryan, L., Kazmierczak, B., Kowal, R., Comolli, J., Hauser, A., and Engel, J.N. (2000). The arginine finger domain of ExoT contributes to actin cytoskeleton disruption and inhibition of internalization of *Pseudomonas aeruginosa* by epithelial cells and macrophages. *Infection and Immunity* **68**, 7100–7113.

Goehring, U.M., Schmidt, G., Pederson, K.J., Aktories, K., and Barbieri, J.T. (1999). The N-terminal domain of *Pseudomonas aeruginosa* exoenzyme S is a GTPase-activating protein for Rho GTPases. *Journal of Biological Chemistry* **274**, 36,369–36,372.

Goguen, J.D., Walker, W.S., Hatch, T.P., and Yother, J. (1986). Plasmid-determined cytotoxicity in *Yersinia pestis* and *Yersinia pseudotuberculosis*. *Infection and Immunity* **51**, 788–794.

Goosney, D.L., Celli, J., Kenny, B., and Finlay, B.B. (1999). Enteropathogenic *Escherichia coli* inhibitis phagocytosis. *Infection and Immunity* **67**, 490–495.

Greenberg, S., Chang, P., Silverstein, S.C. (1993). Tyrosine phosphorylation is required for Fc receptor-mediated phagocytosis in mouse macrophages. *Journal of Experimental Medicine* **177**, 529–534.

Grutzkau, A., Hanski, C., Hahn, H., and Riecken, E. (1990). Involvement of M cells in the bacterial invasion of Peyer's patches: a common mechanism shared by *Yersinia enterocolitica* and other enteroinvasive bacteria. *Gut* **31**, 1011–1015.

Guan, K. and Dixon, J. (1990). Protein tyrosine phosphatase activity of an essential virulence determinant in *Yersinia*. *Science* **249**, 553–556.

Guex, N. and Pietsch, M.C. (1997). SWISS_MODEL and the Swiss-PdbViewer: An environment for comparative protein modeling. *Electrophoresis* **18**, 2714–2723.

Guex, N., Diemand, A., and Peitsch, M.C. (1999). Protein modelling for all. *Trends in Biochemical Science* **24**, 364–367.

Gumbiner, B. (1996). Cell adhesion: the molecular basis of tissue architecture and morphogenesis. *Cell* **84**, 345–357.

Hackstadt, T. (2000). Redirection of host vesicle trafficking pathways by intracellular parasites. *Traffic* **1**, 93–99.

Hackstadt, T., Scidmore, M.A., and Rockey, D.D. (1995). Lipid metabolism in *Chlamydia trachomatis*-infected cells: directed trafficking of Golgi-derived sphingolipids to the chlamydial inclusion. *Proceedings of the National Academy of Sciences USA* **23**, 4877–4881.

Hackstadt, T., Rockey, D.D., Heinzen, R.A., Scidmore, M.A. (1996). *Chlamydia trachomatis* interrupts an exocytic pathway to acquire endogenously synthesized sphingomyelin in transit from the Golgi apparatus to the plasma membrane. *European Molecular Biology Organisation Journal* **15**, 964–977.

Hahn, H.P. (1997). The type-IV pilus is the major virulence associated adhesin of *Pseudomonas aeruginosa*-a review. *Gene* **192**, 99–108.

Håkansson, S., Galyov, E.E., Rosqvist, R., and Wolf-Watz, H. (1996a). The *Yersinia* YpkA Ser/Thr kinase is translocated and subsequently targeted to the inner surface of the HeLa cell plasma membrane. *Molecular Microbiology* **20**, 593–603.

Håkansson, S., Schesser, K., Persson, C., Galyov, E.E., Rosqvist, R., Homblé, F., and Wolf-Watz, H. (1996b). The YopB protein of *Yersinia pseudotuberculosis* is essential for the translocation of Yop effector proteins across the target cell plasma membrane and displays a contact dependent membrane disrupting activity. *European Molecular Biology Organisation Journal* **15**, 5812–5823.

Hall, A. (1998). Rho GTPases and the actin cytoskeleton. *Science* **279**, 509–514.

Hamid, N., Gustavsson, A., Andersson, K., McGee, K., Persson, C., Rudd, C.E., and Fallman, M. (1999). YopH dephosphorylates Cas and Fyn-binding protein in macrophages. *Microbial Pathogenesis* **27**, 231–242.

Hampton, M.B., Kettle, A.J., and Winterbourn, C.C. (1998). Inside the neutrophil phagosome: oxidants, myeloperoxidase, and bacterial killing. *Blood* **92**, 3007–3017.

Hanski, C., Kutschka, U., Schmoranzer, H., Naumann, M., Stallmach, A., Hahn, H., Menge, H., and Riecken, E. (1989). Immunohistochemical and electron microscopic study of interaction of *Yersinia enterocolitica* serotype O8 with intestinal mucosa during experimental enteritis. *Infection and Immunity* **57**, 673–678.

Hardt, W.D., Chen, L.M., Schuebel, K.E., Bustelo, X.R., and Galán, J.E. (1998). *S. typhimurium* encodes an activator of Rho GTPases that induces membrane ruffling and nuclear responses in host cells. *Cell* **93**, 815–826.

Hartland, E., Green, S., Phillips, W., and Robins-Browne, R. (1994). Essential role of YopD in inhibition of the respiratory burst of macrophages by *Yersinia enterocolitica*. *Infection and Immunity* **62**, 4445–4453.

Henriksson, M.L., Rosqvist, R., Telepnev, M., Wolf-Watz, H., and Hallberg, B. (2000). Ras effector pathway activation by epidermal growth factor is inhibited *in vivo* by exoenzyme S ADP-ribosylation of Ras. *Biochemical Journal* **347**, 217–222.

Hensel, M., Shea, J.E., Gleeson, C., Jones, M.D., Dalton, E., and Holden, D.W. (1995). Simultaneous identification of bacterial virulence genes by negative selection. *Science* **269**, 400–403.

Hensel, M., Shea, J.E., Waterman, S.R., Mundy, R., Nikolaus, T., Banks, G., Vazquez-Torres, A., Gleeson, C., Fang, F.C., and Holden, D.W. (1998). Genes encoding putative effector proteins of the type III secretion system of *Salmonella* pathogenicity island 2 are required for bacterial virulence and proliferation in macrophages. *Molecular Microbiology* **30**, 163–174.

Holmström, A., Pettersson, J., Rosqvist, R., Håkansson, S., Tafazoli, F., Fällman, M., Magnusson, K.-E., Wolf-Watz, H., and Forsberg, Å. (1997). YopK of *Yersinia pseudotuberculosis* controls translocation of Yop effectors across the eukaryotic cell membrane. *Molecular Microbiology* **24**, 73–91.

Holmström, A., Olsson, J., Cherepanov, P., Maier, E., Nordfelth, R., Pettersson, J., Benz, R., Wolf-Watz, H., and Forsberg, Å. (2001). LcrV is a channel size determining component of the Yop effector translocon of *Yersinia*. *Molecular Microbiology* **39**, 620–632.

Honda, H., Oda, H., Nakamoto, T., Honda, Z., Sakai, R., Suzuki, T., Saito, T., Nakamura, K., Nakao, K., Ishikawa, T., Katsuki, M., Yazaki, Y., and Hirai, H. (1998). Cardiovascular anomaly, impaired actin bundling and resistance to Src-induced transformation in mice lacking p130Cas *Nature Genetics* **19**, 361–365.

Honda, H., Nakamoto, T., Sakai, R., and Hirai, H. (1999). p130(Cas), an assembling molecule of actin filaments, promotes cell movement, cell migration, and cell spreading in fibroblasts. *Biochemical Biophysical Research Communications* **262**, 25–30.

Hsia, R.-C., Pannekoek, Y., Ingerowski, E., and Bavoil, P. (1997). Type III secretion genes identify a putative virulence locus of *Chlamydia*. *Molecular Microbiology* **25**, 351–359.

Hueck, C.J. (1998). Type III protein secretion systems in bacterial pathogens of animals and plants. *Molecular Microbiology Reviews* **62**, 379–433.

Hunter, A.J., Ottoson, N., Boerth, N., Koretzky, G.A., and Shimizu, Y. (2000). Cutting edge: a novel function for the SLAP-130/FYB adapter protein in beta 1 integrin signaling and T lymphocyte migration. *Journal of Immunology* **164**, 1143–1147.

Ilic, D., Furuta, Y., Kanazawa, S., Takeda, N., Sobue, K., Nakatsuji, N., Nomura, S., Fujimoto, J., Okada, M., and Yamamoto, T. (1995). Reduced cell motility and enhanced focal adhesion contact formation in cells from FAK-deficient mice. *Nature* **377**, 539–544.

Ilic, D., Kanazawa, S., Furuta, Y., Yamamoto, T., and Aizawa, S. (1996). Impairment of mobility in endodermal cells by FAK deficiency. *Experimental Cell Research* **222**, 298–303.

Iriarte, M. and Cornelis, G.R. (1998). YopT, a novel *Yersinia* Yop effector protein, affects the cytoskeleton of host cells. *Molecular Microbiology* **29**, 915–929.

Iriarte, M., Sory, M.P., Boland, A., Boyd, A.P., Mills, S.D., Lambermont, I., and Cornelis, G.R. (1998). TyeA, a protein involved in control of Yop release and in translocation of *Yersinia* Yop effectors. *European Molecular Biology Organisation Journal* **17**, 1907–1918.

Isberg, R. and Leong, J. (1990). Multiple beta 1 chain integrins are receptors for invasin, a protein that promotes bacterial penetration into mammalian cells. *Cell* **60**, 861–871.

Jücker, M., McKenna, K., da Silva, A., Rudd, C., and Feldman, R. (1997). The Fes protein-tyrosine kinase phosphorylates a subset of macrophage proteins that are involved in cell adhesion and cell-cell signaling. *Journal of Biological Chemistry* **272**, 2104–2109.

Juris, S.J., Rudolph, A.E., Huddler, D., Orth, K., and Dixon, J.E. (2000). A distinctive role for the *Yersinia* protein kinase: actin binding, kinase activation, and cytoskeleton disruption. *Proceedings of the National Academy of Sciences USA* **97**, 9431–9436.

Kaniga, K., Uralil, J., Bliska, J.B., and Galán, J.E. (1996). A secreted protein tyrosine phosphatase with modular effector domains in the bacterial pathogen *Salmonella typhimurium*. *Molecular Microbiology* **21**, 633–641.

Kazmierczak, B.I., Jou, T.S., Mostov, K., and Engel, J.N. (2001). RhoGTPase activity modulates *Pseudomonas aeruginosa* internalization by epithelial cells. *Cellular Microbiology* **3**, 85–98.

Kenny, B., DeVinney, R., Stein, M., Reinscheid, D.J., Frey, E.A., and Finlay, B.B. (1997). Enteropathogenic *E. coli* (EPEC) transfers its receptor for intimate adherence into mammalian cells. *Cell* **91**, 511–520.

Klemke, R., Leng, J., Molander, R., Brooks, P., Vuori, K., and Cheresh, D. (1998). CAS/Crk coupling serves as a "molecular switch" for induction of cell migration. *Journal of Cell Biology* **140**, 961–972.

Knutton, S., Rosenshine, I., Pallen, M.J., Nisan, I., Neves, B.C., Bain, C., Wolff, C., Dougan, G., and Frankel, G. (1998). A novel EspA-associated surface organelle of enteropathogenic *Escherichia coli*. *Infection and Immunity* **57**, 2166–2176.

Koornhof, H.J., Smego, R.A., Jr., and Nicol, M. (1999). Yersiniosis. II: The pathogenesis of *Yersinia* infections. *European Journal of Clinical Microbiology and Infectious Disease* **18**, 87–112.

Kounnas, M.Z., Morris, R.E., Thompson, M.R., Fitzgerald, D.J., Strickland, D.K., and Salinger, C.B. (1992). The α2-macroglobulin receptor/low density lipoprotein receptor-related protein binds and internalises *Pseudomonas* exotoxin A. *Journal of Biological Chemistry* **267**, 12,420–12,423.

Krall, R., Schmidt, G., Aktories, K., and Barbieri, J.T. (2000). *Pseudomonas aeruginosa* ExoT is a Rho GTPase-activating protein. *Infection and Immunity* **68**, 6066–6068.

Krause, M., Sechi, A.S., Konradt, M., Monner, D., Gertler, F.B., and Wehland, J. (2000). Fyn-binding protein (Fyb)/SLP-76-associated protein (SLAP), Ena/vasodilator-stimulated phosphoprotein (VASP) proteins and the Arp2/3 complex link T cell receptor (TCR) signaling to the actin cytoskeleton. *Journal of Cell Biology* **149**, 181–194.

Kubori, T., Matsushima, Y., Nakamura, D., Uralil, J., Lara-Tejero, M., Sukhan, A., Galán, J.E., and Aizawa, S.I. (1998). Supramolecular structure of the *Salmonella typhimurium* type III protein secretion system. *Science* **280**, 602–605.

Kubori, T., Sukhan, A., Aizawa, S.I., and Galán, J.E. (2000). Molecular characterization and assembly of the needle complex of the *Salmonella typhimurium* type III protein secretion system. *Proceedings of the National Academy of Sciences USA* **97**, 10,225–10,230.

Leung, K.Y., Reisner, B.S., and Straley, S.C. (1990). YopM inhibits platelet aggregation and is necessary for virulence of *Yersinia pestis* in mice. *Infection and Immunity* **58**, 3262–3271.

Lian, C. and Pai, C. (1985). Inhibition of human neutrophil chemiluminescence by plasmid-mediated outer membrane proteins of *Yersinia enterocolitica*. *Infection and Immunity* **49**, 145–151.

Lin, T., Yurochko, A., Kornberg, L., Morris, J., Walker, J., Haskill, S., and Juliano, R. (1994). The role of protein tyrosine phosphorylation in integrin-mediated gene induction in monocytes. *Journal of Cell Biology* **126**, 1585–1593.

Liu, S., Yahr, T.L., Frank, D.W., and Barbieri, J.T. (1997). Biochemical relationships between the 53-kilodalton (Exo53) and 49-kilodalton (ExoS) forms of exoenzyme S of *Pseudomonas aeruginosa*. *Journal of Bacteriology* **179**, 1609–1613.

Lloyd, S.A., Norman, M., Rosqvist, R., and Wolf-Watz, H. (2001). *Yersinia* YopE is targeted for type III secretion by N-terminal, not mRNA, signals. *Molecular Microbiology* **39**, 520–531.

Magae, J., Nagi, T., Takaku, K., Kataoka, T., Koshino, H., Uramoto, M., and Nagai, K. (1994). Screening for specific inhibitors of phagocytosis of thioglycollate-elicited macrophages. *Bioscience Biotechnology and Biochemistry* **58**, 104–107.

Mandell, G.L., Bennett, J.E., and Dolin, R. (1995). In *Principles and Practise of Infectious Diseases,* pp. 1980–2002. New York: Churchill Livingstone.

Marches, O., Nougayrede, J.P., Boullier, S., Mainil, J., Charlier, G., Raymond, I., Pohl, P., Boury, M., De Rycke, J., Milon, A., and Oswald. E. (2000). Role of tir and intimin in the virulence of rabbit enteropathogenic *Escherichia coli* serotype O103:H2. *Infection and Immunity* **68**, 2171–2182.

Marra, A. and Isberg, R. (1997). Invasin-dependent and invasin-independent pathways for translocation of *Yersinia pseudotuberculosis* across the Peyer's patch intestinal epithelium. *Infection and Immunity* **65**, 3412–3421.

Mauro, L. and Dixon, J. (1994). 'Zip codes' direct intracellular protein tyrosine phosphatases to the correct cellular 'address'. *Trends in Biochemical Sciences* **19**, 151–155.

McGuffie, E.M., Frank, D.W., Vincent, T.S., and Olson, J.C. (1998). Modification of Ras in eukaryotic cells by *Pseudomonas aeruginosa* exoenzyme S. *Infection and Immunity* **66**, 2607–2613.

Montagna, L.G., Ivanov, M.I., and Bliska, J.B. (2001). Identification of residues in the amino-terminal domain of the *Yersinia* tyrosine phosphatase that are critical for substrate recognition. *Journal of Biological Chemistry* **276**, 5005–5011.

Morrissette, N., Gold, E., and Aderem, A. (1999). The macrophage – a cell for all seasons. *Trends in Cell Biology* **9**, 199–201.

Moulder, J.W. (1991). Interaction of chlamydiae and host cells in vitro. *Microbiology Reviews* **55**, 143–190.

Neyt, C. and Cornelis, G.R. (1999). Insertion of a Yop translocation pore into the macrophage plasma membrane by *Yersinia enterocolitica*: requirement for translocators YopB and YopD, but not LcrG. *Molecular Microbiology* **33**, 971–981.

Nhieu, G.T. and Sansonetti, P.J. (1999). Mechanism of *Shigella* entry into epithelial cells. *Current Opinion in Microbiology* **2**, 51–55.

Nicas, T.I. and Iglewski, B.H. (1985). The contribution of exoproducts to virulence of *Pseudomonas aeruginosa*. *Canadian Journal of Microbiology* **31**, 387–392.

Nilles, M.L., Williams, A.W., Skrzypek, E., and Straley, S.C. (1997). *Yersinia pestis* LcrV forms a stable complex with LcrG and may have a secretion-related regulatory role in the low Ca^{2+} response. *Journal of Bacteriology* **179**, 1307–1316.

Norris, F.A., Wilson, M.P., Wallis, T.S., Galyov, E.E., and Majerus, P.W. (1998). SopB, a protein required for virulence of *Salmonella dublin*, is an inositol phosphate phosphatase. *Proceedings of the National Academy of Sciences USA* **95**, 14057–14059.

Ochman, H., Sonsini, F.C., Solomon, F., and Groisman, E.A. (1996). Identification of a pathogenicity island required for *Salmonella* survival in host cells. *Proceedings of the National Academy of Sciences USA* **93**, 7800–7804.

O'Neill, G.M., Fashena, S.J., and Golemis, E.A. (2000). Integrin signalling: a new Cas(t) of characters enters the stage. *Trends in Cell Biology* **10**, 111–119.

Orth, K., Xu, Z., Mudgett, M.B., Bao, Z.Q., Palmer, L.E., Bliska, J.B., Mangel, W.F., Staskawicz, B., and Dixon, J.E. (2000). Disruption of signalling by *Yersinia* effector YopJ, a ubiquitin-like protein protease. *Science* **290**, 1594–1597.

Palmer, L.E., Hobbie, S., Galán, J.E., and Bliska, J.B. (1998). YopJ of *Yersinia pseudotuberculosis* is required for the inhibition of macrophage TNFα production and downregulation of the MAP kinases p38 and JNK. *Molecular Microbiology* **27**, 953–965.

Parsot, C. and Sansonetti, P. (1996). Invasion and the pathogenesis of *Shigella* infections *Current Topics in Microbiology and Immunology* **209**, 25–42.

Pederson, K.J. and Barbieri, J.T. (1998). Intracellular expression of the ADP-ribosyltransferase domain of *Pseudomonas* exoenzyme S is cytotoxic to eukaryotic cells. *Molecular Microbiology* **30**, 751–759.

Pederson, K.J., Vallis, A.J., Aktories, K., Frank, D.W., and Barbieri, J.T. (1999). The amino-terminal domain of *Pseudomonas aeruginosa* ExoS disrupts actin filaments via small-molecular-weight GTP-binding proteins. *Molecular Microbiology* **32**, 393–401.

Peitsch, M.C. (1995). Protein modeling by e-mail. *Bio/Technology* **13**, 658–660.

Persson, C., Nordfelth, R., Holmström, A., Håkansson, S., Rosqvist, R., and Wolf-Watz, H. (1995). Cell-surface-bound *Yersinia* translocate the protein tyrosine phosphatase YopH by a polarized mechanism into the target cell. *Molecular Microbiology* **18**, 135–150.

Persson, C., Carballeira, N., Wolf-Watz, H., and Fällman, M. (1997). The PTPase YopH inhibits uptake of *Yersinia*, tyrosine phosphorylation of p130Cas and FAK, and the associated accumulation of these proteins in peripheral focal adhesions. *European Molecular Biology Organisation Journal* **16**, 2307–2318.

Persson, C., Nordfeldt, R., Andersson, K., Forsberg, Å., Wolf-Watz, H., and Fällman, M. (1999). Localisation of the *Yersinia* PTPase to focal complexes is an important virulence mechanism. *Molecular Microbiology* **33**, 828–838.

Pettersson, J., Holmström, A., Hill, J., Leary, S., Frithz-Lindsten, E., von Euler-Matell, A., Carlsson, E., Titball, R., Forsberg, Å., and Wolf-Watz, H. (1999). The V-antigen of *Yersinia* is surface exposed before target cell contact and involved in virulence protein translocation. *Molecular Microbiology* **32**, 961–976.

Plano, G.V., Day, J.B., and Ferracci, F. (2001). Type III export: new uses for an old pathway. *Molecular Microbiology* **40**, 284–293.

Polte, T. and Hanks, S. (1995). Interaction between focal adhesion kinase and Crk-associated tyrosine kinase substrate p130Cas. *Proceedings of the National Academy of Sciences USA* **92**, 10,678–10,682.

Rabinovitch, M. (1995). Professional and non-professional phagocytes: an introduction. *Trends in Cell Biology* **5**, 85–87.

Ramaro, N., Gray-Owen, S.D., Backert, S., and Meyer, T.F. (2000). *Helicobacter pylori* inhibits phagocytosis by professional phagocytes involving the type IV secretion components. *Molecular Microbiology* **37**, 1389–1404.

Rosenshine, I., Duronio, V., and Finlay, B. (1992). Tyrosine protein kinase inhibitors block invasin-promoted bacterial uptake by epithelial cells. *Infection and Immunity* **60**, 2211–2217.

Rosqvist, R. and Wolf-Watz, H. (1986). Virulence plasmid-associated HeLa cell induced cytotoxicity of *Yersinia pseudotuberculosis*. *Microbial Pathogenesis* **1**, 229–240.

Rosqvist, R., Bölin, I., and Wolf-Watz, H. (1988). Inhibition of phagocytosis in *Yersinia pseudotuberculosis*: a virulence plasmid-encoded ability involving the Yop2b protein. *Infection and Immunity* **56**, 2139–2143.

Rosqvist, R., Forsberg, A., Rimpilainen, M., Bergman, T., and Wolf-Watz, H. (1990). The cytotoxic protein YopE of *Yersinia* obstructs the primary host defence. *Molecular Microbiology* **4**, 657–667.

Rosqvist, R., Forsberg, A., and Wolf-Watz, H. (1991). Intracellular targeting of the *Yersinia* YopE cytotoxin in mammalian cells induces actin microfilament disruption. *Infection and Immunity* **59**, 4562–4569.

Rosqvist, R., Magnusson, K., and Wolf-Watz, H. (1994). Target cell contact triggers expression and polarized transfer of *Yersinia* YopE cytotoxin into mammalian cells. *European Molecular Biology Organisation Journal* **13**, 964–972.

Rosqvist, R., Håkansson, S., Forsberg, A., and Wolf-Watz, H. (1995). Functional conservation of the secretion and translocation machinery for virulence proteins of yersiniae, salmonellae and shigellae. *European Molecular Biology Organisation Journal* **14**, 4187–4195.

Rossier, O., Wengelnik, K., Hahn, K., and Bonas, U. (1999). The *Xanthomonas* type III system secretes proteins from plant and mammalian pathogens. *Proceedings of the National Academy of Sciences USA* **96**, 9368–9373.

Ruckdeschel, K., Roggenkamp, A., Schubert, S., and Heesemann, J. (1996). Differential contribution of *Yersinia enterocolitica* virulence factors to evasion of microbicidal action of neutrophils. *Infection and Immunity* **64**, 724–733.

Ruckdeschel, K., Machold, I., Roggenkamp, A., Schubert, S., Pierre, J., Zumbihl, R., Liautard, J., Heesemann, J., and Rouot, B. (1997). *Yersinia enterocolitica* promotes deactivation of macrophage mitogen-activated protein kinases, extracellular signal-regulated kinase-1/2, p38, and c-Jun NH2-terminal kinase. Correlation with its inhibitory effect on tumor necrosis factor-alpha production. *Journal of Biological Chemistry* **272**, 15,920–15,927.

Sawa, T., Yahr, T.L., Ohara, M., Kurahashi, K., Gropper, M.A., Wiener-Kronish, J.P., and Frank, D.W. (1999). Active and passive immunization with the *Pseudomonas* V antigen protects against type III intoxication and lung injury. *Nature Medicine* **5**, 392–398.

Scheffzek, K., Reza, M., and Wittinghofer, A. (1998). GTP-ase activating proteins: helping hands complement an active site. *Trends in Biochemistry* **23**, 257–262.

Schesser, K., Frithz-Lindsten, E., and Wolf-Watz, H. (1996). Delineation and mutational analysis of the *Yersinia pseudotuberculosis* YopE domains which mediate translocation across bacterial and eukaryotic cellular membranes. *Journal of Bacteriology* **178**, 7227–7233.

Schesser, K., Spiik, A.-K., Dukuzumuremyi, J.-M., Neurath, M.F., Pettersson, S., and Wolf-Watz, H. (1998). The *yopJ* locus is required for *Yersinia*-mediated inhibition of NFκB activation and cytokine expression: YopJ contains a eukaryotic SH2-like domain that is essential for its repressive activity. *Molecular Microbiology* **28**, 1067–1079.

Schmidt, G., Sehr, P., Wilm, M., Selzer, J., Mann, M., and Aktories, K. (1997). Gln-63 of Rho is deamidated by *Escherichia coli* cytotoxic necrotoxing factor-1. *Nature* **387**, 725–729.

Scidmore, M.A. and Hackstadt, T. (2001). Mammalian 14-3-3β associates with the *Chlamydia trachomatis* inclusion membrane via its interaction with IncG. *Molecular Microbiology* **39**, 1638–1650.

Simonet, M., Richard, S., and Berche, P. (1990). Electron microscopic evidence for *in vivo* extracellular localization of *Yersinia pseudotuberculosis* harboring the pYV plasmid. *Infection and Immunity* **58**, 841–845.

Skrzypek, E., Cowan, C., Straley, S.C. (1998). Targeting of the *Yersinia pestis* YopM protein into HeLa cells and intracellular trafficking to the nucleus. *Molecular Microbiology* **30**, 1051–1065.

Sory, M.P. and Cornelis, G.R. (1994). Translocation of a hybrid YopE-adenylate cyclase from *Yersinia enterocolitica* into HeLa cells. *Molecular Microbiology* **14**, 583–594.

Sory, M., Boland, A., Lambermont, I., and Cornelis, G.R. (1995). Identification of the YopE and YopH domains required for secretion and internalization into the cytosol of macrophages, using the cyaA gene fusion approach. *Proceedings of the National Academy of Sciences USA* **92**, 11,998–12,002.

Stebbins, E.C., and Galán, J.E. (2000). Modulation of host signaling by a bacterial mimic: structure of the *Salmonella* effector SptP bound to Rac1. *Molecular Cell* **6**, 1449–1460.

Straley, S.C. and Bowmer, W.S. (1986). Virulence genes regulated at the transcriptional level by Ca^{2+} in *Yersinia pestis* inclue structural genes for outer membrane proteins. *Infection and Immunity* **51**, 445–454.

Subtil, A., Blocker, A., and Dautry-Varsat, A. (2000). Type III secretion system in *Chlamydia*: identified members and candidates. *Microbes and Infection* **2**, 367–369.

Subtil, A., Parsot, C., and Dautry-Varsat, A. (2001). Secretion of predicted Inc proteins of *Chlamydia pneumoniae* by heterologous type III machinery. *Molecular Microbiology* **39**, 792–800.

Sundin, C., Henriksson, M.L., Hallberg, B., Forsberg, Å., and Frithz-Lindsten, E. (2001). Exoenzyme T of *Pseudomonas aeruginosa* elicits cytotoxicity without interfering with Ras signal transduction. *Cellular Microbiology* **3**, 237–246.

Tacket, C.O., Sztein, M.B., Losonsky, G., Abe, A., Finlay, B.B., McNamara, B.P., Fantry, G.T., James, S.P., Nataro, J.P., Levine, M.M., and Donnenberg, M.S. (2000). Role of EspB in experimental human enteropathogenic *Escherichia coli* infection. *Infection and Immunity* **68**, 3689–3695.

Tjelle, T.E., Lovdal, T., and Berg, T.T. (2000). Phagosome dynamics and function. *Bioessays* **22**, 255–263.

Tonks, N. and Neel, B. (1996). From form to function: signaling by protein tyrosine phosphatases. *Cell* **87**, 365–368.

Vallance, B.A. and Finlay, B.B. (2000). Exploitation of host cells by enteropathogenic *Escherichia coli*. *Proceedings of the National Academy of Sciences USA* **97**, 8799–8806.

Vallis, A.J., Finck-Barbancon, V., Yahr, T.L., and Frank, D.W. (1999). Biological effects of *Pseudomonas aeruginosa* type III-secreted proteins on CHO cells. *Infection and Immunity* **67**, 2040–2044.

Van Aelst, L. and D'Souza-Schorey, C. (1997). Rho GTPases and signalling networks. *Genes and Development* **11**, 2295–2322.

Vazquez-Torres, A., Jones-Carson, J., Mastroeni, P., Ischiropoulos, H., and Fang, F.C. (2000a). Antimicrobial actions of the NADPH phagocyte oxidase and inducible nitric oxide synthase in experimental salmonellosis. I. Effects on microbial killing by activated peritoneal macrophages in vitro. *Journal of Experimental Medicine* **192**, 227–236.

Vazquez-Torres, A., Xu, Y., Jones-Carson, J., Holden, D.W., Lucia, S.M., Dinauer, M.-C., Mastroeni, P., and Fang, F.C. (2000b). *Salmonella* pathogenicity island 2-dependent evasion of the phagocyte NADPH oxidase. *Science* **287**, 1655–1658.

Vincent, T.S., Fraylick, J.E., McGuffie, E.M., and Olson, J.C. (1999). ADP-ribosylation of oncogenic Ras proteins by *Pseudomonas aeruginosa* exoenzyme S *in vivo*. *Molecular Microbiology* **32**, 1054–1064.

Visser, L., Annema, A., and van Furth, R. (1995). Role of Yops in inhibition of phagocytosis and killing of opsonized *Yersinia enterocolitica* by human granulocytes. *Infection and Immunity* **63**, 2570–2575.

von Pawel-Rammingen, U., Telepnev, M.V., Schmidt, G., Aktories, K., Wolf-Watz, H., and Rosqvist, R. (2000). GAP activity of the *Yersinia* YopE cytotoxin specifically targets the rho pathway: a mechanism for disruption of actin microfilament structure. *Molecular Microbiology* **36**, 737–748.

Wattiau, P. and Cornelis, G.R. (1993). SycE, a chaperone-like protein of *Yersinia enterocolitica* involved in the secretion of YopE. *Molecular Microbiology* **8**, 123–131.

Wattiau, P., Woestyn, S., and Cornelis, G.R. (1996). Customized secretion chaperones in pathogenic bacteria. *Molecular Microbiology* **20**, 255–262.

Weidow, C.L., Black, D.S., Bliska, J.B., and Bouton, A.H. (2000). CAS/Crk signalling mediates uptake of *Yersinia* into human epithelial cells. *Cellular Microbiology* **2**, 549–560.

Woestyn, S., Sory, M.P., Boland, A., Lequenne, O., and Cornelis, G.R. (1996). The cytosolic SycE and SycH chaperones of *Yersinia* protect the region of YopE and YopH involved in translocation across eukaryotic cell membranes. *Molecular Microbiology* **20**, 1261–1271.

Würtele, M., Wolf, E., Pederson, K., Buchwald, G., Ahmadian, M., Barbieri, J.T., and Wittinghofer, A. (2001). How the *Pseudomonas aeruginosa* ExoS toxin downregulates Rac. *Nature Structural Biology* **8**, 23–26.

Yahr, T.L., Hovey, A.K., Kulich, S.M., and Frank, D.W. (1995). Transcriptional analysis of the *Pseudomonas aeruginosa* exotoxin S structural gene. *Journal of Bacteriology* **177**, 1169–1178.

Yahr, T.L., Goranson, J., and Frank, D.W. (1996). Exoenzyme S of *Pseudomonas aeruginosa* secreted by a type III secretion pathway. *Molecular Microbiology* **22**, 991–1003.

Yahr, T.L., Mende-Mueller, L.M., Friese, M.B., and Frank, D.W. (1997). Identification of type III secreted products of the *Pseudomonas aeruginosa* exoenzyme S regulon. *Journal of Bacteriology* **179**, 7165–7168.

Yahr, T.L., Vallis, A.J., Hancock, M.K., Barbieri, J.T., and Frank, D.W. (1998). ExoY, an adenylate cyclase secreted by the *Pseudomonas aeruginosa* type III system. *Proceedings of the National Academy of Sciences USA* **95**, 13,899–13,904.

Zhang, Z., Clemens, J., Schubert, H., Stuckey, J., Fischer, M., Hume, D., Saper, M., and Dixon, J. (1992). Expression, purification, and physicochemical characterization of a recombinant *Yersinia* protein tyrosine phosphatase. *Journal of Biological Chemistry* **267**, 23,759–23,766.

Zhou, D., Chen, L.M., Hernandez, L., Shears, S.B., and Galán, J.E. (2001). A *Salmonella* inositol polyphosphatase acts in conjunction with other bacterial effectors to promote host cell actin cytoskeleton rearrangements and bacterial internalization. *Molecular Microbiology* **39**, 248–259.

Zumbihl, R., Aepfelbacher, M., Andor, A., Jacobi, C.A., Ruckdeschel, K., Rouot, B., and Heesemann, J. (1999). The cytotoxin YopT of *Yersinia enterocolitica* induces modification and cellular redistribution of the small GTP-binding protein RhoA. *Journal of Biological Chemistry* **274**, 29,289–29,293.

CHAPTER 8

Bacterial superantigens and immune evasion

John Fraser, Vickery Arcus, Ted Baker, and Thomas Proft

8.1 INTRODUCTION

Vertebrates and microbes live together in a precarious balancing act. *Staphylococcus aureus* and *Streptococcus pyogenes* are Gram-positive commensal bacteria that inhabit the human skin, nose, and upper respiratory tract (Wannamaker and Schlievert 1988) and for the most part live an unremarkable, symbiotic existence with humans. Both organisms produce superantigens (SAGs) Table 8.1 that simultaneously bind to the T-cell Receptor (TcR) and the major histocompatibility class II (MHC-II) antigens – two molecules central to host immunity – bringing them together to cause profound T-cell activation (Schlievert, 1993; Marrack and Kappler, 1990; Kotzin et al., 1993; Fleischer, 1994; Acha Orbea and MacDonald, 1995). Any T cell bearing a reactive TcR β-chain becomes a target for a SAG and with only sixty-five TcR β-chain genes resident in the human genome (Rowen et al., 1996), any individual SAG stimulates at least 1–2% of peripheral T cells and often more than this. Superantigen activation produces toxic levels of the pro-inflammatory cytokines Interleukin (IL)-1β, tumour necrosis factor (TNF)-α, and interleukin-2 (IL-2) (see Chapter 10 for more details on cytokines), which can lead to the potentially lethal condition known as toxic shock. SAGs are not limited to *S. aureus* and *S. pyogenes*. Versions of SAGs have also been found in a number of other organisms and all cross-link TcR and MHC class II to overstimulate T-lymphocytes.

Although a great deal is known about the structure and mode of action of the bacterial SAGs, little is known about how they act to enhance the survival of bacteria and how they might disrupt the host immune responses to other antigens. With the completion of both the staphylococcal and streptococcal genomes, one thing is readily apparent – SAGs are a large family of microbial

Table 8.1. *Biochemical, functional properties, and disease association of human SAGs*

Sag	MW	Microbe	Crystal structure	Zinc binding	MHC-II binding $\alpha\|\beta$ chain	Human TcR Vβ specificity	Disease*
SEA	27.1	S. aureus	+	+	+|+	1.1, 5.3, 6.3, 6.4, 6.9, 7.3, 7.4, 9.1, 23.1	SFP
SEB	28.4	S. aureus	+		+|–	1.1, 3.2, 6.4, 15.1	SFP
SEC1	27.5	S. aureus	–		+|–	3.2, 6.4, 6.9, 15.1	SFP
SEC2	27.6	S. aureus	+		+|–	12, 13, 14, 15, 17, 20	SFP
SEC3	27.6	S. aureus	+		+|–	5.1	SFP
SED	26.9	S. aureus	+	+	+|?	1.1, 5.3, 6.9, 7.4, 8.1, 12.1	SFP
SEE	26.8	S. aureus	+	+	+|+	5.1, 6.3, 6.4, 6.9, 8.1	SFP
SEG	27.0	S. aureus	–	?	?|?	?	?
SEH	25.6	S. aureus	+	?	?|?	?	TSS
SEI	24.9	S. aureus	–	?	?|?	?	?
TSST	21.9	S. aureus	+	–	+|–	2.1, 8.1	TSS
SPE-A	26.0	S. pyogenes	–	–	+|–	2.1, 12.2, 14.1, 15.1	TSLS, Scarlet fever
SPE-C	24.4	S. pyogenes	+	+	–|+	2.1, 3.2, 12.5, 15.1	TSLS
SPE-G	24.6	S. pyogenes	–	+	?|+	2.1, 4.1, 6.9, 9.1, 12.3	?
SPE-H	23.6	S. pyogenes	–	+	?|–	2.1, 7.3, 9.1, 23.1	?

SSA	26.9	S. pyogenes	—	—	?|?	1, 3, 15, 17, 19	TSLS
SMEZ	24.3	S. pyogenes	—	+	—|+	2.1, 4.1, 7.3, 8.1	KS?RF?
SMEZ-2	24.1	S. pyogenes	+	+	—|+	4.1, 8.1	KS?RF?
SPE-I	25.9	S. pyogenes	—	+	—|+	2.1, 8.1	?
SPE-J	24.6	S. pyogenes	—	+	—|+	18, 9.1, 5.3	?
YPM	21.0	Y. pseudotuberculosis	—	?		3, 9, 13.1, 13.2	TSLS?KS?
MAM	25.2	M. arthritidis	—	+	?|?	6, 8	Arthritis?
CMV	?	Cytomegalovirus	—	?	?|?	12	?
EBV	?	Epstein Barr Virus	—	?	?	13	?

Notes: The list of SAGs known to affect humans continues to grow. There are currently 35 confirmed SAGs produced by *S. aureus* and *S. pyogenes* and are linked to or are directly responsible for a number of diseases. Not all are shown here. Note that every SAG has a different TcR Vβ profile. T cells bearing Vβ2 and Vβ8 are most commonly targeted by *S. pyogenes* SAGs. The MAM and YPM SAGs are not related to the other bacterial SAGs. The CMV SAG has yet to be isolated and is only known for its ability to enhance HIV replication in human Vβ12 T cells. There are number of SAGs that have yet to have their TcR Vβ profiles mapped. Those Vβs underlined are the predominant responders. *SFP – Staphylococcal food poisoning, TSS – Toxic Shock Syndrome, TSLS – Toxic Shock Like Syndrome, RF – Rheumatic fever, KS – Kawasaki's Syndrome.

virulence factors clustered within pathogenicity islands – a clear indication of their importance to microbial defense and their similarity in function and repertoire diversity to the immune system of higher vertebrates. This chapter provides an overview of the SAG family, their structure function and diverse mechanisms of the staphylococcal and streptococcal SAG family, and their role in gram-positive mediated disease including thoughts on their role in immune evasion.

8.2 FEATURES OF SUPERANTIGENS

8.2.1 Superantigens stimulate T cells via the Vβ-region of TcR

The superantigens (SAGs) of *S. aureus* and *S. pyogenes* are single-chain globular proteins of molecular mass 23–28 kD secreted to different extents by different strains. They represent some of the most potent bioactive molecules discovered. Addition of femtomolar amounts of any purified recombinant SAG to cultured human peripheral blood lymphocytes results in profound activation and proliferation within three days (Li et al., 1999). Common to all those T cells proliferating in response to any one SAG is that they share a limited repertoire of TcR β-chains. This Vβ conservation is the classic "footprint" of a superantigen. In contrast, normal peptide antigens stimulate no more than 0.0001–0.001% of naïve T cells because binding to peptide:MHC complexes requires surface complementarity to the combined TcR α/β-chain surface generated by the germ-line recombination mechanisms of the T-cell receptor locus (Davis et al., 1998). If nothing else, the extraordinary potency of SAGs, equivalent to only a few SAG molecules per MHC class II antigen presenting cell (APC), highlights the ability of the immune system to recognise and respond to infinitesimally small numbers of antigens. Thus, one of the key questions about SAGs is what features endow them with the ability to stimulate so many T cells at such low concentrations?

Both human and mouse T cells respond to SAGs, but human T cells are significantly more sensitive (Marrack and Kappler, 1990). Some strains of mice are more responsive to some SAGs than others and certain SAGs do not stimulate murine T cells at all. For example, the streptococcal SAG (streptococcal pyrogenic exotoxin C or SPE-C), a potent activator of human T cells, fails to activate murine T cells from any mouse strain even though it binds as well to murine I-E molecules as it does to human class II (Li et al., 1997). SAGs ligate the β-chain side of TcR coded entirely by the β-chain V-region segment. There is no involvement from either the TcR α-chain or those hypervariable regions of the TcR formed from the junction of V-D-J gene segment recombination (i.e., the CDR3 loop regions of the antigen binding surface) (Li et al., 1998). The human TcR β-chain gene locus contains approximately sixty-five V-region germ-line segments (Rowen et al., 1996) and each SAG usually binds more than 1 Vβ domain (Fraser et al., 2000). Mice are different in their responses to SAGs, because certain Vβs have been deleted because of the expression of endogenous viral SAGs in the thymus (Acha Orbea and MacDonald, 1995; discussed below). Humans have little bias in their peripheral TcR Vβ repertoire so that there is little difference in either the extent or the range of Vβs stimulated between different healthy individuals (Kotzin et al., 1993). Staphylococcal Enterotoxin A (SEA), for example, stimulates approximately 20% of human T cells bearing Vβ1, 5.3, 6.6, 6.4, 6.9, 7.4 9.1, and 23, whereas staphylococcal enterotoxin E (SEE), which differs in amino acid sequence from SEA by only 20%, stimulates a different but overlapping group of T-cells bearing Vβ1, 5.1, 6.3, 6.4, 6.9, 7.4, 8.1. Toxic shock syndrome toxin (TSST), on the other hand, almost exclusively binds human Vβ2 bearing cells. These profiles are the same for different individuals (Hudson et al., 1993). No one has yet found a SAG that stimulates all peripheral T cells or one that binds to the Vα domain of the TcR.

8.3 SAGs REQUIRE MHC CLASS II FOR FUNCTION

One feature common to all SAGs, is their dependence on APCs expressing MHC class molecules. SAGs must first bind to the distal region of MHC-II before they stimulate T cells and removal of class II expressing APCs from *in vitro* culture renders soluble SAGs inactive (Fraser, 1989). However, MHC class II for most SAGs (there are some exceptions), does not itself contribute other than to provide a scaffold for the SAG to interact with the TcR alone (Li et al., 1999). Immobilized SAGs bound to plastic for example, can stimulate the same T cells, albeit at much higher concentrations. In addition, CD8 T cells are equally as reactive to SAGs as CD4 T cells, even though CD8 T cells are normally restricted by MHC class I molecules (Fleischer and Schrezenmeier, 1988). This indicates that SAG binding of TcR does not depend on either CD4 or CD8 molecules that are normally critical coreceptors for peptide recognition but does not exclude their involvement either. More recent studies of crystal complexes of SAgs such as SPE-C. (Li et al., 2001) and SEH (Hakansson et al., 2000) bound to MHC class II, indicate that the SAG masks much of the peptide:MHC class II surface that would normally interact with TcR. Not all SAGs bind in this orientation however and staphylococcal enterotoxin B (SEB), which binds on the other side of MHC class II, requires continued interaction between MHC class II and TcR for stability (Seth et al., 1994; discussed in more detail below).

Why then must SAGs bind to MHC class II first before they bind TcR? The strict dependence on MHC class II appears to be critical for optimal potency presumably through the coordinated expression and clustering of adhesion molecules and TcR oligomerisation into RAFTS in the surface of the APC (Davis et al., 1998; Bromley et al., 2001).

BACTERIAL SUPERANTIGENS

8.3.1 The staphylococcal and streptococcal SAG family

The best-studied SAGs are the family of staphylococcal enterotoxins (SE) and streptococcal pyrogenic exotoxins (SPE) secreted by *Staphylococcus aureus* and *Streptococcus pyogenes* commonly found on the skin, the nose, and upper respiratory tract of humans (Schlievert, 1993; Marrack and Kappler, 1990; Kotzin et al., 1993; Fleischer, 1994; Li et al., 1999; Fraser et al., 2000). These toxins are potent virulence factors in Gram-positive-mediated disease but their exact function in the life cycle of either *S. aureus* or *S. pyogenes* still remains uncertain.

For many years only six staphylococcal (SEA, SEB, SEC, SED, SEE, and TSST) and two streptococcal (SPE-A and SPE-C) toxins were known. The highly expressed SPE-B was for many years thought to be a SAG, but has since been shown, through expression of the recombinant protein, to be a potent cysteine protease. (Gerlach et al., 1994). Preparations of SPE-B were contaminated with the highly potent SAG SME-Z (Proft et al., 1999). The number of SAGs has increased considerably in recent years with the availability of both *S. aureus* and *S. pyogenes* genome sequencing databases (Ferretti et al., 2001; Kuroda et al., 2001). The number of confirmed staphylococcal and streptococcal SAGs now stands at 35 (Fig. 8.1) but grows steadily larger as other strains are mined for SAG homologues.

Similarity in amino acid sequences between SAG family members varies between 20 and 90% (Fig. 8.1), and it is clear they have all evolved from a common ancestral gene. TSST is the clear outlier of the family and represents an intriguing window on the evolution of the SAG family especially considering that a whole new family of genes with none of the typical SAG activity has been found. These genes are similar to TSST sequence (see discussion of the SET sequences below; Williams et al., 2000; Kuroda et al., 2001). Many of the SAG genes reside on mobile genetic elements that have increased their

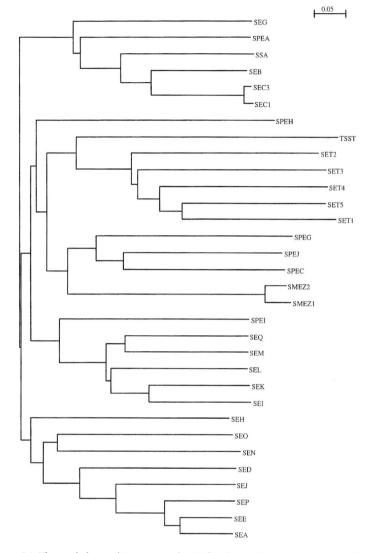

Figure 8.1. The staphylococcal/streptococcal SAG family tree. The current number of staphylococcal and streptococcal SAGs stands at thirty-five, although this is likely to be an underestimate of the total. This tree has been constructed using the program Clustal X (Thompson et al., 1997) and is based on a multiple amino acids sequence alignment from all SAG- and SAG-like sequences present in GenBank. A new group of SAG-like gene products have been included (SET1-15) and are most similar in sequence to TSST but they do not display the classical SAG activity. A few of these SAGs such as SMEZ show considerable allelic polymorphism and are expressed at high frequency in the general Group A streptococcal population. Only SMEZ-1 and SMEZ-2 are shown but there are in fact twenty-five alleles of smez identified.

horizontal transfer. The staphylococcal enterotoxin A gene, for example, resides on a variable genetic element (Betley et al., 1984), whereas the *spe-a* gene from *S. pyogenes* is located on streptococcal phage T12 (Weeks and Ferretti, 1986) and is found in about 30% of all streptococcal isolates. The gene for streptococcal pyrogenic exotoxin C (SPE-C) is also phage encoded. (Goshorn and Schlievert, 1989).

Despite the variation in primary amino acid sequence throughout the SAG family, the SAG 3-D structure remains highly conserved (discussed below). Almost all strains of *S. aureus* and *S. pyogenes* secrete at least one SAG, and many strains produce multiple SAGs.

8.4 THE STAPHYLOCOCCAL GENOME AND NEW SAG-LIKE MOLECULES

The complete genome sequence of *S. aureus* methicillin resistant strain N315 identified a much larger array of SAG and SAG-like genes residing in at least seven pathogenicity islands (Kuroda et al., 2001). In the N315 strain, three toxic shock syndrome toxin pathogenicity islands SaPln1, Sapl1, and SaPlbov were identified containing the *tst* gene upstream of a cluster of genes with unknown function that are in turn upstream of an intergrase gene. Two new pathogenicity islands, SAPln2 and SaPlm2, were also identified and contain ten SET genes (SET6-15). The SET genes are related to the prototype SAGs (Williams et al., 2000) by sequence similarity only, with the closest SAG gene being TSST (Fig. 8.1), but so far experiments have failed to determine their function. They do not appear to be classical SAGs. They are secreted proteins but do not activate T cells or bind selectively to MHC class II. In addition, the 3-D structure of one SET is considerably different from the prototype SAG structure (Fraser, unpublished). In addition to the TSST containing pathogenicity islands, a fourth enterotoxin pathogenicity island called SaPln3 has also been identified in the N315 genome. SaPIn3 contains two distinct gene clusters with five new SAG genes downstream of a serine protease gene cluster. The extraordinary array of SAG and SAG-like genes found within a single highly pathogenic, methicillin-resistant strain of *S. aureus* reflects the importance these genes must play in the survival of the organism.

8.5 ALLELIC VARIATION OF SAGs IN *STREPTOCOCCUS PYOGENES*

Species of pathogenic bacteria are characterised by evolutionary distinct genotypes resulting in diversity in gene content and sequence. This allelic variation is based on the high mutation rates and DNA recombination frequencies in bacteria and is an important strategy to counteract the diversity of their biggest "enemy," the cells of the host immune response. Allelic variation in SAGs was first described for the *S. pyogenes* SAGs: SPE-A, SPE-C, and SSA. However, the respective genes possess only small numbers of polymorphic sites and allelic variants (1 to 3).

In contrast, the analysis of the *smez* gene locus revealed a far more extensive allelic variation. Thus far, twenty-four *smez* alleles were identified containing sixty-two polymorphic sites (10% of the gene) and many more alleles are expected to exist. Neutralisation experiments showed that blood samples from healthy donors contain protective antibodies against distinctive sets of SMEZ variants, indicating a significant antigenic variation against humoral responses. Despite the sequence variation, recombinant forms of SMEZ variants did not show any differences in function, such as MHC class II binding and TcR Vβ specificity (Proft et al., 2000). Antigenic variation is a widespread bacterial defense strategy to avoid a specific Th2 response of the host immune system and usually involves important virulence factors that are essential for bacterial infection. SMEZ, which is also the most potent of all known SAGs, is therefore believed to play a unique role within the SAG family of *Streptococcus pyogenes*.

8.6 SAGs FROM OTHER BACTERIA

Unrelated SAGs genes have also been isolated from *Mycoplasma arthritidis.* (Cole and Atkin, 1991) and *Yersinia pseudotuberculosis.* (Ito et al., 1995) and both have been confirmed by testing their recombinant products in T-cell stimulation assays. *M. arthritidis* is a rodent pathogen causing chronic arthritis. Although the *M. arthritidis* mitogen (MAM) has been known for many years, research on this SAG has been hindered by difficulties in its isolation and stability. The MAM gene was finally cloned in 1996 and did not show any significant homology to the staphylococcal or streptococcal SAGs (Knudtson et al., 1997). MAM is a single 25kD protein that preferentially binds to murine I-E and human HLA-DR molecules with DR4, DR7, and DR12 subtypes presenting MAM most efficiently. Little is known about its structure and mechanism of action, but a recent study on the TcRs that respond to MAM indicates that it is likely to contact not just the germ-line encoded TcR β chain but also the CDR3 region that is generated through somatic mutation (Hodtsev et al., 1998). This interesting finding reveals for the first time that some SAGs may have a much more restricted TcR repertoire and thus may not leave the obvious Vβ "footprint" seen for the prototypic staphylococcal and streptococcal SAGs. This raises the question of how many more semi-restricted bacterial SAGs exist that can stimulate T cells in an unrestricted fashion without producing the visible signs of massive T-cell activation, cytokine release, and shock that accompany acute poisoning by one of the prototype staphylococcal or streptococcal SAGs. That such unrestricted SAGs could cause alterations in immune system function is clear but the exact nature of the pathology they could induce is not.

The *Y. pseudotuberculosis* derived mitogen (YPM) is a small 21 kD protein that specifically targets human Vβ13.1 and 13.2 bearing TcRs (Ito et al., 1995). Like MAM, it shows no significant homology to other SAGs and its role in the pathogenicity of *Y. pseudotuberculosis* is unclear.

8.7 VIRAL SUPERANTIGENS

8.7.1 Mouse mammary tumor virus

The other major SAG group is the product of an endogenous mouse retrovirus. The viral (v)SAGs are type II membrane proteins with no sequence similarity to the bacterial SAGs. They were first discovered in 1974 by Hillyard Festenstein and were referred to as minor lymphocyte stimulating (mls) antigens. They were known to generate strong T-cell proliferative responses between strains of mice matched at all MHC loci (Festenstein and Kimura, 1988). The T-cell response to mls antigens was identical to the response to bacterial SAGs with expansion of unique Vβ subsets. In 1991, the mls antigens were shown by several groups to be the products of the endogenous murine retrovirus mouse mammary tumor virus (MMTV). So far, MMTV is the only virus confirmed to express SAG genes. The gene for the MMTV SAG resides in the 3' Long Terminal Repeat of the retroviral genome and resides in all laboratory and many wild-type strains of mice as an integrated pro-viral form (Acha-Orbea and MacDonald, 1995; Winslow et al., 1992). Infectious MMTV is present at very high viral titres in the mammary tissue and breast milk of only a few strains of mice. The vSAG molecule is an essential component of the life cycle of the virus; providing efficient viral replication in newly infected gut B-cells by recruiting Vβ mediated T-cell "help" to promote B cell proliferation and viral replication. Integration of the vSAG into the mouse genome provides mice with a defense against MMTV infection. Expression of the endogenous vSAG genes in thymic stromal cells deletes reactive T cells during development thus removing T-cell "help" at the source and preventing wild-type viral infection in the gut. The net effect, as we humans have observed, is the absence of Vβ bearing T-cell populations in the peripheral T-cell repertoire of all mice (Acha-Orbea and MacDonald, 1995). The structure of the viral SAGs has not been determined despite many attempts to express the gene in heterologous systems. The protein sequence is highly conserved among all strains of MMTV except in the last 29–32 C-terminal residues that are highly variable and confer the Vβ specificity. The 45 kD protein is extensively glycosylated and is only weakly expressed in infected B cells. Glycosylation is important for efficient expression, and Grigg et al. have shown that the SAG protein traffics independently of MHC-II in B cells (Grigg et al., 1998). Evidence also suggests that proteolytic cleavage to release an 18 kD C-terminal fragment is required for efficient binding to MHC-II molecules once the viral SAG reaches the cell surface (Winslow et al., 1992).

8.7.2 Other viral SAGs

So far, mice are the only animals known for certain to express endogenous viral SAGs. Despite extensive investigations, there is no evidence for Vβ specific T-cell deletions or skewing in the peripheral T-cell repertoire in humans, which would suggest a similar viral superantigen affects the human T-cell repertoire. There have been some tantalizing preliminary studies suggesting skewed Vβ responses to some viral infection however (listed below).

8.7.3 Cytomegalovirus (CMV)

Evidence for a cytomegalovirus-encoded SAG has come from the unorthodox observation that human immunodeficiency virus (HIV) replicated preferentially in Vβ12 bearing T cells by a factor of up to 100-fold (Dobrescu et al., 1995). This was true both *in vivo* (where T cells isolated from normal individuals display a similar pattern of differential permissiveness to HIV) and *in vitro* (where Vβ12-bearing T cells may act as a viral reservoir). This selectivity was attributed to a Vβ-selecting element that activates these T cells, allowing HIV to replicate in them. Furthermore, this selectivity was MHC-II-dependent but not MHC-restricted and required the participation of non-T cells. Such an element has the hallmark of a Vβ12-specific SAG. The possibility that a ubiquitous virus such as CMV may be responsible for this element was tested. It was shown that the Vβ12-selective HIV replication depended on non-T cells that came from CMV-positive individuals. In addition, a monocytic cell line was able to promote this phenomenon only when the cells themselves were infected with CMV and were in direct contact with the T cells, thus ruling out the of a possibility of a soluble mediator (Dobrescu et al., 1995).

8.7.4 Epstein Barr Virus (EBV)

Evidence for an EBV encoded SAG arose from observations that EBV infected human B-cell lines induced into the lytic cycle with a B-cell mitogen, selectively stimulate Vβ13 bearing T cells. This Vβ13 specific T-cell activation remains the only tantalizing evidence so far of an EBV encoded SAG (Sutkowski et al., 1996).

8.8 SUPERANTIGENS AND DISEASE

8.8.1 Staphylococcal and Streptococcal SAGs in the general population

Staphylococcal and streptococcal expressing strains are extremely common in the general population, and all adults show evidence of exposure to at least one, and in most cases, multiple, SAGs. Exposure to SAGs occurs very early in life with sero-conversion evident in many infants (Reiser et al., 1988). Interestingly, not all sero-conversion leads to effective neutralizing antibodies with some individuals displaying high titers of anti-SAG antibody but no neutralizing response. The absence of a neutralizing antibody is generally regarded as a critical determinant in individual susceptibility to the effects of particular SAGs. For example, in the TSST-mediated toxic shock syndrome, no patients display a neutralizing response to TSST in either the acute or convalescent stages of the disease (Vergeront et al., 1983).

8.8.2 Staphylococcal food poisoning

The SAGs were first identified as staphylococcal enterotoxins by the late Merlin S. Bergdoll and colleagues at the Food Research Institute in Wisconsin as the causative agents in staphylococcal food poisoning – a common outcome for people with the misfortune of having eating food contaminated with *S. aureus*. Ingestion of microgram amounts of an enterotoxin induces violent vomiting and diarrhoea within 4–6 hours. The patient usually makes a full recovery within 24 hours. The violent enteric response is presumably a means to rid the body of these agents before they enter the bloodstream. The exact mechanism by which SAGs induce the vomiting response remains unclear. Several studies attempting to relate T-cell mitogenicity with enterotoxicity have produced confusing results. Mutants engineered to be defective in T-cell activation do not consistently ablate the enterotoxic effect of a SAG, leading some to speculate that another separate gut receptor might be involved that is separate from the immune recognition receptors MHC-II or TcR (Harris et al., 1993).

The severity of enterotoxaemia is dependent on the extent of contamination, the amount ingested, and the sensitivity of the individual. The SAG most commonly found in food poisoning outbreaks is staphylococcal enterotoxin B (SEB). The SEB gene is located on the staphylococcal chromosome and is expressed at high levels by many strains of *S. aureus*. Some strains isolated from food poisoning outbreaks produce SEB at levels approaching 1 mg/ml in culture. Thus, only minimal bacterial growth on the food is required before sufficient toxin is produced to cause food poisoning. Another SAG commonly associated with food poisoning is staphylococcal enterotoxin A (SEA). Unlike SEB, the SEA gene is located on a phage and is expressed at a 1000 times lower concentration (around 1–5 μg/ml) (Kokan and Bergdoll, 1987). All other staphylococcal SAGs except TSST have been shown to be enterotoxic. Because of their structural stability, the staphylococcal enterotoxins resist both elevated cooking temperatures as well as the acidic conditions of the stomach. One possible reason why TSST producing strains are not associated with food poisoning is because TSST is less stable and is denatured in the low pH conditions found in the stomach. The streptococcal SAGs, although structurally very similar, are not enterotoxic for the simply reason that streptococci do not grow well on food.

8.8.3 Toxic Shock Syndrome

Toxic shock is a relatively rare consequence of staphylococcal or streptococcal infection. It is a rapid onset syndrome characterized by fever, vomiting, rash, desquamation, hypotension, microvascular leakage, and multiorgan failure leading to death. Toxic shock syndrome was first formally described by Todd in the late 1970s during an outbreak of toxic shock in young-to-middle aged women using a new extended wear tampon marketed under the brand-name Rely (Todd and Tishant, 1978). This high absorbency tampon provided optimal conditions for the florid intra-vaginal growth of toxic shock syndrome toxin (TSST-1)-producing strains of *S. aureus*, which in turn lead to high systemic levels of TSST in the blood. TSST-producing strains of *S. aureus* are widely distributed within the community. The disease however is not limited to TSST or to *S. aureus*. Toxic shock can also arise from streptococcal wound infections. Onset of symptoms can be extremely rapid and usually begins with a localized rash around the wound followed by fever and rigors then vomiting, hypotension, unconsciousness, and death (Schlievert, 1993).

Toxic shock is not caused by the SAG itself, but by the sudden release of cytokines following SAG T-cell activation. In an elegant study by Fleischer and colleagues, production of cytokines in the mouse spleen was examined following intravenous injection of purified SEB. Within 30 minutes, a dramatic burst of TNF-α, INF-γ, and IL2 expression by the PAL (periarteriolar lymphocytes) region of the spleen was observed followed by a second burst of TNF-α expression by the B cells and resident APCs of the spleen (Bette et al., 1993). SAG mutants defective in TcR binding were no longer lethal in mice sensitized with the liver toxin D-galactosamine, which indicates that the shock is dependent upon T-cell activation (Marrack et al., 1990).

Links between disease severity and the expression of a particular SAG have been tenuous at best, except in the case of TSST and toxic shock syndrome. The strongest association is between toxic shock syndrome and TSST-1 where 50% of all *S. aureus* strains isolated from toxic shock patients in the USA produce TSST-1 (Musser et al., 1990). Moreover, there is a strong correlation between the disease susceptibility and absence of neutralizing antibodies to TSST-1 (Vergeront et al., 1983). However, this figure does not absolutely imply that TSST-1 is the causative agent because the prevalence of TSST-1 producing strains in the general community is high.

8.8.4 Scarlet Fever

A less severe form of toxic shock is scarlet fever, commonly attributed to *S. pyogenes* strains producing the scarlatina toxin (SAG pyrogenic exotoxin A – SPE-A) infecting the head and neck region. This acute onset disease is characterized by a fever, rash, and the characteristic purple tongue and is confined to infants and adolescents (Schlievert, 1993). Prior to the antibiotic era, this was a serious childhood disease but is now effectively treated with penicillin. Scarlet fever has been commonly associated with the expression of streptococcal pyrogenic exotoxin A (SPE-A), but SPE-A is found in 30% of all streptococcal isolates so it is not surprising that this association has arisen. Other studies have identified SPE-C as the principal culprit (Hoebe et al., 2000). With the discovery of many more streptococcal SAGs following the completion of a streptococcal genome sequencing project (Ferretti et al., 2001), it is likely that other previously unknown SAGs also contribute to this disease including the most potent of all SAGs yet discovered, SMEZ. Many streptococcal isolates express more than one SAG and the combination of SAGs combined with relative levels of expression may be a stronger indicator of this disease.

8.8.5 Streptococcal throat and rheumatic fever

The most common site of streptococcal infection is the throat. Painful swelling and inflammation of the tonsils is a result of lymphocyte proliferation and expansion in response to *S. pyogenes* infection. Group A streptococcal infection remains 100% sensitive to penicillin. Repeated streptococcal throat infections can lead to the more serious complication of rheumatic fever and poststreptococcal glomerulonephritis. The cause of this disease remains a mystery. One possible cause is the generation of autoantibodies following repeated streptococcal antigenic or SAG challenge. Some studies have suggested the presence of cross-reactive antibodies to heart valve antigens but these antibodies are low affinity and are also present in non-RF patients (Jones et al., 2000).

8.8.6 Kawasaki's disease

This is an acute febrile disease of children aged about 5 years. Without treatment, it can be fatal. In Japan and the United States, it has become one of the most common causes of acquired heart disease in children. Intravenous immunoglobulin therapy is highly effective when given early suggesting that the agent is a toxin that is neutralized by naturally occurring antitoxin antibodies. The etiological agent is not known but several lines of evidence suggest that it might be SAG mediated. The symptoms of the disease resemble toxic shock syndrome and several studies have shown that there is a selective expansion of Vβ2 subset of T cells (Leung et al., 1995). In one study of sixteen patients, thirteen were found to have TSST +ve isolates of *S. aureus* in their pharynx, rectum, or groin (Leung et al., 1993). Further investigations however have not been able to substantiate the connection with *S. aureus* TSST strains, and there is still debate over whether SAGs really are the culprit in Kawasaki's disease. In another article, a link between the *Yersinia pseudotuberculosis* SAG, YPM, and Kawasaki's disease has been suggested (Konishi et al., 1997), although this link also has yet to be firmly established.

8.8.7 SAGs in autoimmune diseases

It has been proposed that SAGs derived from bacteria, viruses, or mycoplasma might contribute to the pathogenesis of autoimmune disease by activating T cells specific for self-antigen (Marrack and Kappler, 1990; Kotzin et al., 1993). Early reports observed Vβ-specific enrichment of T cells in common diseases such as arthritis. These created much excitement. Paliard and coworkers (1991) for example analysed the TcR β-chain profiles of synovial T cells and found a selective expansion of human Vβ14 expressing T cells. Using an animal model of multiple sclerosis, experimental autoimmune encephelomyelitis (EAE), it has been shown that administration of SEB to mice recovering from EAE triggered a rapid relapse of the disease. This effect was due to direct stimulation of the Vβ3 positive autoreactive MBP peptide specific T cells that initially caused the brain inflammation. This was clear evidence that SAGs could, under the right conditions, break the tolerance or suppression of autoreactive T-cell clones and induce a state of autoimmune disease (Brocke et al., 1994; Racke et al., 1994).

Despite these early reports, there has been little clear evidence directly linking a SAG to any autoimmune disease. Nevertheless, there is sufficient evidence to suggest that SAGs can and do play a secondary role in autoimmune activation through the relatively indiscriminate stimulation of autoreactive T cells.

8.8.8 Crohn's disease associated SAG from *Pseudomonas fluoresceins*

A suspected SAG has been identified from a Crohn's disease associated microbial gene I2 isolated from a CD disease lesion in murine intestine. Although still unconfirmed, the recombinant gene product stimulated mVβ5-bearing T cells, required MHC class II APC for activity and did not require antigen processing for activity (Dalwadi et al., 2001).

8.9 THE STRUCTURE AND FUNCTION OF THE STAPHYLOCOCCAL AND STREPTOCOCCAL SUPERANTIGENS

The crystal structures of SEA (Schad et al., 1995), SEB (Swaminatham et al., 1992), SEC2 (Papageorgiou et al., 1995), SEE (Swaminatham et al., 1995), TSST (Prasad et al., 1993), SEH (Hakansson et al., 2000), SPEC (Roussel et al., 1997), and SMEZ-2 (Arcus et al., 2000) reveal a common core fold based upon two globular domains, a C-terminal domain of the β-grasp motif and a smaller N-terminal pseudo β-barrel domain (Fig. 8.2). The crystal structures of SEB/HLA-DR1, TSST/DR1, SPE-C/DR1, and SEH/DR1 complexes reveal very different mechanisms of binding (Li et al., 2001; Kim et al., 1994; Jardetsky et al., 1994; Petersson et al., 2001). Staphylococcal enterotoxin B (SEB) has a single, low-affinity MHC-II binding site located in the smaller N-terminal domain. Residues in this region bind to a hydrophobic groove located in the distal region of the invariant β1 domain of HLA-DR (Fig. 8.2). The region of SEB that binds TcR is a shallow groove between the two domains. SEB contacts both the complementarity determining region 2 (CDR2) and the fourth hypervariable loop (HV4) of the TcR β-chain (Fields et al., 1996). Thus, SEB can be likened most simply to a wedge that fits snugly

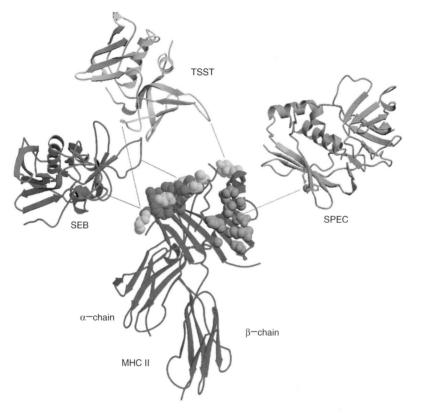

Figure 8.2. **Different modes of binding to MHC-II molecules**. At least three separate binding orientations have been found for bacterial SAGs. The extracellular domain of MHC-II is shown in blue and the MHC-II residues with which SEB interacts are shown as space-filling spheres in red. SEB has been moved by a simple translation away from its position on the human HLA-DR1 molecule (from the crystal structure (Jardetzky et al., 1994)). TSST forms a complex with MHC-II using the same face of the toxin (when compared to SEB) but in a different orientation. TSST has also been moved up from its position by a simple translation from the crystal structure (Kim et al., 1994). Residues on MHC-II that interact with TSST upon complex formation include those that interact with SEB (red spheres) and, in addition, those shown as yellow spheres. Note that TSST bridges one end of the peptide binding groove and interacts with two residues of the β-chain of MHC-II. The orientation of SPE-C (grey) on MHC-II is mediated by zinc and His-81 on the MHC-II β-chain (shown as grey spheres). SPE-C binds across the peptide groove and excludes contact between MHC-class II and TcR (Li et al., 2001). SPE-C only binds in this orientation. SEA, on the other hand, binds in both this orientation and in the SEB orientation also.

between the TcR and MHC-II while the TcR and MHC-II are bound together, although the "wedge" analogy is not entirely accurate. A recent comparison of several crystal structures reveals that SEB is likely to tip and rotate the TcR away from its normal position on the MHC-II molecule so that entirely new contacts must be accommodated from those determined during thymic selection (Li et al., 1998).

The crystal structure of TSST/DR1 reveals a different mode of binding from SEB (Kim et al., 1994) (Fig. 8.2). Although TSST binds to the same region of DRβ1 domain as SEB, it covers most of the top of MHC-II preventing any further contact with the TcR. TSST also makes contact with residues from the bound peptide that limits the number of MHC-II molecules that TSST can bind and also raises the intriguing possibility that different peptides might regulate the potency of SAG-mediated T-cell activation.

A third type of MHC-II binding occurs with a subset of the bacterial SAGs that contain a high affinity zinc site in the C-terminal domain on the opposite side of the molecule to the low affinity MHC-II β-chain binding site. Members of this group include SEA, SEE, SED, SSA SEH, SPE-C, and SME-Z. The zinc atom forms a stable coordination complex with a conserved histidine (His81) in the polymorphic β-chain of MHC-II so that these SAGs can bind the polymorphic β-chain. Recent crystal structures of SPE-C (Li et al., 2001) and SEH (Petersson et al., 2001) bound to MHC class II show that this class of SAGs bind across the top of the MHC class II and make significant contacts with peptide residues also. The predominant interaction results from the formation of a tetravalent zinc coordination complex between three residues in the SAG and His81 located in the α-helix of the β-chain of MHC class II. This provides an interesting view of how this subclass of SAGs has evolved to overcome the polymorphism that exists on the β-chain of MHC class II by developing a metal coordination complex.

A fourth mechanism is found with the subset of SAGs, SEA, and SEE. These molecules have both an high affinity zinc mediated β-chain binding site and a low affinity β-chain site similar to SEB (Hudson et al., 1995; Tiedemann and Fraser, 1996). This allows SEA to bind both sides of MHC class II and to cross-link MHC-II on the APC surface (Fig. 8.3). This leads to TNFα and IL-1 gene transcription in the APC (Tiedemann and Fraser, 1996; Mehindate et al., 1995). It is not clear how APC activation increases the potency of SEA, but one current theory is that cross-linking induces localized regions on the APC surface that possess high concentrations of MHC-II, SEA, and adhesion molecules such as ICAM-1 (intracellular adhesion molecule) and CD80 and CD86, which are surface ligands for T-cell co-stimulation. These activation clusters might be required for efficient triggering of T cells and studies using

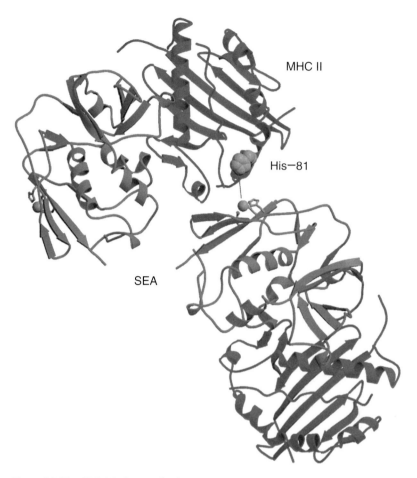

Figure 8.3. The SAG SEA has two binding sites to cross-link MHC-II. The Superantigen SEA is displayed in its dual orientation to MHC class II. One orientation is identical to the SEB site to the MHC class II β-chain whereas the other much higher affinity zinc mediated binding orientation is to the MHC class II β-chain. This picture is looking down from above with the distal domains of MHC-II β and β-chains shown in blue. On the left of the figure, an SEA molecule (red) is bound via the generic β-chain binding site in the same fashion as SEB. On the right, another SEA molecule is bound to the same MHC-II molecule via a zinc bridge between His81 on the MHC-II β-chain and His187, His225, and Asp227 in SEA. This interaction is about 100 times stronger than the first and therefore occurs first. There is some cooperation between the two SEA molecules. Once bound, it is obvious to see that cross-linking of another MHC-II molecules can occur via both bound SEA molecules. This image is only inferred from mutational data. There is yet no crystal structure of an SEA/MHC-II complex.

high-resolution confocal microscopy confirm the formation of MHC/TcR supramolecular structures during T-cell activation (Monks et al., 1998). It is likely that this subgroup of SAGs has evolved a second MHC-II binding site to increase the efficiency at which they concentrate on the APC surface and to cross-link MHC-II molecules into activation clusters.

Yet another variation in SAG binding is seen with the streptococcal SAG SPE-C, which has dispensed with MHC-II β-chain binding altogether in favor of a zinc-mediated β-chain binding site. Moreover, the crystal structure of SPE-C reveals a dimer structure formed through an interface where the low-affinity N-terminal binding is normally located (Li et al., 1997; Roussel et al., 1997). The discovery of three new streptococcal SAGs indicates that SPE-C belongs to a subgroup containing SMEZ-1, SMEZ-2, SPE-G, and SPE-H that all appear to bind exclusively through only a single zinc mediated binding to the MHC-II β-chain (Proft et al., 1999; Proft et al., 2000).

8.10 HOW DO SAGs BIND TO THE TcR?

One of the more intriguing questions about SAGs is how they manage to ligate so many different TcR molecules and whether SAG/TcR binding is any stronger or more stable than normal TcR/MHC-peptides, which might explain how SAGs activate T cells at concentrations many orders of magnitude below their dissociation binding constants for MHC-II. These questions are not easily answered, simply because soluble forms of TcR are hard to produce. Moreover, solution studies are generally not regarded as representing the complex multicomponent interactions that occur on the cell surface. Mariuzza and colleagues have succeeded in crystallizing SEC3 with a soluble form of the murine Vβ8 TcR (Fields et al., 1996). This complex shows that the SAG makes multiple contacts with residues from the CDR2, the third framework (FR3) region and the fourth hypervariable loop (HV4) of the TcR β-chain of murine Vβ8. What is most significant to the function is that individual contacts between SAG side-chain residues are predominantly made to atoms in the β-carbon backbone of the TcR, not to atoms of the side-chains, so that side-chain amino acid variation in TcR is less likely to affect the affinity of binding. The affinity of binding between SEC3 and Vβ8 has been calculated at 3.5 μM and mutations of SEC3 residues in the TcR binding site indicate that the contribution to binding energy is spread evenly across 5 or 6 amino acids that are generally conserved among other SAGs. Leder et al. (1998) performed a careful mutational analysis of this region in an attempt to correlate binding affinity to T-cell activation. By mutating individual residues in the TcR binding site, then comparing TcR binding affinity to potency, they were able to show a proportional relationship between binding affinity and potency. Their results were intriguing from the point of view of comparison to normal peptide recognition in which even the smallest drop in affinity results in a complete loss in T-cell stimulation. In contrast, SEC3 withstands an affinity decrease of over 60-fold without loss of T-cell stimulation. This difference perhaps reflects the different mechanisms of binding and ligation. Whereas peptide-MHC complexes trigger TcRs in a sequential fashion (Valitutti et al., 1995) and only transiently associate with TcR, SAGs may bind more tightly and induce much more stable surface complexes (Proft and Fraser, 1998).

8.11 SAG STIMULATION OF T CELLS

T cells must convert trivial affinity differences between self and nonself peptide-MHC into a functional T-cell response. This task is made more difficult because the number of relevant peptide–MHC complexes as a proportion of the total peptide–MHC molecules evenly distributed across the entire APC surface is likely to be very small. The TcR must therefore bind multiple peptide–MHC complexes in a serial fashion (Valitutti et al., 1995) so that binding must be transitory, yet of sufficient duration and frequency to produce a series of net positive intracellular signals. Full agonist signaling in T cells appears to be regulated by the "dwell time" of TcR:peptide–MHC complexes and binding studies of soluble TcR and peptide–MHC reveal that the half-life determined by the off-rate of the TcR:peptide–MHC complex most closely predicts the functional response by the T cell. Full agonist peptide–MHC complexes have the slowest off-rates while partial agonist and antagonist pep-MHC complexes have intermediate and high solution off-rates respectively (Davis et al., 1998).

Unlike peptide that are irreversibly bound within MHC, SAGs only bind MHC-II in a transitory fashion. SEB for example has an affinity toward HLA-DR molecules in the nM range yet activates at femtomolar concentrations – a full six orders of magnitude difference. How SAGs achieve such profoundly positive T-cell signals at concentrations many orders of magnitude below their dissociation constants for MHC class II remains a compelling question. Peptides with even a modest affinity drop toward TcR are antagonists to T cells, yet SAGs never display antagonist activity, even when their affinity for TcR is reduced by mutagenesis by several orders of magnitude (Leder et al., 1998). Clearly, there are differences in the way SAGs bind to and ligate TcRs compared to peptide, which ensures that the outcome of SAG ligation is always activation and proliferation of the T cell.

One mechanism SAGs have employed to enhance their activity at very low concentrations is to possess two separate binding sites for the MHC-II receptor. SEA is one such SAG that has two cooperative binding sites for MHC-II, one on the β-chain side and another zinc-dependent site to the β-chain side of MHC-II (Hudson et al., 1995; Tiedemann et al., 1995). SEA binds and cross-links MHC-II on the APC surface. There is strong cooperative binding of SEA to MHC-II observed at very low concentrations of SEA, well below its calculated dissociation constant for either binding site, thus explaining the ability of this particular SAG to remain bound to MHC class II at very low concentrations.

8.12 WHAT IS THE FUNCTION OF SAGs?

The simple question as to why SAGs are produced, and what benefit they are to the organisms, has been very difficult to answer. There are no effective models to examine survival rates of *S. aureus* or *S. pyogenes* strains devoid of all SAGs because it is now apparent that strains produce multiple SAGs. Once such study recently completed compared two isogenic strains of *S. pyogenes* devoid of the predominant SAG SMEZ and lacking both SPE-A and SPE-C. The knockout strain no longer produced any detectable T-cell mitogenic activity but was not cleared any faster in a mouse infection model than the isogenic strain producing SMEZ (measuring the clearance rate of infection in mice has shown little difference between SAG-producing and nonproducing strains; Unnikrishnan et al., 2001). Despite the inability to show that loss of SAG activity results in a loss of virulence for the strain, the simple fact that both *S. aureus* and *S. pyogenes* have retained so many different SAGs and the degree of variation in the *smez* locus suggests they provide an important survival advantage to the microbe. One possible function of SAGs might be to promote Th1-like immune responses that effectively suppress Th2-driven antibody responses that confers protection against Gram-positive organisms. Macrophages are generally regarded as the most important defense mechanisms against Gram-positive organisms, and one mediator of macrophage recognition is via bound immune complexes on the surface of the bacterium. The expression of protein A on *S. aureus* and protein G on *S. pyogenes* is another effective mechanism to sequester antibodies to the surface of the pathogen in an unfavorable orientation, thus rendering Fc receptor mediated and complement mediated recognition defective.

Another possible function for SAGs might be to promote a localized inflammatory response that isolates the growing colony and provides increased blood flow and nutrients to prolonging survival. At present, these roles remain speculation.

8.13 SAGs CAUSE PERIPHERAL T-CELL DELETION

SAG activation serves as an excellent model to examine the process of peripheral deletion of T cells because of the high proportion of responder cells. In animals injected with purified SEB, there is an immediate Vβ8-specific T-cell expansion followed by significant peripheral T-cell deletion of Vβ+ T-cells. T cells stimulated by a SAG are refractory to further antigenic stimulation and are rapidly removed from the periphery by a process of Fas/FasL mediated deletion. Recent experiments using Fas and FasL homozygous knockout mice indicate that FasL is predominantly expressed on nonlymphoid tissue such as liver, lung, and small intestine rather than lymphoid tissue (Bonfoco et al., 1998). Moreover, the induction on nonlymphoid FasL is not inhibited by cyclosporin, which normally inhibits the induction of FasL on T cells. This suggests a separate pathway for activation of nonlymphoid cells to express FasL. Thus, one function of bacterial SAGs might be to promote the deletion of T cells that contribute help to B-lymphocytes in the immune defense against the bacteria.

8.14 CONCLUSION

As a family of molecules, SAGs display a remarkable degree of variation in structure and function around the common goal of bringing the TcR and MHC class II molecules together. The recent completion of several staphylococcal and streptococcal genome projects has highlighted the extraordinary size and diversity of this family of molecules. As tools, they have provided remarkable insight into the mechanisms of T-cell activation and the extraordinary sensitivity of T-cell antigen recognition. Because of their ubiquitous expression and widespread carriage of both staphylococci and streptococci, it is clear that our immune systems are under constant challenge from these extremely powerful toxins and is therefore likely to influence our response to other challenges. What benefit they serve the bacteria remains a mystery but it is clear that their primary function is as a defense mechanism against host immune responses – similar to a bacterial immune system.

REFERENCES

Acha Orbea, H. and MacDonald, H.R. (1995). Superantigens of mouse mammary tumor virus. *Annual Review of Immunology* **13**, 459–486.

Arcus V., Proft, T., Sigrell, J.A., Baker, H.M., Fraser, J.D., and Baker, E.N. (2000). Conservation and variation in superantigen structure and activity highlighted by 3-dimensional structures of two new superantigens from *Streptococcus pyogenes. Journal of Molecular Biology* **299**, 157–168.

Betley, M.J., Lofdahl, S., Kreiswirth, B.N., Bergdoll, M.S., and Novick, P. (1984). Staphylococcal enterotoxin A gene is associated with a variable genetic element. *Proceedings of the National Academy of Sciences USA* **81**, 5179–5183.

Bette, M., Schafer, M.K., van Rooijen, N., Weihe, E., and Fleischer, B. (1993). Distribution and kinetics of superantigen-induced cytokine gene expression in mouse spleen. *Journal of Experimental Medicine* **178**, 1531–1539.

Bonfoco, E., Stuart, P.M., Brunner, T., Lin, T., Griffith, T.S., Gao, Y., Nakajima, H., Henkart, P.A., Ferguson, T.A., and Green, D.R. (1998). Inducible non-lymphoid expression of Fas ligand is responsible for superantigen-induced peripheral deletion of T cells. *Immunity* **9**, 711–720.

Brocke, S., Veromaa, T., Weissman, I.L., Gijbels, K., and Steinman, L. (1994). Infection and multiple sclerosis: a possible role for superantigens? *Trends in Microbiology* **2**, 250–254.

Bromley, S.K., Burack, W.R., Johnson, K.G., Somersalo, K., Sims, T.N., Sumen, C., Davis, M.M., Shaw, A.S., Allen, P.M., and Dustin, M.L. (2001). The immunological synapse. *Annual Review of Immunology* **19**, 375–396.

Cole, B.C. and Atkin, C.L. (1991). The *Mycoplasma arthritidis* T-cell mitogen, MAM: a model superantigen. *Immunology Today* **12**, 271–276.

Dalwadi, H., Wei, B., Kronenberg, M., Sutton, C.L., and Braun, J. (2001). The Crohn's disease-associated bacterial protein I2 is a novel enteric T cell superantigen. *Immunity* **15**, 149–158.

Davis, M.M., Boniface, J.J., Reich, Z., Lyons, D., Hampl, J., Arden, B., and Chien, Y.H. (1998). Ligand recognition by alpha-beta T cell receptors. *Annual Review of Immunology* **16**, 523–544.

Dobrescu, D., Ursea, B., Pope, M., Asch, A.S., and Posnett, D.N. (1995). Enhanced HIV-1 replication in V beta 12 T cells due to human cytomegalovirus in monocytes: evidence for a putative herpesvirus superantigen. *Cell* **82**, 753–763.

Ferretti, J.J., McShan, W.M., Ajdic, D., Savic, D.J., Savic, G., Lyon, K., Primeaux, C., Sezate, S., Suvorov, A.N., Kenton, S., Lai, H.S., Lin, S.P., Qian, Y.D., Jia, H.G., Najar, F.Z., Ren, Q., Zhu, H., Song, L., White, J., Yuan, X.L., Clifton, S.W., Roe, B.A., and McLaughlin, R. (2001). Complete genome sequence of an M1 strain of *Streptococcus pyogenes. Proceedings of the National Academy of Sciences USA* **98**, 4658–4663.

Festenstein, H. and Kimura, S. (1988). The Mls system: past and present. *Journal of Immunogenetics* **15**, 183–196.

Fields, B.A., Malchiodi, E.L., Li, H., Ysern, X., Stauffacher, C.V., Schlievert, P.M., Karjalainen, K., and Mariuzza, R.A. (1996). Crystal structure of a

T-cell receptor beta-chain complexed with a superantigen. *Nature* **384**, 188–192.

Fleischer, B. (1994). Superantigens. *Apmis* **102**, 3–12.

Fleischer, B. and Schrezenmeier, H. (1988). T cell stimulation by staphylococcal enterotoxins. Clonally variable response and requirement for major histocompatibility complex class II molecules on accessory or target cells. *Journal of Experimental Medicine* **167**, 1697–1707.

Fraser, J.D. (1989). High-affinity binding of staphylococcal enterotoxins A and B to HLA-DR. *Nature* **339**, 221–223.

Fraser, J.D., Arcus, V., Kong, P., Baker, E.N., and Proft, T.P. (2000). Superantigens – powerful modifiers of the immune system. *Molecular Medicine Today* **6**, 125–135.

Gerlach, D., Reichardt, W., Fleischer, B., and Schmidt, K.H. (1994). Separation of mitogenic and pyrogenic activities from so-called erythrogenic toxin type B (Streptococcal proteinase). *International Journal of Medical Microbiology, Virology, Parasitology, and Infectious Diseases* **280**, 507–514.

Goshorn, S.C. and Schlievert, P.M. (1989). Bacteriophage association of streptococcal pyrogenic exotoxin type C. *Journal of Bacteriology* **171**, 3068–3073.

Grigg, M.E., McMahon, C.W., Morkowski, S., Rudensky, A.Y., and Pullen, A.M. (1998). Mtv-1 superantigen traffics independently of major histocompatibility complex class II directly to the B-cell surface by the exocytic pathway. *Journal of Virology* **72**, 2577–2588.

Hakansson, M., Petersson, K., Nilsson, H., Forsberg, G., Bjork, P., Antonsson, P., and Svensson, L.A. (2000). The crystal structure of staphylococcal enterotoxin H: Implications for binding properties to MHC class II and TcR molecules. *Journal of Molecular Biology* **302**, 527–537.

Harris, T.O., Grossman, D., Kappler, J.W., Marrack, P., Rich, R.R., and Betley, M.J. (1993). Lack of complete correlation between emetic and T-cell stimulatory activities of staphylococcal enterotoxins. *Infection and Immunity* **61**, 3175–3183.

Hodtsev, A.S., Choi, Y., Spanopoulou, E., and Posnett, D.N. (1998). Mycoplasma superantigen is a CDR3-dependent ligand for the T cell antigen receptor. *Journal of Experimental Medicine* **187**, 319–327.

Hoebe, C.J., Wagenvoort, J.H., and Schellekens, J.F. (2000). An outbreak of scarlet fever, impetigo and pharyngitis caused by the same *Streptococcus pyogenes* type T4M4 in a primary school. *Nederlands Tijdschrift voor Geneeskunde* **144**, 2148–2152.

Hudson, K.R., Robinson, H., and Fraser, J.D. (1993). Two adjacent residues in staphylococcal enterotoxins A and E determine T cell receptor V beta specificity. *Journal of Experimental Medicine* **177**, 175–184.

Hudson, K.R., Tiedemann, R.E., Urban, R.G., Lowe, S.C., Strominger, J.L., and Fraser, J.D. (1995). Staphylococcal enterotoxin A has two cooperative binding sites on major histocompatibility complex class II. *Journal of Experimental Medicine* **182**, 711–720.

Ito, Y., Abe, J., Yoshino, K., Takeda, T., and Kohsaka, T. (1995). Sequence analysis of the gene for a novel superantigen produced by *Yersinia pseudotuberculosis* and expression of the recombinant protein. *Journal of Immunology* **154**, 5896–5906.

Jardetzky, T.S., Brown, J.H., Gorga, J.C., Stern, L.J., Urban, R.G., Chi, Y.I., Stauffacher, C., Strominger, J.L., and Wiley, D.C. (1994). Three-dimensional structure of a human class II histocompatibility molecule complexed with superantigen. *Nature* **368**, 711–718.

Jones, K.F., Whitehead, S.S., Cunningham, M.W., and Fischetti, V.A. (2000). Reactivity of rheumatic fever and scarlet fever patients' sera with group A streptococcal M protein, cardiac myosin, and cardiac tropomyosin: a retrospective study. *Infection and Immunity* **68**, 7132–7136.

Kim, J., Urban, R., Strominger, J.L., and Wiley, D. (1994). Toxic shock syndrome toxin-1 complexed with a major histocompatibility molecule HLA-DR1. *Science* **266**, 1870–1874.

Knudtson, K.L., Manohar, M., Joyner, D.E., Ahmed, E.A., and Cole, B.C. (1997). Expression of the superantigen *Mycoplasma arthritidis* mitogen in *Escherichia coli* and characterization of the recombinant protein. *Infection and Immunity* **65**, 4965–4971.

Kokan, N.P. and Bergdoll, M.S. (1987). Detection of low-enterotoxin-producing *Staphylococcus aureus* strains. *Applied and Environmental Microbiology* **53**, 2675–2676.

Konishi, N., Baba, K., Abe, J., Maruko, T., Waki, K., Takeda, N., and Tanaka, M. (1997). A case of Kawasaki's disease with coronary aneurysms documenting *Yersinia pseudotuberculosis* infection. *Acta Paediatrics* **86**, 661–664.

Kotzin, B.L., Leung, D.Y., Kappler, J., and Marrack, P. (1993). Superantigens and their potential role in human disease. *Advances in Immunology* **54**, 99–166.

Kuroda, M., Ohta, T., Uchiyama, I., Baba, T., Yuzawa, H., Kobayashi, I., Cui, L.Z., Oguchi, A., Aoki, K., Nagai, Y., Lian, J.Q., Ito, T., Kanamori, M., Matsumaru, H., Maruyama, A., Murakami, H., Hosoyama, A., Mizutani Ui, Y., Takahashi, N.K., Sawano, T., Inoue, R., Kaito, C., Sekimizu, K., Hirakawa, H., Kuhara, S., Hiramatsu, K. et al. (2001). Whole genome sequencing of methicillin-resistant *Staphylococcus aureus*. *Lancet* **357**, 1225–1240.

Leder, L., Llera, A., Lavoie, P.M., Lebedeva, M.I., Li, H., Sekaly, R.P., Bohach, G.A.,

Gahr, P.J., Schlievert, P.M., Karjalainen, K., and Mariuzza, R.A. (1998). A mutational analysis of the binding of staphylococcal enterotoxins B and C3 to the T cell receptor beta chain and major histocompatibility complex class II. *Journal of Experimental Medicine* **187**, 823–833.

Leung, D.Y., Meissner, H.C., Fulton, D.R., Murray, D.L., Kotzin, B.L., and Schlievert, P.M. (1993). Toxic shock syndrome toxin-secreting *Staphylococcus aureus* in Kawasaki syndrome. *Lancet* **342**, 1385–1388.

Leung, D.Y., Meissner, C., Fulton, D., and Schlievert, P.M. (1995). The potential role of bacterial superantigens in the pathogenesis of Kawasaki syndrome. *Journal of Clinical Immunology* **15**, 11S–17S.

Li, P.L., Tiedemann, R.E., Moffat, S.L., and Fraser, J.D. (1997). The superantigen streptococcal pyrogenic exotoxin C (SPE-C) exhibits a novel mode of action. *Journal of Experimental Medicine* **186**, 375–383.

Li, H., Llera, A., Tsuchiya, D., Leder, L., Ysern, X., Schlievert, P.M., Karjalainen, K., and Mariuzza, R.A. (1998). Three-dimensional structure of the complex between a T cell receptor beta chain and the superantigen staphylococcal enterotoxin B. *Immunity* **9**, 807–816.

Li, H.M., Llera, A., Malchiodi, E.L., and Mariuzza, R.A. (1999). The structural basis of T cell activation by superantigens. *Annual Review of Immunology* **17**, 435–466.

Li, Y.L., Li, H.M., Dimasi, N., McCormick, J.K., Martin, R., Schuck, P., Schlievert, P.M., and Mariuzza, R.A. (2001). Crystal structure of a superantigen bound to the high-affinity, zinc-dependent site on MHC class II. *Immunity* **14**, 93–103.

Marrack, P. and Kappler, J. (1990). The Staphylococcal enterotoxins and their relatives. *Science* **248**, 705–711.

Marrack, P., Blackman, M., Kushnir, E., and Kappler, J. (1990). The toxicity of staphylococcal enterotoxin B in mice is mediated by T cells. *Journal of Experimental Medicine* **171**, 455–464.

Mehindate, K., Thibodeau, J., Dohlsten, M., Kalland, T., Sekaly, R.P., and Mourad, W. (1995). Cross-linking of major histocompatibility complex class II molecules by staphylococcal enterotoxin A superantigen is a requirement for inflammatory cytokine gene expression. *Journal of Experimental Medicine* **182**, 1573–1577.

Monks, C.R., Freiberg, B.A., Kupfer, H., Sciaky, N., and Kupfer, A. (1998). Three-dimensional segregation of supramolecular activation clusters in T cells. *Nature* **395**, 82–86.

Musser, J.M., Schlievert, P.M., Chow, A.W., Ewan, P., Kreiswirth, B.N., Rosdahl, V.T., Naidu, A.S., Witte, W., and Selander, R.K. (1990). A single clone of

Staphylococcus aureus causes the majority of cases of toxic shock syndrome. *Proceedings of the National Academy of Sciences USA* **87**, 225–229.

Paliard, X., West, S.G., Lafferty, J.A., Clements, J.R., Kappler, J.W., Marrack, P., and Kotzin, B.L. (1991). Evidence for the effects of a superantigen in rheumatoid arthritis. *Science* **253**, 325–329.

Papageorgiou, A.C., Acharya, K.R., Shapiro, R., Passalacqua, E.F., Brehm, R.D., and Tranter, H.S. (1995). Crystal structure of the superantigen enterotoxin C2 from *Staphylococcus aureus* reveals a zinc-binding site. *Structure* **3**, 769–779.

Petersson, K., Hakansson, M., Nilsson, H., Forsberg, G., Svensson, L.A., Liljas, A., and Walse, B. (2001). Crystal structure of a superantigen bound to MHC class II displays zinc and peptide dependence. *European Molecular Biology Organisation Journal* **20**, 3306–3312.

Prasad, G.S., Earhart, C.A., Murray, D.L., Novick, R.P., Schlievert, P.M., and Ohlendorf, D.H. (1993). Structure of toxic shock syndrome toxin 1. *Biochemistry* **32**, 13,761–13,766.

Proft, T. and Fraser, J.D. (1998). Superantigens: just like peptides only different. *Journal of Experimental Medicine* **187**, 819–821.

Proft, T., Moffatt, S.L., Berkahn, C.J., and Fraser, J.D. (1999). Identification and characterization of novel superantigens from *Streptococcus pyogenes*. *Journal of Experimental Medicine* **189**, 89–102.

Proft, T., Moffatt, S.L., Weller, K.D., Paterson, A., Martin, D., and Fraser, J.D. (2000). The streptococcal superantigen SMEZ exhibits wide allelic variation, mosaic structure, and significant antigenic variation. *Journal of Experimental Medicine* **191**, 1765–1776.

Racke, M., Quigley, L., Cannella, B., Raine, C.S., McFarlin, D.E., and Scott, D.E. (1994). Superantigen modulation of experimental allergic encephalomyelitis: activation of anergy determines outcome. *Journal of Immunology* **152**, 2051–2059.

Reiser, R.F., Jacobson, J.A., Kasworm, E.M., and Bergdoll, M.S. (1988). Staphylococcal enterotoxin antibodies in pediatric patients from Utah. *Journal of Infectious Diseases* **158**, 1105–1108.

Roussel, A., Anderson, B.F., Baker, H.M., Fraser, J.D., and Baker E.N. (1997). Crystal structure of the streptococcal superantigen SPE-C: dimerization and zinc binding suggest a novel mode of interaction with MHC class II molecules. *Nature Structural Biology* **4**, 635–643.

Rowen, L., Koop, B.F., and Hood, L. (1996). The complete 685 kilobase DNA sequence of the human beta T cell receptor locus. *Science* **272**, 1755–1762.

Schad, E.M., Zaitseva, I., Zaitsev, V.N., Dohlsten, M., Kalland, T., Schlievert, P.M., Ohlendorf, D.H., and Svensson, L.A. (1995). Crystal structure of the superantigen staphylococcal enterotoxin type A. *European Molecular Biology Organisation Journal* **14**, 3292–3301.

Schlievert, P.M. (1993). Role of superantigens in human disease. *Journal of Infectious Diseases* **167**, 997–1002.

Seth, A., Stern, L.J., Ottenhoff, T.H., Engel, I., Owen, M.J., Lamb, J.R., Klausner, R.D., and Wiley, D.C. (1994). Binary and ternary complexes between T-cell receptor, class II MHC and superantigen in vitro. *Nature* **369**, 324–327.

Sutkowski, N., Palkama, T., Ciurli, C., Sekaly, R.-P., Thorley-Lawson, D.A., and Huber, B. (1996). An Epstein-Barr virus-associated superantigen. *Journal of Experimental Medicine* **184**, 971–980.

Swaminathan, S., Furey, W., Pletcher, J., and Sax, M. (1992). Crystal structure of Staphylococcal enterotoxin B, a superantigen. *Nature* **359**, 801–806.

Swaminathan, S., Furey, W., Pletcher, J., and Sax, M. (1995). Residues defining V beta specificity in staphylococcal enterotoxins. *Nature Structural Biology* **2**, 680.

Thompson, J.D., Gibson, T.J., Plewniak, F., Jeanmougin, F., and Higgins, D.G. (1997). The ClustalX windows interface: Flexible strategies for multiple sequence alignment aided by quality anlysis tools. *Nucleic Acids Research* **24**, 4876–4882.

Tiedemann, R.E., Urban, R.J., Strominger, J.L., and Fraser, J.D. (1995). Isolation of HLA-DR1.(staphylococcal enterotoxin A)2 trimers in solution. *Proceedings of the National Academy of Sciences USA* **92**, 12,156–12,159.

Tiedemann, R.E. and Fraser, J.D. (1996). Cross-linking of MHC class II molecules by staphylococcal enterotoxin A is essential for antigen-presenting cell and T cell activation. *Journal of Immunology* **157**, 3958–3966.

Todd, J. and Tishant, D. (1978). Toxic Shock syndrome associated with phage-group-1 Staphylococci. *Lancet* **2**, 1116–1118.

Unnikrishnan, M., Altman, D., Proft, T., Wahid, F., Cohen, J., Fraser, J.D., and Sriskandan, S. (2002). The bacterial superantigen SMEZ is the major immunoreactive agent of *Streptococcus pyogenes, submitted*.

Valitutti, S., Muller, S., Cella, M., Padovan, E., and Lanzavecchia, A. (1995). Serial triggering of many T-cell receptors by a few peptide-MHC complexes. *Nature* **375**, 148–151.

Vergeront, J.M., Stolz, S.J., Crass, B.A., Nelson, D.B., Davis, J.P., and Bergdoll, M.S. (1983). Prevalence of serum antibody to staphylococcal enterotoxin F among Wisconsin residents: implication for toxic shock syndrome. *Journal of Infectious Diseases* **142**, 692–698.

Wannamaker, L.W. and Schlievert, P.M. (1988). Exotoxins of group A Streptococci. In Bacterial toxins. ed. C.M. Hardegree and A.T. Tu. pp. 267–295. New York: Marcel Dekker.

Weeks, C.R. and Ferretti, J.J. (1986). Nucleotide sequence of the type A Streptococcal exotoxin (Erythrogenic toxin) gene from *Streptococcus pyrogenes* Bacteriophage T12. *Infection and Immunity* **52**, 144–150.

Williams, R.J., Ward, J.M., Henderson, B., Poole, S., O'Hara, B.P., Wilson, M., and Nair, S.P. (2000). Identification of a novel gene cluster encoding staphylococcal exotoxin-like proteins: Characterization of the prototypic gene and its protein product, SET1. *Infection and Immunity* **68**, 4407–4415.

Winslow, G.M., Scherer, M.T., Kappler, J.W., and Marrack, P. (1992). Detection and biochemical characterization of the mouse mammary tumor virus 7 superantigen (Mls-1a). *Cell* **71**, 719–730.

CHAPTER 9

Bacterial quorum sensing signalling molecules as immune modulators

David Pritchard, Doreen Hooi, Eleanor Watson, Sek Chow, Gary Telford,
Barrie Bycroft, Siriram Chhabra, Christopher Harty, Miguel Camara,
Stephen Diggle, and Paul Williams

9.1 INTRODUCTION

For many pathogens, the outcome of the interaction between host and
bacterium is strongly influenced by bacterial population size. Coupling the
production of virulence determinants with cell population density ensures
that the host lacks sufficient time to mount an effective defence against con-
solidated attack. Such a strategy depends on the ability of an individual bacterial cell to sense other members of the same species and, in response, differen-
tially express specific sets of genes. Such bacterial cell-to-cell communication
or "quorum sensing" describes the phenomenon whereby the accumulation
of a diffusible, low molecular weight signal molecule (sometimes referred to
as a "pheromone" or "autoinducer") enables individual bacterial cells to sense
when the minimal population unit or "quorum" of bacteria has been achieved
for a concerted population response to be initiated. Quorum sensing thus constitutes a mechanism for multicellular behaviour in prokaryotes and is now
known to control many different aspects of bacterial physiology including the
production of virulence determinants in animal, fish, and plant pathogens.
A number of chemically distinct quorum sensing signal molecules (QSMs)
have been described of which the *N*-acylhomoserine lactone (AHL) family
in Gram-negative bacteria has been the most intensively investigated (for
reviews see Salmond et al., 1995; Fuqua et al., 1996; Dunny and Winans,

Abbreviations and nomenclature.

AHLs are *N*-acyl homoserine lactones. QSM is a collective abbreviation for quorum sensing
signal molecules. IL = interleukin; NK = natural killer cell; PBMC = human peripheral
blood mononuclear cell; SLPI = secreted leucocyte proteinase inhibitor; T cell = thymus
derived lymphocyte; Th = T helper cell; TNF = tumour necrosis factor.

1999; Williams et al., 2000; Withers et al., 2001). The acyl groups of the naturally occurring AHLs identified to date range from 4 to 14 carbons in length and may be saturated or unsaturated with or without a C3 substituent (usually hydroxy or oxo; see Fig. 9.1).

In contrast, many Gram-positive bacteria employ posttranslationally modified peptides processed from larger precursors as QSMs. In *Staphylococcus aureus*, for example, peptide thiolactone QSMs (Fig. 9.1d) control the expression of cell wall colonization factors and exotoxins (Mayville et al., 1999). More recently, a quorum sensing system common to both Gram-negative and Gram-positive bacteria has been described, in which as yet chemically uncharacterized QSM is produced via the product of the *luxS* gene (Surette et al., 1999). In entero-virulent *E. coli* and *Shigella flexneri*, LuxS activity has been implicated in the enhancement of type III secretion (see Chapters 7 and 12) and in the regulation of virulence respectively (Sperandio et al., 1999; Day and Maurelli, 2001).

While QSMs clearly function in the regulation of prokaryotic gene expression, their production in host tissues during bacterial infection (Williams et al., 2000) raises the possibility that they may impact on eukaryotic signalling systems and in particular those employed by the immune system to control infection. This research field is largely unexplored at the cellular or molecular levels, yet early indications from immunological experiments indicate that complex interactions may be occurring. Before embarking on a detailed description of these data, it is necessary to understand the types of immune effector likely to be encountered by pathogens deploying QSMs to regulate virulence gene expression.

9.2 COMPARTMENTALISED IMMUNE RESPONSES TO PATHOGENIC ORGANISMS

Given the nature of parasitism and pathogenesis, which in essence constitutes an arms race between host and pathogen, it is not surprising that infecting organisms have evolved multiple molecular strategies to subvert the host's immune response (Wilson et al., 1998). The strategies involved are varied, and include, as discussed in other chapters: complement avoidance, immunoglobulin degradation, the modulation of cytokine networks, and the production of super-antigens and toxins, illustrating the complexity of the molecular interface between the host and infectious organisms.

To aid our understanding of this complexity, it is possible to subdivide the immune system into sectors considered important for the control

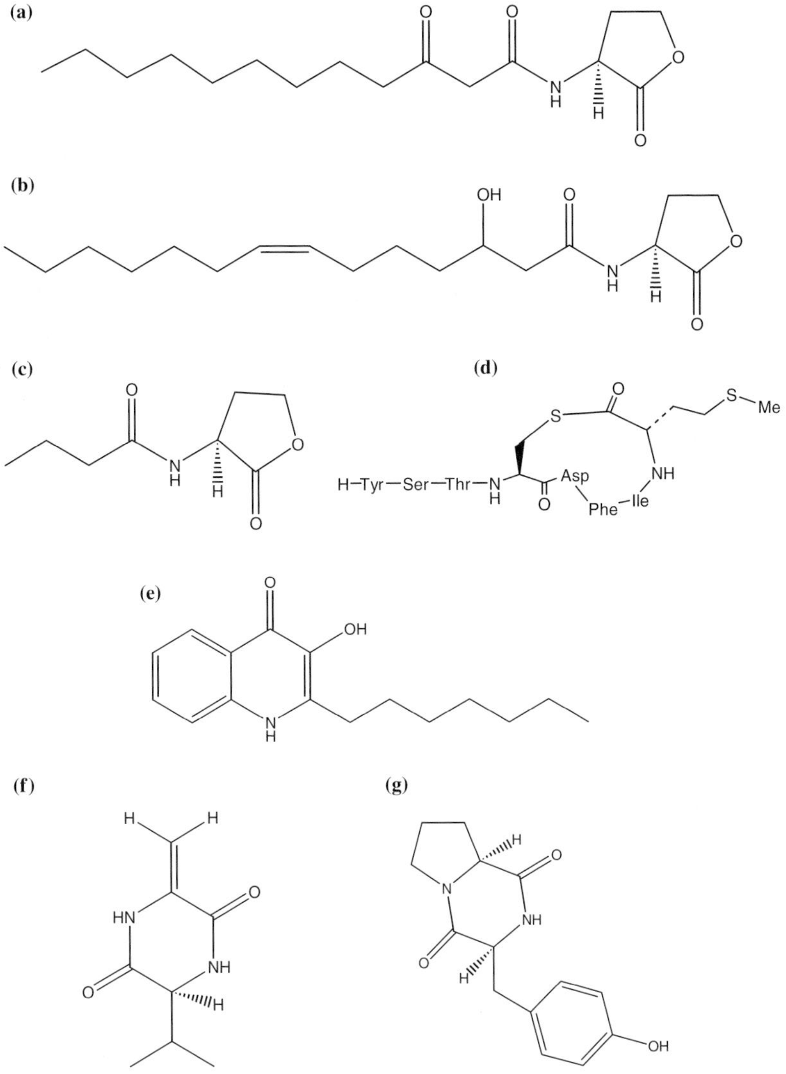

Figure 9.1. Structures of some representative bacterial QSMs.
(a) *N*-(3-oxododecanoyl)-L-homoserine lactone (3O-C12-HSL); (b) *N*-(3-hydroxy-7 *cis*-tetradecenoyl)-L-homoserine lactone (3-hydroxy-C14:1-HSL;
(c) *N*-butanoyl-L-homoserine lactone (C4-HSL); (d) group I *Staphylococcus aureus* cyclic peptide thiolactone; (e) 2-heptyl-3-hydroxy-4-quinolone; (f) cyclo(ΔAla-L-Val); and
(g) cyclo(L-Pro-L-Tyr).

of bacterial infections in general. Although it could be considered unwise to overcompartmentalise the immune system, which is a highly interactive physiological mechanism, this has been done in recent years to simplify and delineate immunological networks governing effector mechanisms controlling viral, bacterial, and protozoan pathogens, as opposed to the larger and possibly more sophisticated helminth parasites. On this basis it has been suggested, for example, that bacterial pathogens have evolved strategies to subvert T helper 1 (Th1) responses, which in the main are antibacterial (although, inevitably, there are exceptions to this rule such as *Helicobacter pylori* (Hatzifoti et al., 2000), whereas parasitic helminths have evolved strategies to subvert antiparasitic T helper 2 (Th2) responses. In this way, the parasite or pathogen subverts the very arm of the immune response that evolved to combat the type of infectious organism involved (Pritchard and Brown, 2001), thus imparting necessary economy to the immune system.

Such compartmentalised co-evolution between pathogen and host is perhaps inevitable given the differing types of molecular stimuli experienced during the infectious process. In many bacterial infections, cell wall materials and membrane proteins [for example, lipopolysaccharide (LPS), lipoteichoic acid, and lipoproteins] induce a cytokine profile in host macrophages and NK cells (IL-12, TNFα, IFNγ), which tends to promote the Th1 phenotype. This type of response to surface antigens has, as indicated above, evolved to be host protective, supporting the maturation of an activated and bactericidal reticulo-endothelial system, leading bacteria to co-evolve coincident counterimmune strategies (above). The end result in many of these situations is programmed cell death or apoptosis of the host cell (Balcewicz-Sablinska et al., 1998; Bayles et al., 1998; Chen et al., 1996; Cornelis, 1998; Mills et al., 1997; Wesson et al., 1998), and bacterial survival, propagation, and transmission (see Chapters 7 and 12 for further discussion of apoptosis).

The recent demonstration of the immune regulatory potential of AHLs adds a chemical immune-evasive element to a predominantly macromolecular bacterial armoury (Wilson et al., 1998). In contrast to bacteria, the larger helminth parasites produce secretions, which are rich in proteinases and lectins, often into mucosal layers, inducing a cytokine milieu conducive to the establishment of an allergic response. Consequently, the immune-evasion armoury of helminths is probably distinct from that used by bacteria in that it is potentially antiallergic in nature (Pritchard and Brown, 2001).

A final consideration in this section is that each compartment is claimed to be counter-regulatory, a property currently under exploitation to produce vaccines against immunological diseases by stimulating antibacterial

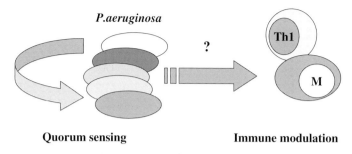

P.aeruginosa

Th1

M

Quorum sensing   Immune modulation

Figure 9.2. Immune modulation by QSMs. Selectivity or serendipity?. Th1, helper T cell
type I; M. monocyte/macrophage.

responses and cytokines which down-regulate allergy (Rook, 2000; Rook
et al., 2000). Additionally, the controlled stimulation of pro-allergic immunological networks could alleviate the many autoimmune conditions driven by
hypersensitive Th1 responses.

In this chapter, data are presented that indicate that some bacteria may
have evolved a sophisticated dual purpose system, termed quorum sensing,
which upon infection of an appropriate host serves to coordinate virulence
gene expression via bacterial cell-to-cell communication and to down-regulate
appropriate immune effector mechanisms (Fig. 9.2). The converse argument
is that the discovery of immune modulation by QSMs is totally serendipitous,
bearing little relevance to the host–pathogen interface.

9.3 BACTERIAL QUORUM SENSING SIGNALLING MOLECULES AND IMMUNE SUPPRESSION

9.3.1 Quorum sensing networks in *Pseudomonas aeruginosa*

The Gram-negative bacterium *Pseudomonas aeruginosa* is an opportunistic human pathogen responsible for infections in immuno-compromised
hosts and capable of chronic colonization of the lungs of individuals with
cystic fibrosis (Smith et al., 1996; Lyczak et al., 2000). An attribute that is
widely regarded as a major contributor to the pathogenesis of *P. aeruginosa*
is its ability to secrete numerous toxic compounds and degradative enzymes
(elastase, LasA protease, phospholipase C, exotoxin A, exoenzyme S, rhamno-
lipid, alginate, pyocyanin, PA-1L, and PA-II-L lectins and cyanide; Nicas, and
Iglewski, 1985; Lyczak et al., 2000; Van Delden and Iglewski, 1998; Winzer
et al., 2000). Many of these enzymes and metabolites are not actively produced
until the late logarithmic phase of growth, when the cell density is high. In
P. aeruginosa, the expression of many exoproducts is cell density-dependent

and is coordinated via AHL-dependent quorum sensing (Latifi et al., 1996; Winson et al., 1995; Pesci et al., 1997; Van Delden and Iglewski, 1998). Two separate quorum sensing circuits (termed *las* and *rhl*, respectively), each of which possesses an AHL synthase (LasI or RhlI) and a sensor-regulator (LasR or RhlR), modulate gene transcription in response to increasing AHL concentrations (Latifi et al., 1995, 1996; Winson et al., 1995; Pesci et al., 1997). The AHLs that signal within the *las* and *rhl* systems are *N*-(3-oxododecanoyl)-L-homoserine lactone (3O-C12-HSL; Fig. 9.1a) and *N*-butanoyl-L-homoserine lactone (C4-HSL; Fig. 9.1c), respectively. Both LasR and RhlR are members of the LuxR family of transcriptional activators, which incorporate AHL- and DNA-binding regions within their amino and carboxyl domains, respectively. A widely accepted model by which LuxR-type proteins are considered to activate gene expression is that the binding of the cognate AHL ligand to the amino-terminal domain of the protein causes a conformational change that leads to exposure of the DNA-binding domain at the C-terminus of the protein (Welch et al., 2000). Thus, activation of LasR and RhlR depends on the binding of 3O-C12-HSL and C4-HSL, respectively. Together, the *las* and *rhl* systems comprise a hierarchical cascade that coordinates the production of virulence factors and stationary phase genes (via the alternative sigma factor, RpoS: Latifi et al., 1996; Pesci and Iglewski, 1997). Individually, each system's sensor-regulator modulates a regulon comprising an overlapping set of genes. However, the *las* system directly regulates the *rhl* system, thus providing overall coordination of quorum sensing and temporal gene expression in response to cell-to-cell communication (Latifi et al., 1996; Pesci and Iglewski, 1997).

In addition to the AHLs, two additional, chemically distinct classes of putative QSMs have recently been described in *P. aeruginosa*. By studying the expression of *lasB* in a *lasR*-negative mutant, Pesci et al. (1999), discovered another *P. aeruginosa* QSM unrelated to the AHLs. This signal molecule (termed PQS for Pseudomonas Quinolone Signal), the synthesis and bioactivity of which is dependent on both the LasRI and RhlRI quorum sensing systems, was chemically characterized as 2-heptyl-3-hydroxy-4-quinolone (Fig. 9.1e), a compound related to the well-known class of 4-quinolone antibiotics (Pesci et al., 1999). A second class of putative QSMs in *P. aeruginosa* was identified by Holden et al. (1999) via their capacity to weakly activate LuxR-dependent AHL biosensors and to inhibit AHL-mediated swarming in *Serratia liquefaciens*. These molecules were chemically characterized as the cyclic dipeptides, cyclo(ΔAla-L-Val) (Fig. 9.1f) and cyclo(L-Pro-L-Tyr) (Fig. 9.1g). Whether these cyclic dipeptides modulate AHL-dependent quorum sensing in the producer organism or in other organisms occupying the same ecological niche or indeed whether they function as diffusible signal molecules *per se* remains to be established. However, such cyclic dipeptides are known to modulate eukaryotic cell function; cyclo(L-Pro-L-Tyr) for example is a plant phytotoxin while other members of this cyclic dipeptide family exhibit a variety of endocrine and central nervous system-related biological activities (Prasad, 1995).

Thus, *P. aeruginosa* exports at least three chemically distinct molecules, each of which could influence the immune response and contribute to the survival of this Gram-negative pathogen in the mammalian host.

9.4 AHLs AND EUKARYOTIC CELLS

Although AHLs have so far largely been considered as effectors of prokaryotic gene expression, they are capable of influencing eukaryotic cell behaviour and potentially modulating disease processes (Telford et al., 1998; Lawrence et al., 1999; Saleh et al., 1999; Gardiner et al., 2001).

In particular, 3O-C12-HSL (but not the short chain 3O-C6-HSL):

- suppresses human and murine lymphocyte proliferation (Fig. 9.3)
- inhibits LPS-induced secretion of TNF α (Fig. 9.4) and IL-12 (Telford et al., 1998)
- induces apoptosis in responsive leucocyte populations (Figs. 9.5 and 9.6)
- promotes IgG1 production by antigen-challenged splenocyte cultures (Telford et al., 1998).

This is, in essence, an upregulation of a Th2 response, which could be considered pro-bacterial if T helper cell contra-suppression is operative. Such apparent selectivity must be fully investigated, given the observation that 3O-C12-HSL induces apoptosis in leucocytes.

This immunological profile obtained suggests a degree of immunological selectivity for suppression of Th1 responses, which might be expected if AHLs evolved to subvert immunity while simultaneously promoting bacterial communication. However, it should be remembered that *P. aeruginosa* is an ubiquitous organism that perhaps is not entirely suited to life in the immune competent human body. Consequently, the discovery of the immune modulatory capacity of some AHLs may be entirely serendipitous.

9.5 QSMs AND IMMUNITY TO *PSEUDOMONAS AERUGINOSA*

Pseudomonas aeruginosa infection occurs when immunological defences are down, when normal lung physiology is genetically impaired, and in

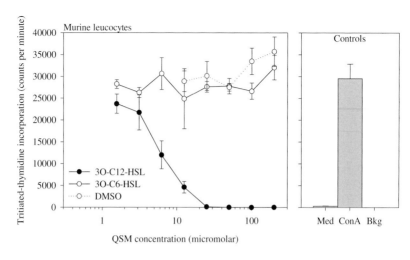

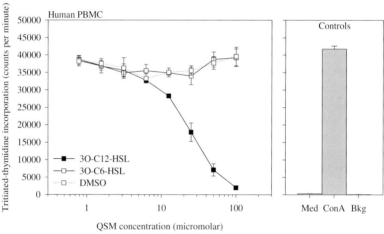

Figure 9.3. Suppression of human and murine leucocyte proliferation by 3O-C12-HSL. 10^5 cells were cultured with 3O-C12-HSL or 3O-C6-HSL and 1 μg/ml Concanavalin A (ConA), a lectin, for 3 days with 0.25 μCi tritiated-thymidine added in the final 18 h of culture. Cells were then harvested onto filters and the incorporated radioisotope counted on a Packard TopCount scintillation counter.

cases of severe tissue damage (Allewelt et al., 2000; Lyczak et al., 2000). Consequently, it is a predominant pathogen in hospital infections (Banerjee et al., 1991; Jarvis and Martone, 1992) and in cystic fibrosis (CF) patients (Singh et al., 2000; Lyczak et al., 2000). It is therefore important to fully understand the immunological events occurring during the initial stages of

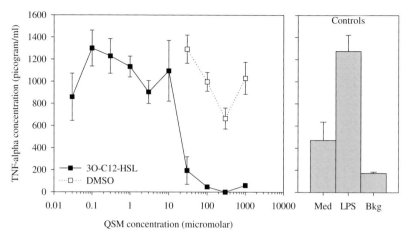

Figure 9.4. Suppression of LPS-induced TNFα secretion by 3O-C12-HSL. 10^5 human peripheral blood mononuclear cells (PBMC) were stimulated with 1 ng/ml of LPS from *E. coli* serotype 055:B5 in the presence of 3O-C12-HSL or 3O-C6-HSL for 24 hours. Levels of TNFα released into the culture supernatants were assayed in a sandwich ELISA. Briefly a mouse anti-human TNFα antibody was immobilised onto a 96-well ELISA plate. After blocking with 1% bovine serum albumin in phosphate buffered saline (PBS), culture supernatants were added to the wells; dilutions of standard human TNFα was included in parallel. The cytokine was captured overnight followed by a wash in PBS containing 0.05% Tween 20 (PBS/Tween). The captured TNFα was detected with the addition of a biotinylated mouse antihuman TNFα antibody. Following a wash in PBS/Tween streptavidin-horseradish peroxidase was added to the wells. After a final wash in PBS/Tween the bound peroxidase was developed for 10 min using tetramethyl-benzidine as substrate and product development was recorded at 450 nm in a Dynex plate reader.

infection and during the establishment of infection if the relevance of QSM-induced immune modulation *in vivo* is to be authenticated.

Immunity to *P. aeruginosa* in model systems appears to be dependent on the functionality of alveolar macrophages, neutrophils, T cells, and cytokines, particularly TNFα, secretion (Amura et al., 1994; Buret et al., 1994; Morissette et al., 1995), which are immunological effectors known to be affected by AHLs, at least *in vitro*. Chemokines, particularly those binding to CXCR2 chemokine receptors, are key in immunological protection against murine *P. aeruginosa* pneumonia (Tsai et al., 2000).

Despite this apparent understanding of the host–pathogen interface, further complex immunological interactions are undoubtedly initiated by

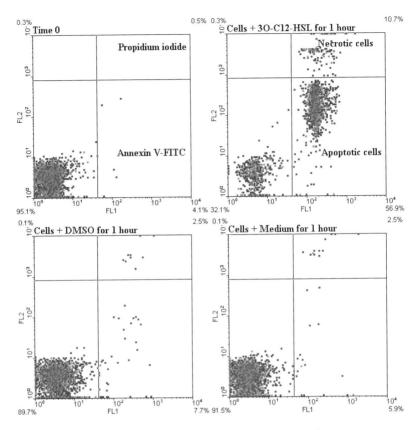

Figure 9.5. 3O-C12-HSL induces apoptosis in murine splenocytes. 10^6 murine leucocytes were incubated with 100 μM 3O-C12-HSL for 1 hour. An aliquot of 10^5 cells was removed for labelling with Annexin V-FITC, which detected apoptotic cells and propidium iodide, which stained for necrotic cells. Analysis was by flow cytometry.

infection, and investigators have also discovered a role for IL-4 (an accepted T helper 2 cytokine) and possibly the upregulation of surfactant protein (Jain-Vora et al., 1998) in immunity; AHLs seem to promote this phenotype *in vitro* (Telford et al., 1998) providing an interesting caveat to the argument that

Figure 9.6. 3O-C12-HSL induces apoptosis in murine peritoneal macrophages.
(A) Macrophages incubated in 3O-C12-HSL exhibit several hallmarks of apoptosis. A normal cell is shown for comparison (i). Cytoplasm pinches off in a process known as blebbing (arrow-ii) until it is completely reduced (iii). Meanwhile endonucleases are activated which cleave chromatin, causing it to condense and line the nuclear membrane. Nuclear morphology in (iv) is typical of a cell undergoing this process. Eventually the

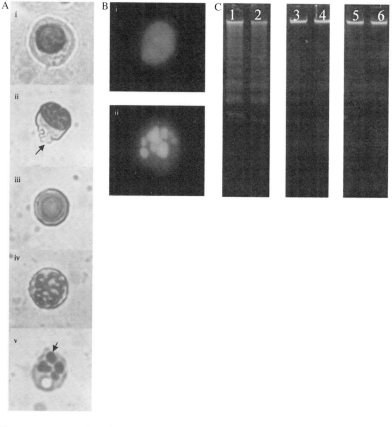

Figure 9.6 (*cont*) nucleus fragments into apoptotic bodies (arrow-v). Macrophages were incubated in 12.5, 25, 50, and 100 μM 3O-C12-HSL and 3O-C6-HSL and CTCM alone for 1, 2, 4, and 6 h. Macrophages were cytospun, fixed with methanol and stained with Giemsa. Examples shown are representative of macrophages found after incubation with 3O-C12-HSL at all concentrations and times. Control and 3O-C6-HSL incubations showed no apoptotic morphology. (B) Representative Hoechst stained cells. Hoechst 33342 binds DNA which allows the determination of nuclear morphology. DNA in the nucleus of normal cells is distributed diffusely (i) and fragmented in apoptotic cells (ii). Macrophages were incubated in 50 μM 3O-C12-HSL and 3O-C6-HSL and CTCM for 3 h, detached and stained. 3O-C12-HSL incubation induced the apoptotic morphology shown and 3O-C6-HSL and CTCM incubated cells appeared normal. (C) Incubation with 3O-C12-HSL yields the characteristic DNA ladder. DNA was extracted from macrophages incubated for 3 h in 50 μM 3O-C12-HSL (lanes 1 and 2) and 50 μM 3O-C6-HSL (lanes 3 and 4) and CTCM (lanes 5 and 6). 12 cultures of 1×10^5 macrophages were detached and pooled allowing sufficient amounts of DNA to be analysed yet ensuring that results reflect previous experiments. DNA was visualised on a 1.8% agarose gel by ethidium bromide staining.

AHLs promote bacterial survival by modulating Th1 function. Interleukin 10 could also be acting as a regulatory cytokine in *Pseudomonas* infection; in mice, it is reported to be a major mediator of sepsis-induced impairment of innate antibacterial defences in the lung (Steinhauser et al., 1999), yet also plays a role in attenuating excessive lung inflammation (Chmiel et al., 1999). This reinforces the belief that we are dealing with a very immunologically complex host–bacterial interaction, where the temporal nature of bacterial stimulus/immune modulation and host immune response warrants dissection.

3O-C12-HSL has been detected directly in the sputum of cystic fibrosis patients experiencing acute exacerbations of *P. aeruginosa* infection, (Williams et al., 2000). In addition, using an experimental approach in which ^{14}C-labelled methionine is incorporated into AHLs via *S*-adenosylmethionine during AHL biosynthesis, Singh et al. (2000) reported that in biofilms and in cystic fibrosis sputa (incubated *ex vivo* for 4 hours to stimulate *de novo* AHL biosynthesis), *P. aeruginosa* produces both C4-HSL and 3O-C12-HSL. Intriguingly, the ratio of C4-HSL to 3-oxo-C12-HSL in the biofilms was found to be converse of that observed during planktonic growth in L-broth, i.e., C4-HSL predominates. This suggests that during chronic infections at mucosal surfaces, *P. aeruginosa* down-regulates the production of the immune regulatory AHL, 3O-C12-HSL (Fig. 9.7, Table 9.1).

Thus, while QSMs are generated *in vivo* in quantities considered to be immune modulatory *in vitro*, it is also important to consider that they are made in combination with a battery of putative immune regulatory molecules such as ExoU cytotoxin, exotoxin A, Opr1, nitrite reductase, outer membrane porin proteins, and the phenazine pigment, pyocyanin (Michalkiewitz et al., 1998, 1999; Muller and Sorrell, 1997; Schümann et al., 1998; Ino et al., 1999; Oishi et al., 1997; Cusumano et al., 1997; Denning et al., 1998). It is not known whether the immune regulatory AHLs would have the opportunity to modify the immune system in the selective manner observed to date when incorporated into such a complex molecular cocktail, which undoubtedly has the capacity to over-extend the local response to bacterial infection. However, it is possible that immune suppression by AHLs during bacterial infection is a relatively early cellular event (Fig. 9.7). Alternatively, given that small uncharged molecules such as 3O-C12-HSL may readily cross membranes and diffuse through tissues much more rapidly than the much larger *P. aeruginosa* exoprotein virulence determinants, which are also likely to be immunogenic and neutralised by antibodies, immune regulatory AHLs may provide the bacterium with an important advance foothold prior to the secretion of diverse tissue damaging exoproducts.

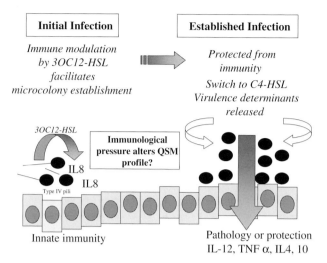

Figure 9.7. The immunological complexity of *Pseudomonas aeruginosa* infection at mucosal surfaces. A theoretical immunological scenario is envisaged where the bacterium progresses from microcolony formation to biofilm establishment aided by a temporal switch in QSM production. Early immunological events would include chemokine, particularly IL8, secretion by compromised epithelium and neutrophil recruitment, which is reported to actually assist microcolony formation (Mathee et al., 2000). 3O-C12-HSL would act at this stage to modulate immunity, innate, and adaptive given the opportunity for multiple repeated exposure, particularly in the hospital environment (Jarvis and Martone, 1992). As microcolony formation proceeds, and the colonial organism becomes protected from immunological effectors, C4-HSL becomes the major QSM (Singh et al., 2000). C4-HSL does not modulate immunity but, in common with 3O-C12-HSL, is responsible for driving the production of a cocktail of virulence determinants from a site of immune privilege, the biofilm. This cocktail, consisting of exotoxins with cytolytic activity, a pyocyanin which can destroy the anti-proteinase defence of the lungs (Britingan et al., 1999), and cytokine inducing secretions serve to hyperactivate the immune system, resulting in clearance of infection or pathology. Bacterial SOD and catalase activities within the biofilm may shield organisms from oxidative attack (Hassett et al., 1999), while the proteinase-antiproteinase balance of the lung may be severely compromised in *P. aeruginosa* infection by a combination of LPS-induced SLPI expression and a1-antiproteinase destruction by pyocyanin (Jin et al., 1998; Britingan et al., 1999).

With reference to Fig. 9.7, we suggest that in chronic *P. aeruginosa* infections, C4-HSL is produced, once a site of immune privilege has been established. C4-HSL is *not* immune modulatory *in vitro* (Fig. 9.8), and the need for the immune modulatory 3O-C12-HSL may be surpassed by biofilm formation.

Table 9.1. *The putative role of AHLs in early and late phase infection with P. aeruginosa*

Initial contact with host 3O-C12-HSL predominant?	Established infection C4-HSL predominant?
Foothold gained by bacterium through the secretion of immune modulatory and pro-apoptotic QSMs, which suppress leucocyte function and orchestrate synthesis of elastase, alkaline proteinases, and exotoxin A.	Once infection is established, a combination of virulence factors, including exotoxins, membrane proteins, porins, pyocyanin and nitrite reductase, lipase, and cyanide induce cytokine and chemokine activities at sites of infection (TNF α, IL2,6,8, GM-CSF).
It should also be noted that IL-8 induction in epithelial cells by QSMs has been reported. This would indicate that QSMs modulate inevitable self-inflicted inflammation.	It is not known whether the QSMs preserve a site immune privilege around the bacterial colony, and whether QSMs and biofilm formation synergise to promote bacterial survival. Bacterial SOD and catalase may constitute an anti-oxidant defence shield.

Notes: Recent data suggest that C4-HSL or BHL is produced in excess of 3O-C12-HSL or OdDHL in CF sputum compared to laboratory broth culture (Singh, 2000). Sputum was collected in PBS in the presence of 2% n-acetylcysteine, then labelled with ^{14}C carboxy-methionine. Labelled HSLs were extracted under acidic conditions, and analysed by HPLC. From broth cultures of 9 clinical isolates from 6 sputums, 6/9 demonstrated a bias toward C4 synthesis. This trend was also seen in biofilm vs broth cultures manufactured under laboratory conditions. The inference is that a switch to predominant C4-HSL synthesis is a marker of biofilm formation. This might suggest that C4-HSL is produced once a site of immune privilege has been established. C4-HSL is not immune modulatory *in vitro*, and the need for the immune modulatory 3O-C12-HSL may be surpassed by biofilm formation. It is also possible that the chemical phenotype of the bacterium may be regulated by the eukaryotic response.

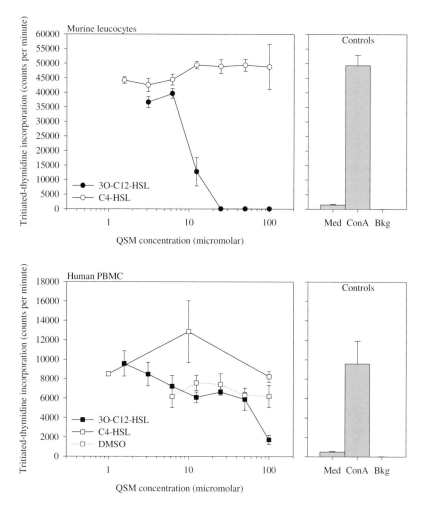

Figure 9.8. The comparative immune suppressive activity of 3O-C12-HSL and C4-HSL. 10^5 murine and human leucocytes were cultured with 3O-C12-HSL or C4-HSL and 1 μg/ml ConA for 3 days with 0.25 μCi tritiated-thymidine added in the final 18 hours of culture. Cells were then harvested onto filters and the incorporated radioisotope counted on a Packard TopCount scintillation counter.

9.6 CONCLUSION

While we await data providing a greater understanding of the role of bacterial QSMs in immune modulation and pathogenesis at mucosal surfaces, immunological experiments conducted in our laboratories indicate that AHLs represent a family of potentially important therapeutic compounds for the alleviation of Th1-driven autoimmune disease, where inappropriate levels of macrophage-derived TNFα and IL-12 are thought to contribute to pathology. On the basis that the AHLs are simple organic molecules readily amenable to chemical synthesis and modification, they constitute an attractive family of potential immunotherapeutic agents. The fact that AHLs function by binding to and activating members of the LuxR family of transcriptional regulator proteins in diverse Gram-negative bacteria supports our belief that eukaryotic receptors exist and mediate the immunological events discovered to date. Furthermore, the existence of a number of chemically distinct QSM families (Fig. 9.1), highly amenable to chemical synthesis, is likely to offer fruitful future opportunities for the discovery of novel immune modulatory agents for the treatment of immunological disease. Finally, an investigation of the role of immunological effectors in controlling or modulating AHL synthesis, secretion, and activity would be worthwhile.

ACKNOWLEDGMENTS

Work in the authors' laboratories has been supported via grants and studentships from the Biotechnology and Biological Sciences Research Council, UK, the Medical Research Council, UK, and the University of Nottingham. Thanks are also extended to Adrian Robins and Alison Galvin for their advice and practical help with flow cytometry, to our brave donor panel, and to our exuberant phlebotomist (Mrs. F. Brown) for PBMCs.

REFERENCES

Allewelt, M., Coleman, F.T., Grout, M., Priebe, G.P., and Pier, G.B. (2000). Acquisition of expression of the *Pseudomonas aeruginosa* ExoU cytotoxin leads to increased bacterial virulence in a murine model of acute pneumonia and systematic spread. *Infection and Immunity* **68**, 3998–4004.

Amura, C.R., Fontan, P.A., Sanjuan, N., and Sordelli, D.O. (1994). The effect of treatment with interleukin-1 and tumour necrosis factor on *Pseudomonas aeruginosa* lung infection in a granulocytopenic mouse model. *Immunology and Immunopathology* **73**, 261–266.

Balcewicz-Sablinska, M.K., Kornfeld, K.J., and Remold, H.G. (1998). Pathogenic *Mycobacterium tuberculosis* evades apoptosis of host macrophages by release of TNF-R2, resulting in inactivation of TNF-alpha. *Journal of Immunology* **161**, 2636–2641.

Banerjee, S.D., Emori, T.G., Culver, D.H., Gaynes, R.P., Jarvis, W.R., Horan, T., Edwards, J.R., Tolson, J., Henderson, T., and Martone, W.J. (1991). Secular trends in nocosomial primary bloodstream infections in the United States. *American Journal of Medicine* **91**, 86S–89S.

Bayles, K.W., Wesson, C.A., Liou, L.E., Fox, L.K., Bohach, G.A., and Trumble,

W.R. (1998). Intracellular *Staphylococcus aureus* escapes the endosome and induces apoptosis in epithelial cells, *Infection and Immunity* **66**, 336–342.

Britingan, B.E., Railsback, M.A., and Cox, C.D. (1999). The *Pseudomonas aeruginosa* secretory product pyocyanin inactivates alpha 1 protease inhibitor: implications for the pathogenesis of cystic fibrosis lung disease. *Infection and Immunity* **67**, 1207–1212.

Buret, A., Dunkley, M.L., Pang, G., Clancy, R.L., and Cripps, A.W. (1994). Pulmonary immunity to *Pseudomonas aeruginosa* in intestinally immunized rats: role of alveolar macrophages, tumor necrosis factor alpha, and interleukin-1∀. *Infection and Immunity* **62**, 5335–5343.

Chen, Y., Smith, M.R., Thirumalai, K., and Zychlinsky, A. (1996). A bacterial invasin induces macrophage apoptosis by binding directly to ICE. *European Molecular Biology Organisation Journal* **15**, 3853–3860.

Chmiel, J.F., Konstan, M.W., Knesebeck, J.E., Hilliard, J.B., Bonfield, T.L., Dawson, D.V., and Berger, M. (1999). IL-10 attenuates excessive inflammation in chronic Pseudomonas infection in mice. *American Journal of Respiratory Critical Care Medicine* **160**, 2040–2047.

Cornelis, G.R. (1998). The *Yersinia* deadly kiss. *Journal of Bacteriology* **180**, 5495–5504.

Cusumano, V., Tufano, M.A., Mancuso, G., Carbone, M., Rossano, F., Fera, M.T., Ciliberti, F.A., Ruocco, E., Merendino, R.A., and Teti, G. (1997). Porins of Pseudomonas aeruginosa induce release of tumor necrosis factor alpha and interleukin-b by human leukocytes. *Infection and Immunity* **65**, 1683–1687.

Day, W.A. Jr. and Maurelli, A.T. (2001). *Shigella flexneri* LuxS quorum-sensing system modulates *virB* expression but is not essential for virulence. *Infection and Immunity* **69**, 15–23.

Denning, G.M., Wollenweber, L.A., Railsback, M.A., Cox, C.D., Stoll, L.L., and Britigan, B.E. (1998). *Pseudomonas* pyocyanin increases interleukin-8 expression by human airway epithelial cells. *Infection and Immunity* **66**, 5777–84.

Dunny, G.M. and Winans, S.C. eds. (1999). *Cell–cell Signaling in Bacteria*. Washington DC: ASM Press.

Fuqua, W.C., Winans, S.C., and Greenberg, E.P. (1996). Census and consensus in bacterial ecosystems: the LuxR-LuxI family of cell quorum-sensing regulators. *Annual Reviews of Microbiology* **50**, 727–751.

Gardiner, S.M., Chhabra, S.R., Harty, C., Williams, P., Pritchard, D.I., Bycroft, B.W., and Bennett, T. (2001). Haemodynamic effects of the bacterial quorum sensing signal molecule N-(3-oxododecanoyl)-L-homoserine lactone in conscious rats. *British Journal of Pharmacology* **133**, 1047–1054.

Hassett, D.J., Ma, J.-F., Elkins, J.-G., McDermott, T.R., Ochsner, U.A., West, S.E.H., Huang, C.-T., Fredericks, J., Burnett, S., Stewart, P.S., McFeters, G., Passador, L., and Iglewski, B.H. (1999). Quorum sensing in *Pseudomonas aeruginosa* controls expression of catalase and superoxide dismutase genes and mediates biofilm susceptibility to hydrogen peroxide. *Molecular Microbiology* **34**, 1082–1093.

Hatzifoti, C., Wren, B.W., and Morrow, W.J.W. (2000). *Helicobacter pylori* vaccine strategies – triggering a gut reaction. *Immunology Today* **21**, 615–619.

Holden, M.T.G., Chhabra, S.R., de Nys, R., Stead, P., Bainton, N.J., Hill, P.J., Manefield, M., Kumar, N., Labatte, M., England, D., Rice, S., Givskov, M., Salmond, G.P.C., Stewart, G.S.A.B., Bycroft, B.W., Kjelleberg, S., and Williams, P. (1999). Quorum sensing cross-talk: isolation and chemical characterization of cyclic dipeptides from *Pseudomonas aeruginosa* and other Gram-negative bacteria. *Molecular Microbiology* **33**, 1254–1266.

Ino, M., Nagase, S., Nagasawa, T., Koyama, A., and Tachibana, S. (1999). The outer membrane protein I of *Pseudomonas aeruginosa* PAO1, a possible pollutant of dialysate in haemodialysis, induces cytokines in mouse bone marrow cells. *Nephron* **82**, 324–330.

Jain-Vora, S., LeVine, A.M., Chroneos, Z., Ross, G.F., Hull, W.M., and Whitsett, J.A. (1998). Interleukin-4 enhances pulmonary clearance of *Pseudomonas aeruginosa*. *Infection and Immunity* **66**, 4229–4236.

Jarvis, W.R. and Martone, W.J. (1992). Predominant pathogens in hospital infections. *Journal of Antimicrobial Chemotherapy* **29**, 19–24.

Jin, F., Nathan, C.F., Radzioch, D., and Ding A. (1998). Lipopolysaccharide-related stimuli induce expression of the secretory leukocyte protease inhibitor, a macrophage-derived lipopolysaccharide inhibitor. *Infection and Immunity* **66**, 2447–2452.

Latifi, A., Winson, M.K., Foglino, M., Bycroft, B.W., Stewart, G.S.A.B., Lazdunski, A., and Williams, P. (1995). Multiple homologues of LuxR and LuxI control expression of virulence determinants and secondary metabolites through quorum sensing in *Pseudomonas aeruginosa* PAO1. *Molecular Microbiology* **17**, 333–343.

Latifi, A., Foglino, M., Tanaka, K., Williams P., and Lazdunski, A. (1996). A hierarchical quorum sensing cascade in *Pseudomonas aeruginosa* links the transcriptional activators LasR and RhlR (VsmR) to expression of the stationary-phase sigma factor RpoS. *Molecular Microbiology* **21**, 1137–1146.

Lawrence, R.N., Dunn, W.R., Bycroft, B.W., Camara, M., Chhabra, S.R., Williams, P., and Wilson, V.G. (1999). The *Pseudomonas aeruginosa* quorum-sensing signal molecule, *N*-(3-oxododecanoyl)-L-homoserine lactone, inhibits porcine arterial smooth muscle contraction. *British Journal of Pharmacology* **128**, 845–848.

Lyczak, J.B., Cannon, C.L., and Pier, G.B. (2000). Establishment of *Pseudomonas aeruginosa* infection: lessons from a versatile opportunist. *Microbes and Infection* **2**, 1051–1060.

Mathee, K., Hentzer, M., Heydorn, A., Høiby, N., Ohman, D.E., Molin, S., and Kharazmi, A. (2000). Microcolony formation of *Pseudomonas aeruginosa* in cystic fibrosis lungs is influenced by infiltrating polymorphonuclear leukocytes. Available at http://view.abstractonline.com.

Mayville, P., Ji, G., Beavis, R., Yang, H., Goger, M., Novick, R.P., and Muir, T.W. (1999). Structure-activity analysis of synthetic autoinducing thiolactone peptides from *Staphylococcus aureus* responsible for virulence. *Proceedings of the National Academy of Sciences USA* **96**, 1218–1223.

Michalkiewicz, J., Stachowski, J., Barth, C., Patzer, J., Dzierzanowska, D., Runowski, D., and Madalinski, K. (1998). Effect of *Pseudomonas aeruginosa* exotoxin A on CD3-induced human T-cell activation, *Immunology Letters* **61**, 79–88.

Michalkiewicz, J., Stachowski, J., Barth, C., Patzer, J., Dzierzanowska, D., and Madalinski, K. (1999). Effect of *Pseudomonas aeruginosa* exotoxin A on IFN-gamma synthesis: expression of costimulatory molecules on monocytes and activity of NK cells. *Immunology Letters* **69**, 359–366.

Mills, S.D., Boland, A., Sory, M.-P., Van Der Smissen, P., Kerbourch, C., Finlay, B.B., and Cornelis, G.R. (1997). *Yersinia enterocolitica* induces apoptosis in macrophages by a process requiring functional type III secretion and translocation mechanisms and involving YopP, presumably acting as an effector protein, *Proceedings of the National Academy of Sciences USA* **94**, 12,638–12,643.

Morisette, C., Skamene, E., and Gervais, F. (1995). Endobronchial inflammation following *Pseudomonas aeruginosa* infection in resistant and susceptible strains of mice. *Infection and Immunity* **63**, 2251–2257.

Muller, M. and Sorrell, T.C. (1997). Modulation of neutrophil superoxide response and intracellular diacylglyceride levels by the bacterial pigment pyocyanin. *Infection and Immunity* **65**, 2483–2487.

Nicas, T.I. and Iglewski, B.H. (1985). The contribution of exoproducts to the virulence of *Pseudomonas aeruginosa*. *Canadian Journal of Microbiology* **31**, 387–392.

Oishi, K., Sar, B., Wada, A., Hidaka, Y., Matsumoto, S., Amano, H., Sonoda, F., Kobayashi, S., Hirayama, T., Nagatake, T., and Matsushima, K. (1997). Nitrite reductase from *Pseudomonas aeruginosa* induces inflammatory cytokines in cultured respiratory cells. *Infection and Immunity* **65**, 2648–2655.

Pesci, E.C. and Iglewski, B.H. (1997). The chain of command in Pseudomonas quorum sensing. *Trends in Microbiology* **5**, 132–134.

Pesci, E.C., Pearson, J.P., Seed, P.C., and Iglewski, B.H. (1997). Regulation of *las* and *rhl* quorum sensing in *Pseudomonas aeruginosa*. *Journal of Bacteriology* **179**, 3127–3132.

Pesci, E.C., Milbank, J.B.J., Pearson, J.P., McKnight, S., Kende, A.S., Greenberg, E.P., and Iglewski, B.H. (1999). Quinolone signaling in the cell-to-cell communication system of *Pseudomonas aeruginosa*. *Proceedings of the National Academy of Sciences USA* **96**, 11,229–11,234.

Prasad, C. (1995). Bioactive cyclic dipeptides. *Peptides* **16**, 151–164.

Pritchard, D.I. and Brown, A. (2001). Is *Necator americanus* approaching a mutualistic symbiotic relationship with humans? *Trends in Parasitology* **17**, 169–172.

Rook, G.A.W. (2000). Clean living increases more than just atopic disease (multiple letters). *Immunology Today* **21**, 249–250.

Rook, G.A.W., Ristori, G., and Salvetti, M. (2000). Bacterial vaccines for the treatment of multiple sclerosis and other autoimmune diseases. *Immunology Today* **21**, 503–508.

Saleh, A., Figarella, C., Kammouni, W., Marchand-Pinatel, S., Lazdunski, A., Tubul, A., Brun, P., and Merten, M.D. (1999). *Pseudomonas aeruginosa* quorum sensing signal molecule N-(3-oxododecanoyl)-L-homoserine lactone inhibits expression of P2Y receptors in cystic fibrosis tracheal gland cells. *Infection and Immunity* **67**, 5076–5082.

Salmond, G.P.C., Bycroft, B.W., Stewart, G.S.A.B., and Williams, P. (1995). The bacterial enigma: cracking the code of cell-cell communication. *Molecular Microbiology* **16**, 615–624.

Schümann, J., Angermüller, S., Bang, R., Lohoff, M., and Tiegs, G. (1998). Acute hepatoxicity of *Pseudomonas aeruginosa* exotoxin A in mice depends on T cells and TNF. *Journal of Immunology* **161**, 5745–5754.

Singh, P.K., Schaefer, A.L. Parsek, R., Moninger, T.O., Welsh, M.J., and Greenberg, E.P. (2000). Quorum-sensing signals indicate that cystic fibrosis lungs are infected with bacterial biofilms. *Nature* **407**, 762–764.

Smith, J.J., Travis, S.M., Greenberg, E.P., and Welsh, M.J. (1996). Cystic fibrosis airway epithelia fail to kill bacteria because of abnormal airway surface fluid. *Cell* **85**, 229–236.

Sperandio, V., Mellies, J.L., Nguyen, W., Shin, S., and Kaper, J.B. (1999). Quorum sensing controls expression of the type III secretion gene transcription and protein secretion in enterohemorrhagic and enteropathogenic *Escherichia coli*. *Proceedings of the National Academy of Sciences USA* **96**, 15,196–15,201.

Steinhauser, M.L., Hogaboam, C.M., Kunkel, S.L., Lukacs, N.W., Strieter, R.M., and Standiford, T.J. (1999). IL-10 is a major mediator of sepsis-induced impairment in lung antibacterial host defense. *Journal of Immunology* **162**, 392–399.

Surrette, M.G., Miller, M.B., and Bassler, B.L. (1999). Quorum-sensing in *Escherichia coli, Salmonella typhimurium,* and *Vibrio harveyi*: A new family of genes responsible for autoinducer production. *Proceedings of the National Academy of Sciences USA* **96**, 1639–1644.

Telford, G., Wheeler, D., Williams, P., Tomkins, P.T., Appleby, P., Sewell, H., Stewart, G.S.A.B., Bycroft, B.W., and Pritchard, D.I. (1998). The *Pseudomonas aeruginosa* quorum sensing signal molecule, N-(3-oxododecanoyl)-L-homoserine lactone has immunomodulatory activity. *Infection and Immunity* **66**, 36–42.

Tsai, W.C., Strieter, R.M., Mehrad, B., Newstead, M.W., Zeng, X., and Standiford, T.J. (2000). CXC Chemokine receptor CXCR2 is essential for protective innate host response in murine *Pseudomonas aeruginosa* pneumonia. *Infection and Immunity* **68**, 4289–4296.

Van Delden, C. and Iglewski, B.H. (1998). Cell-to-cell signaling and *Pseudomonas aeruginosa* infections. *Emerging Infectious Diseases* **4**, 551–560.

Welch, M., Todd, D.E., Whitehead, N.A. McGowan, S.J., Bycroft, B.W., and Salmond, G.P.C. (2000). N-acylhomoserine lactone binding to CarR receptor determines quorum-sensing specificity in *Erwinia*. *European Molecular Biology Organisation Journal* **19**, 631–641.

Wesson, C.A., Liou, L.E., Todd, K.M., Bohach, G.A., Trumble, W.R., and Bayles, K.W. (1998). *Staphylococcus aureus* Agr and Sar global regulators influence internalization and induction of apoptosis. *Infection and Immunity* **66**, 5238–5243.

Williams, P., Camara, M., Hardman, A., Swift, S., Milton, D., Hope, V.J., Winzer, K., Middleton, B., Pritchard, D.I., and Bycroft, B.W. (2000). Quorum sensing and the population dependent control of virulence. *Philosophical Transactions of the Royal Society of London Series B – Biological Science* **355**, 667–680.

Wilson, M., Seymour, R., and Henderson, B. (1998). Bacterial perturbation of cytokine networks. *Infection and Immunity* **66**, 2401–2409.

Winson, M.K., Camara, M., Latifi, A., Foglino, M., Chhabra, S.R., Daykin, M., Bally, M., Chapon, V., Salmond, G.P.C., Bycroft, B.W., Lazdunski, A., Stewart, G.S.A.B., and Williams, P. (1995). Multiple N-acyl-L-homoserine lactone signal molecules regulate production of virulence determinants and secondary metabolites in *Pseudomonas aeruginosa*. *Proceedings of the National Academy of Sciences USA* **92**, 9427–9431.

Winzer, K., Falconer, C., Garber, N.C., Diggle, S.P., Camara, M., and Williams, P. (2000). The *Pseudomonas aeruginosa* lectins PA-IL and PA-IIL are controlled by quorum sensing and by RpoS. *Journal of Bacteriology* **182**, 6401–6411.

Withers, H., Swift, S., and Williams, P. (2001). Quorum sensing as an integral component of gene regulatory networks in Gram-negative bacteria. *Current Opinions in Microbiology* **4**, 186–193.

CHAPTER 10

Microbial modulation of cytokine networks

Brian Henderson and Robert M. Seymour

10.1 INTRODUCTION

Inflammation is a paradoxical process. This protective mechanism, whose absence spells prolonged illness or death, is also the cause of an enormous amount of morbidity worldwide with many idiopathic chronic inflammatory states, including asthma, autoimmune diseases (rheumatoid arthritis, multiple sclerosis, etc.), psoriasis, and inflammatory bowel disease, still awaiting a cure. The signs of inflammation were defined by the Roman encyclopaedist, Celsus, almost two millennia ago, and the humoral and cellular factors that drive inflammation have been under scrutiny since the middle of the nineteenth century. However, it was not until the 1950s that clues emerged as to how the enormously complex inflammatory/immune response, with its multiple cells and mediators (discussed in other chapters in this volume), was integrated and controlled. In the United States, the study of endotoxin-induced pyrexia (reviewed by Dinarello, 1989) and in the United Kingdom, the study of viral "interference" (reviewed by Gresser, 1997), led to the discovery of polypeptides with potent effects on cell behaviour. These proteins, interleukin (IL)-1 and interferon (IFN)α, respectively, were the forerunners of the enormous lists of proteins now known as cytokines that we recognise as inducing, and suppressing, inflammation (see Horst Ibelgauft's website, COPE, for a crash course in cytokines and Table 10.1). Cytokines can now be defined on the basis of their structural biology, or subdivided according to their historical naming/function, as in Table 10.1. Over the past thirty years it has become established that cytokines control the interactions among the many cell populations that constitute the mammalian immune and inflammatory responses (Meager, 1998; Mantovani et al., 2000;

Table 10.1. *The cytokines defined by function*

Cytokine family	Examples	Functions
Interleukins	IL-1, IL-2, IL-10, IL-18	Regulators of inflammatory responses (IL-1, IL-10, IL-12, IL-18), lymphocyte growth factors (e.g. IL-2, IL-4, IL-6, IL-7, IL-9, IL-17), chemokines (IL-8)
Necrotic cytokines	TNF family	Involved in regulation of inflammation and of apoptosis
Interferons	IFN α, β, γ	Antiviral and immune modulating
Colony-stimulating factors	IL-3, G-CSF, M-CSF	Growth and differentiation factors for myeloid cells
Growth factors	TGFβ, FGF, PDGF	Growth factors for mesenchymal and epithelial cells
Chemokines	IL-8, MCP, RANTES	Chemotactic cytokines

Balkwill, 2000; Oppenheim and Feldmann, 2000; Callard et al., 2001). In this chapter we will consider how bacteria interact with their hosts in terms of stimulating or inhibiting the synthesis and action of cytokines. These bacteria-host/host-bacteria interactions have co-evolved to maximise both the survival of the bacterium and of the host species. As a preface to this discussion, the reader will be introduced to the concept of the cytokine network. It has to be emphasised that cytokines differ from endocrine hormones in that in any given tissue most, if not all, cells can generate, and respond to, multiple cytokines. Thus, the basic level of control of cytokines is not the individual cytokine, but the network of cytokines generated in any particular tissue or group of cells in response to a particular stimulus or group of stimuli.

10.2 CYTOKINE NETWORKS

A familiar and vital physiological system involves the release of corticotrophin-releasing factor (CRF) by the hypothalamus, which stimulates the synthesis and release of adrenocorticotrophic hormone (ACTH) by the anterior pituitary. In turn, ACTH stimulates cortisol release by the adrenal cortex. Cortisol has a negative feedback effect both on the hypothalamus and pituitary (Fig. 10.1a). Compare this simple negative feedback loop between

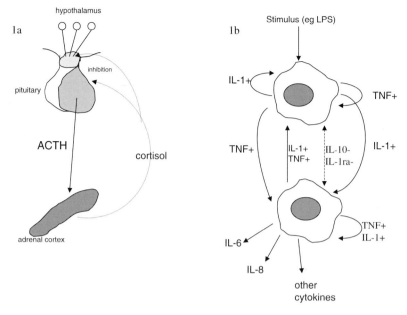

Figure 10.1. Comparison between endocrine hormone control, in this case the control of release of the glucocorticoid, cortisol (1a), and the control of cytokine synthesis. In the case of the former, signals acting at the level of the hypothalamus cause the release of corticotrophin-releasing factor (CRF), which acts on the pituitary to stimulate the release of adrenocorticotrophic hormone (ACTH). This enters the blood and binds to adrenocortical cells, which are stimulated to release cortisol. The cortisol can then act at the level of the pituitary and hypothalamus on which tissue it has a direct negative feedback effect inhibiting CRF and ACTH synthesis. In Fig. 10.1b the effect of stimulating a cell (e.g., macrophage) with a bacterial pro-inflammatory signal (e.g., LPS) is schematically shown. This causes the release of the early response cytokines IL-1β and TNFα. These cytokines can act on nearby cells in a paracrine manner. They can also act in an autocrine fashion to stimulate the producing cell. These early response cytokines can also act to induce the production of additional cytokines including regulatory proteins such as IL-10 and IL-1 ra (receptor antagonist). These proteins are believed to act to down-regulate pro-inflammatory cytokine networks. In addition, the early response cytokines can induce the formation of so-called secondary cytokines such as IL-6, IL-8 etc., which can act to promulgate the inflammatory response. Note that this diagram is an enormous simplification of what the real situation is believed to be. Many other cytokines will be involved and the role of receptor turnover and shedding has not been included.

these three tissues with the interactions between two cells exchanging information in the form of cytokines as a result of stimulation with a bacterial agonist (Fig. 10.1b). The example chosen is the synthesis and release of the so-called early response cytokines, IL-1, and tumour necrosis factor (TNF)α in response to a bacterial agonist. This results in a "network" of cytokines, produced by paracrine and autocrine interactions, including TNFα, IL-1ra, IL-6, IL-8, and IL-10. It is these cytokines that control the behaviour of the cells involved in bacterial infections. In simplistic terms, the key to the homeostatic control of inflammation resulting from bacterial infection is the ability to switch on the appropriate level and temporal pattern of production of specific cytokines and to then be able to switch off these cytokines in an appropriate temporal manner. It must be emphasised that cytokine networks are dynamic entities with both spatial and temporal manifestations.

More generally, a (local) cytokine network consists of a localised collection of cells of one or more types, which communicate with each other by means of various cytokines. Producer cells release a cytokine signal into the extracellular environment, and receiver cells detect the signal by ligand binding to cytokine-specific, high affinity, cell-surface receptors, which in turn transmit a signal via intracellular signalling pathways to the nucleus. The receiver cells may then respond by producing further cytokines, which may be the same, or different, to that received. Thus, cells can act as both producers and receivers of information carried by cytokine signals. In particular, a cell can act on itself by autocrine signalling. Different species of cell may respond, more or less strongly, to different sets of cytokines, depending on the type and density of cell-surface receptors they express.

This picture is complicated by additional features such as cytokine-stimulated cells responding by up-regulating or down-regulating the expression of cytokine receptors rather than the cytokines themselves. Cells can also react by producing soluble receptors or by expressing decoy receptors, both of which act to absorb information-carrying cytokine molecules without transmitting a signal. In addition, cytokines can act as receptor antagonists (e.g., IL-1ra), which do not carry information, but instead act by blocking signal-transduction pathways otherwise open to signal-carrying cytokines.

Networks of the form described above may be conceived more abstractly as consisting of a set of "cells" (of various species), pairwise connected by one or more binary relations, which determine the information traffic (signals carried by cytokines) between them. Multiple signals (carried by different cytokines) between pairs of cells and auto-signalling from a cell to itself are also possible. Signals can be *activating*, stimulating the receiving cell to up-regulate the production of some molecule(s), or *inhibiting*, inducing the down-regulation of some molecule(s). The same signal can, in principle, be both activating for some molecules and inhibiting for others in the same cell (Wilson et al., 1998; Seymour and Henderson, 2001). This concept can be seen visually in Fig. 10.2 which shows a simple cytokine network involving the two

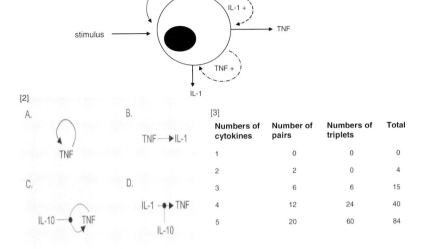

Figure 10.2. A simple cytokine network is shown produced by the activation of the cell (a monocyte) by a stimulus (e.g., LPS) [1]. The induction of IL-1 and TNF synthesis and release of these cytokines can result in the feedback of these cytokines onto the producing cell. The IL-1 can, for example, activate the producing cell to make TNF or it can even induce IL-1 synthesis. In [2] the various modes of interaction between cytokines and cells are shown schematically. These include (A) autoinduction; (B) cross-induction; (C) inhibition of autoinduction; and (D) inhibition of cross-induction. In [3] the rapid complexification of cytokine networks as a response to the numbers of cytokines involved in shown in a table. Here the relationship is shown between the number of cytokines in a network and the calculated number of interactions that can occur in a binary and ternary manner. Higher order interactions have been omitted. Thus with only four cytokines up to forty different interactions can occur.

early response cytokines, IL-1 and TNF, and a major anti-inflammatory cytokine, IL-10. The way these cytokines can interact with the cell to control each other's synthesis and the enormous complexity of the potential interactions that can occur in such networks is shown schematically in the diagram (Fig. 10.2).

Thus, each cell in the network exists in a "soup" of cytokines, carrying various activating and inhibiting signals, some of which may reinforce, and some cancel or reduce, the effect of others, and which may act in different ways on different cells. The cytokine network is *pro-inflammatory* if the net effect is to provoke an inflammatory response, and is *anti-inflammatory* if the net effect is to suppress an inflammatory response. Furthermore, this cytokine soup (network) is not static, but changes dynamically as the various

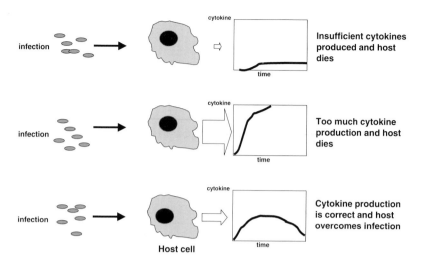

Figure 10.3. The kinetics and dynamics of cytokine network induction in infection. Failure to mount an appropriate cytokine network response by either under- or over-producing cytokines can result in tissue or systemic pathology and even death.

responses unfold in time. Even the cells involved may change dynamically through: (i) induced cell proliferation, (ii) cell maturation (for example monocytes maturing into macrophages or dendritic cell precursor into mature dendritic cells – see Chapter 1), (iii) cell death, or (iv) through migration into or out of the site of cytokine network activity. It is envisaged that such cytokine networks are the main controlling agents in the inflammation and immune reactions that occur in infections. The dynamical relationships within a cytokine network, invoked in response to an initiating external event, will control the induction, perpetuation, and the possible eventual collapse of the network and the consequent cellular events that it controls. However, it is also possible that, once induced, a "homeostatic" state of dynamic stability may be maintained more-or-less indefinitely between cells participating in the network, even after the primary initiating event has passed. This may be the situation in idiopathic diseases such as rheumatoid arthritis, which is now believed to be initiated by some infectious agent.

10.3 BACTERIA AND CYTOKINE NETWORKS

The host cytokine response to a bacterial infection can take one of three dynamic states (Fig. 10.3). Failure to produce sufficient "cytokine" or, more correctly, the optimal network of cytokines, will inevitably lead to

Table 10.2. *Classes of bacterial components able to induce cytokine synthesis*

Lipopolysaccharides	Oligosaccharides	Lipoarabinomannans
Peptidoglycans	Oligopeptides	Molecular chaperones
Lipopeptides	Lipoproteins	Lipids
Glycolipids	Glycoproteins	Fimbriae
Superantigens	Adhesins	Phospholipases
Exotoxins	CpG DNA	Proteinases

Note: NB Each class of bacterial molecule contains individual molecules with a range of potencies.

overwhelming infection and death. If, on the other hand, the host generates a network "rich" in pro-inflammatory cytokines or fails to produce sufficient inhibitory cytokines, then, again, morbidity and mortality can ensue. Good examples are the conditions of septic and toxic shock. We have some limited insight into the problems of cytokine network formation in pathology. What we clearly lack is an understanding of how infecting organisms and the host combine to produce the optimum cytokine networks to enable bacterial infections to be overcome and for the host return to normality without any notable sequelae.

The obvious starting point is to understand how bacteria stimulate cytokine synthesis. It is now established that many bacterial components stimulate cytokine synthesis. Most attention has focused on the pro-inflammatory cytokine-inducing actions of lipopolysaccharide (LPS) from Gram-negative bacteria and peptidoglycan and lipoteichoic acid from Gram-positive organisms (Henderson et al., 1998). We now know that these components bind to, among others, the Toll-like receptors (TLRs) (Akira et al., 2001) described briefly in Chapter 1 to induce the early response-type cytokine network shown diagrammatically in Fig. 10.1. Here the two key pro-inflammatory cytokines, IL-1 and TNFα, are produced. However, these bacterial cell wall components are only the tip of the iceberg of cytokine-inducing bacterial components (Table 10.2). It is now clear that bacterial molecules of all chemical classes are able to induce cytokine synthesis (reviewed by Henderson et al., 1996a, 1996b; Henderson et al., 1998). Perhaps the most active cytokine inducers are bacterial exotoxins (Henderson et al., 1997) such as the superantigens described in Chapter 8. The induction and inhibition of cytokine synthesis by enterotoxins is discussed in Chapter 11. There is evidence that some of these varied cytokine-inducing molecules induce cytokine networks that differ from those induced by LPS and peptidoglycan and that are generally thought of as the prototypic network. For example, the cytolethal distending toxin from the oral bacterium, *Actinobacillus actinomycetemcomitans,* induces human monocytes to secrete IL-1β but not TNFα (Akifusa et al., 2001). Thus, each bacterium may generate dozens of cytokine-inducing molecules that can interact with the host to create tissue pathology. It has been proposed that these cytokine-inducing molecules be classified as a novel form of bacterial virulence factor and the term modulin has been coined, as these molecules, by inducing cells to produce autocrine-acting cytokines, modulate cell behaviour (Henderson et al., 1996a, 1996b; Henderson and Wilson, 1998; Henderson et al., 1998).

10.4 THE NORMAL MICROBIOTA AND THE COMMENSAL PARADOX

Ninety percent of the cells in the average human are not eukaryotic but are, in fact, bacteria (Tanner, 1995). This is the normal human microbiota, which is very complex, containing at least 1,000 different bacterial species (Relman and Falkow, 2001). These bacteria can make intimate contact with our mucosal cells but do appear not provoke an inflammatory response. As all bacteria studied produce cytokine-inducing pro-inflammatory molecules and must be constantly releasing them, the finding that the mucosae harbouring the microflora are noninflamed is a paradox, which has been termed the commensal paradox (Henderson et al., 1996a, 1996b, 1998; Henderson and Wilson, 1998; Wilson et al., 1998). Arguably, the failure of the mucosae to become inflamed in response to the adjacent normal microbiota is the most widespread example of evasion of the inflammatory defence mechanism in multicellular organisms. One solution to this paradox is that either some, or possibly all, of the members of the microbiota produce molecules with the ability to block the synthesis of pro-inflammatory cytokines or pro-inflammatory cytokine networks. Possible clues as to the cytokine networks involved has come from the phenotypes of transgenic mice with inactivated cytokine genes. Mice lacking IL-2 (Sadlack et al., 1993) or IL-10 (Kuhn et al., 1993) develop entero/colitis. However, transgenic knockout mice raised under gnotobiotic conditions fail to develop inflammation of the gut (Contractor et al., 1998; Sellon et al., 1998) revealing that the removal of IL-2 or IL-10 from local cytokine networks in the gut generates an inflammatory response to members of the normal gut microbiota. Thus, a plausible hypothesis is that under normal conditions the gut microbiota signals to the gut to induce a noninflammatory cytokine network. This is not so implausible as it is known

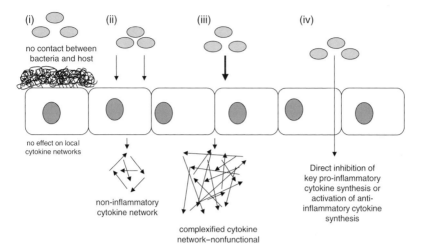

(i) no contact between bacteria and host

no effect on local cytokine networks

non-inflammatory cytokine network

complexified cytokine network–nonfunctional

Direct inhibition of key pro-inflammatory cytokine synthesis or activation of anti-inflammatory cytokine synthesis

Figure 10.4. Four theoretical states of interaction of bacteria with host cells and the consequent evasion of host cytokine-driven responses. In (i) and (ii), we have the two possible forms of interaction between the normal bacterial flora and mucosal surfaces. In (i) there is no interaction either because the presence of the mucus barrier or through selective desensitisation of the epithelial cells of the mucosa. An alternative hypothesis is that there is constant crosstalk between the bacteria and the mucosal (and even submucosal) cells resulting in the formation of cytokine networks which do not evince an inflammatory state. In (iii) the bacteria under consideration are pathogens and the release of multiple cytokine-inducing components is proposed to complexify protective cytokine networks thus rendering them nonfunctional. In (iv) the effect of selective inhibitors of pro-inflammatory cytokine networks deactivates inflammatory defences.

that the gut organism, *Bacteroides thetaiotamicron,* interacts with intestinal epithalial cells to control their cell surface glycosylation (Hooper et al., 2001).

10.5 BACTERIAL EVASION OF PROTECTIVE PRO-INFLAMMATORY CYTOKINE NETWORKS

To survive within its human host, we propose the hypothesis that a bacterium must either: (i) fail to interact with host cells and therefore not affect local cytokine networks; (ii) interact with host cells and induce a noninflammatory cytokine network; (iii) generate a complexified network that is essentially nonfunctional or selectively suppresses protective pro-inflammatory networks; or (iv) synthesise components that selectively inhibit the synthesis or action of key protective pro-inflammatory cytokines (Fig. 10.4). It is envisaged that members of the normal microbiota exist within categories (i) and (ii). The failure to induce cytokine networks implies a lack of communication between bacteria and host or an unresponsiveness of host epithelium. It has recently been argued that intestinal epithelial cells lack TLR-4 and MD-2 and that this renders them resistant to the effects of LPS released by the bacteria microflora (Abreu et al., 2001). In contrast, it has been reported that flagellin from *Salmonella* spp can activate intestinal epithelial cells in a TLR4-independent fashion (Eaves-Pyles et al., 2001). However, as bacteria must constantly be releasing a range of cytokine-inducing components (TLR-dependent and TLR-independent) onto mucosal surfaces it would seen more sensible that these surfaces compensate for this by some active form of active cytokine network control. However, the molecules required to support this hypothesis have not yet been identified.

The finding that bacteria can produce a wide range of cytokine-inducing molecules suggests another strategy for evading host cytokine network defence. If we follow Charles Janeway's train of thought about pathogen-associated molecular patterns (PAMPs) and pattern recognition receptors (PRRs; Janeway, 1992, as discussed in Chapter 1), then we have a situation in which the host recognises bacteria through evolutionarily conserved molecules (PAMPs) such as LPS, CpG DNA, peptidoglycan, and molecular chaperones. This recognition induces an appropriate induction of protective host cytokine networks whose kinetics and dynamics is controlled by the levels of these PAMPs. However, if bacteria produce additional cytokine-inducing molecules that do not signal through the PRRs it could lead to dysequilibrium (complexification) of these protective cytokine networks and therefore to the evasion of the protective host inflammatory/immune system. It is of interest that many, if not all, of the bacterial exotoxins have potent cytokine-inducing actions in addition to their more accepted functions as cell toxins. Some toxins are more active cytokine inducers than they are toxins (Henderson et al., 1997). It must be emphasised that there is no direct experimental evidence to support the proposal that bacteria can induce complexified networks of cytokines. However, the concept is a testable hypothesis. The final mechanism will be dealt with in the next section.

10.6 MICROBIAL INHIBITION OF PRO-INFLAMMATORY CYTOKINE NETWORKS

During the past fifteen to twenty years, we have seen the identification of a large number of viral gene products with homology to human cytokines and cytokine receptors and some proteins with the capacity to affect cytokine gene transcription and transcriptional control. Viruses such as vaccinia and Epstein Barr encode proteins with homology to human cytokines such as epidermal growth factor (Brown et al., 1985), IL-6 (Moore et al., 1996) and

Table 10.3. *Examples of bacterial exotoxins that inhibit cytokine synthesis*

Toxin	Cytokines inhibited
Cholera toxin	TNFα, IL-2, IL-12 (not IL-6)
E. coli heat-labile toxin	IL-4
Anthrax oedema toxin	TNF(stimulates IL-1)
Pseudomonas aeruginosa exotoxin A	IL-1, TNF, lymphotoxin, IFNγ
Botulinum toxin type D	TNF
Lymphostatin	IL-2, IL-4, IL-5, IFNγ

IL-17 (Yao et al., 1995). These cytokine-like proteins have been termed virokines. In addition, many viruses are now established to encode soluble forms of cytokine receptors including those binding to IL-1, TNFα, IFNα/β, IFNγ, and chemokines (Kotwal, 1999). Chemokine receptor-like proteins are also encoded by various viruses (Kotwal, 2000). These receptor-like proteins have been termed viroceptors. The role of these various viral gene products, which have been "borrowed" from the host, are assumed to aid the virus in infecting the host and evading cytokine-driven immune responses. Indeed, there are experimental animal studies confirming that various virokines can inhibit inflammatory processes *in vitro*, and this led to the suggestion that such viral proteins could be utilised as anti-inflammatory agents (Henderson, 2000).

To date, analysis of the growing number of bacterial genomes that have been fully sequenced has failed to identify bacterial equivalents of the virokines and viroceptors, although proteins with homology, for example, to human NRAMP have been identified (see Chapter 3 for details). However, there is evidence that various bacterial proteins can act to modulate cytokine synthesis and some limited evidence exists for the existence of cytokine receptors on some common pathogens.

10.6.1 Bacterial proteins or mechanisms inhibiting cytokine synthesis

It is now established that bacteria can inhibit macrophage cytokine synthesis via type III secretion systems; the mechanism of NF-κB inhibition via such systems is described in detail in Chapter 12. A number of bacterial toxins have been reported to selectively inhibit cytokine synthesis (Table 10.3). The mechanism of action of cholera toxin, *E. coli* heat labile toxin, and lymphostatin is described in detail in Chapter 11.

The beginning of our understanding of the role of type III secretion and of exotoxins in controlling cytokine synthesis is likely to be the tip of the iceberg of mechanisms utilised by bacteria to control host cytokine networks. A growing number of reports have appeared in recent years emphasising that bacteria can modulate the production of a number of cytokines. The ability of non-virulent *Salmonella* strains to inhibit NF-κB by a direct cell contact mechanism has been reported. The mechanism of action is at the level of the polyubiquitination of the IκB – a process required to enable this protein to be degraded by the proteasome. In some as yet unexplained manner, the IκB in cells in contact with these nonvirulent *Salmonella* is not degraded and thus the NF-κB cannot enter the nucleus (Neish et al., 2000). Intracellular bacteria appear to have the ability to modulate cytokine synthesis or cytokine signalling to enable persistent infection of cells. Evidence is now appearing that bacteria that normally exist within host cells such as *Mycobacterium tuberculosis* and *Chlamydia trachomatis* have additional mechanisms of controlling cytokine signalling systems – but this early-phase work will not be discussed in this chapter.

A number of reports have shown that bacteria including enteropathogenic *E. coli* (Klapproth et al., 1995), *A. actinomycetemcomitans* (Kurita-Ochiai and Ochiai, 1996), *Brucella* spp (Caron et al., 1996), *Salmonella typhimurium* (Matsui, 1996), and *Myobacterium ulcerans* (Pahlevan et al., 1999) produce proteins able to inhibit the synthesis of specific cytokines. In the case of enteropathogenic *E. coli,* the active protein has been identified as novel toxin termed lymphostatin. The discovery and mechanism of this toxin is described in detail in Chapter 11. In the case of *Brucella* spp., the use of isogenic mutants has identified the major outer membrane protein, (Omp)25, as the cytokine inhibitor (Jubier-Maurin et al., 2001). In addition, mannosylated lipoarabinomannans from mycobacteria have been demonstrated to inhibit IL-12 production by human dendritic cells (Nigou et al., 2001). Thus, it is slowly starting to become clear that bacteria can exert down-regulatory effects on pro-inflammatory cytokine networks by direct cell binding (through type III systems) or by the release of particular molecules or indeed while living within cells. Much more work is now needed to identify the populations of pro- and anti-inflammatory cytokine regulating proteins produced by individual bacterial species in order to determine how bacteria regulate their abilities to control host cytokine networks.

10.6.2 Bacterial proteinases and cytokine inactivation/activation

Bacteria secrete a range of proteinases that can play a role in nutrition, pathogenesis, or immune evasion. One example of proteinases involved in evading humoral immune responses is the IgA proteinases described by Mogens Kilian in Chapter 5. *Pseudomonas aeruginosa*, an environmental organism that can cause human pathology produces two major proteinases – an alkaline proteinase and an elastase. Both proteinases can inactivate IFNγ and TNFα but were unable to affect the biological activity of IL-1α or IL-1β. The activity of the alkaline proteinase were unaffected by serum while that of the elastase was reduced in the presence of serum (Horvat et al., 1989; Parmely et al., 1990). *Legionella pneumophila* produces a metalloproteinase with haemolytic and cytotoxic actions. In addition to these activities, the metalloproteinase also cleaves IL-2 and removes CD4 from human T cells inhibiting T cell proliferation (Mintz et al., 1993). A number of bacteria have been reported to release proteinases with the ability to cleave the IL-6 receptor from human monocyctes. In this way, these bacterial proteinases mimic the action of natural host proteinases (sheddases) evolved to catalyse the release of cytokine receptors from cells as part of the control of cytokine networks (Mullberg et al., 1997). One of the consequences of this is that bystander cells, which normally lack IL-6 receptors, can become responsive to this cytokine (Vollmer et al., 1996). In contrast to the cytokine-inhibitory effects of these bacterial proteinases, the extracellular cysteine proteinase of *Streptococcus pyogenes*, also known as streptococcal pyrogenic exotoxin B (SPE B), cleaves biologically inactive pro-IL-1β to produce an active form of this mature cytokine (Kapur et al., 1993).

Periodontal disease is a very common chronic bacterially driven disease in which bacterial biofilms play a major role in pathology. Examination of two of the major bacterial pathogens, *A. actinomycetemcomitans* and *Porphyromonas gingivalis*, for the presence of cytokine-degrading activity revealed that the culture media supporting the latter organism was able to proteolyse IL-1β, IL-6, and the anti-inflammatory cytokine IL-1ra (receptor antagonist). This proteolytic breakdown of cytokines could be produced even in the presence of a large excess of serum proteins (Fletcher et al., 1997). It was also shown that biofilms of *P. gingivalis* were able to degrade cytokines (Fletcher et al., 1998). *P. gingivalis* produces two major cysteine proteinase types – the Arg-gingipains and the Lys-gingipains. The purified gingipains have been reported to inactivate TNFα (Calkins et al., 1998), IL-6 (Banbula et al., 1999), and IL-8 (Mikolajczyk-Pawlinska et al., 1998; Zhang et al., 1999). Of interest was the finding that the initial attack of soluble gingipains on IL-8 produced a cleavage product with enhanced bioactivity (Mikolajczyk-Pawlinska et al., 1998). These findings seem to suggest that the proteinases acts as inhibitors of pro-inflammatory cytokine networks. However, to complicate the picture it has been established that an internal fragment of the adhesin domain of

Lys-gingipain is a potent inducer of cytokine synthesis (Sharp et al., 1998) and that the arginine-specific cysteine proteinase (RgpB) can stimulate oral epithelial cells to secrete IL-6 by activation of protease-activated receptors (PARs) (Lourbakos et al., 2001).

Much more work is needed to identify the overall effect of the proteinases of bacterial pathogens on the workings of pro- or anti-inflammatory cytokine networks. However, the available data have revealed that proteases can have profound effects on such networks.

10.6.3 Bacterial receptors for cyokines

Cytokines are extremely potent molecules and only small amounts have to be produced at sites of bacterial infection to maximally activate host defence cells. In consequence, therefore, only small amounts of cytokine "inhibitors" would have to be produced to negate the activity of any particular cytokine. This weakness in cytokine network design has been exploited by viruses. As explained above, viruses have pirated host genes encoding cytokine receptors to produce soluble forms of cytokine receptors (viroceptors) that prevent the formation of defensive cytokine networks. This would be a sensible strategy for bacteria to evolve. However, is there any evidence for this?

In 1991 it was reported that freshly isolated strains of virulent *E. coli*, but not avirulent strains or virulent strains that had been laboratory-passaged, specifically bound IL-1. Binding isotherms suggested that each bacterium contained 20–40,000 IL-1 receptors, which is a much greater number than that present on mammalian cells (Porat et al., 1991). Intriguingly, these workers reported that IL-1 acted as a growth factor (the levels of IL-1 were too low for this to be a nutrient effect) for virulent *E. coli*. If true, the ability to bind (neutralise) and utilise key defence cytokines as growth factors would be an extremely useful evasion mechanism. Luo and coworkers (1993) examined a number of bacterial species: *E. coli*, *Shigella flexneri*, and *S. typhimurium* and reported that they had specific receptors for TNFα. The identity of these putative receptors remains a mystery. Perhaps the most intriguing bacterial "cytokine receptor" is the glycolytic enzyme, glyceraldehyde 3-phosphate dehydrogenase (GAP), which is found on the surface of *M. tuberculosis* and *M. avium* and which binds the human cytokine, epidermal growth factor, with high affinity. Binding of this cytokine acts as a growth factor for the mycobacteria (Bermudez et al., 1996; Parker and Bermudez, 2000). The mechanism of transduction of this eukaryotic growth factor signal in bacteria has not been identified. Other workers have reported that mammalian cytokines such as IL-2 and granulocyte-macrophage colony-stimulating factor (GM-CSF) are growth factors for certain bacterial species (Denis et al., 1991a, 1991b). This ability of signalling molecules from one superkingdom (*Eukarya*) to reach across and act within another (*Prokarya*) has recently been reported in the other direction. As described by Pritchard and coworkers in Chapter 9, acyl homoserine lactones, the mediators of quorum sensing, can interact with leukocytes and interfere with cytokine signalling.

While these reports are fascinating, apart from the identification of the mycobacterial receptor as GAP, no further information on these putative bacterial cytokine receptors has been forthcoming. A recent report slightly extends this area of study. This was the finding that a cyclic peptide produced by *Streptomyces* spp. inhibits the binding of platelet-derived growth factor to its receptor (Toki et al., 2001). However, the possibility that bacteria, or at least pathogenic strains of individual bacteria, express proteins that can mimic the action of host cytokine receptors is a fascinating hypothesis and one that requires further study. The problem is that such hunting of potential needles in virtual haystacks is difficult to fund either by granting agencies or within the biopharmaceutical industry.

10.7 CONCLUSION

This chapter is one of four that describe diverse mechanisms by which bacteria can interfere with cytokine signalling. In order to make sense of how bacteria interact with host defence systems, it must be emphasised that cytokines are the controlling and integrating signals of inflammation and immunity and that cytokines exhibit the behaviour of network phenomena. Thus, we are dealing not with individual cytokines but with the network behaviour of cytokines both spatially and temporally. The interaction of bacteria with the host sets up another putative network – that of the various molecules emanating from bacteria that can interfere with cytokine network control. Thus, understanding how any one bacterium, under a single set of circumstances, can give rise to a protective or nonprotective cytokine network is essential if we are to be able to make sense of bacterial infection. This information is also essential to understand why the enormous number of bacteria that constitute our normal microbiota do not act as a constant source of inflammation. One powerful tool that the authors are using is the combination of mathematical modelling in close combination with cell-based studies of cytokine synthesis to assess the validity of models. This approach is now being used to determine how networks of cytokine-inducing bacterial molecules (modulins) interact with the networks of cytokines produced in response to these modulins. We are confident that this approach allows the complexities of cytokine signalling in response to bacterial infection to be teased apart and allows the cytokine evasion mechanisms utilised by bacteria to be elucidated.

ACKNOWLEDGMENTS

We acknowledge the financial support of the Biotechnology and Biology Science Research Council (BBSRC) [grant number 31/MM109754].

REFERENCES

Abreu, M.T., Vora, P., Faure, E., Thomas, L.S., Arnold, E.T., and Arditi, M. (2001). Decreased expression of Toll-like receptor-4 and MD-2 correlates with intestinal cell protection against dysregulated proinflammatory gene expression in response to bacterial lipopolysaccharide. *Journal of Immunology* **167**, 1609–1616.

Akifusa, S., Poole, S., Lewthwaite, J., Henderson, B., and Nair, S.P. (2001). Recombinant *Actinobacillus actinomycetemcomitans* cytolethal distending toxin. proteins are required to interact to inhibit human cell cycle progression and to stimulate human leukocyte cytokine synthesis. *Infection and Immunity* **69**, 5925–5930.

Akira, S., Takeda, K., and Kaisho, T. (2001). Toll-like receptors: critical proteins linking innate and acquired immunity. *Nature Immunology* **2**, 675–680.

Balkwill, F., ed (2000). *The Cytokine Network*. Oxford: Oxford University Press.

Banbula, A., Bugno, M., Kuster, A., Heinrich, P.C., Travis, J., and Potempa, J. (1999). Rapid and efficient inactivation of IL-6 by gingipains: lysine- and arginine-specific proteinases from *Porphyromonas gingivalis*. *Biochemical and Biophysical Research Communications* **261**, 598–602.

Bermudez, L.E., Petrofsky, M., and Shelton, K. (1996). Epidermal growth factor-binding protein in *Mycobacterium avium* and *Mycobacterium tuberculosis:* a possible role in the mechanism of infection. *Infection and Immunity* **64**, 2917–2922.

Brown, J.P., Twardzik, D.R., Marquadt, H., and Todaro, G.J. (1985). Vaccinia virus encodes a polypeptide homologous to epidermal growth factor and transforming growth factor. *Nature* **313**, 491–492.

Calkins, C.C., Platt, K., Potempa, J., and Travis, J. (1998). Inactivation of tumor necrosis factor-α by proteinases (gingipains) from the periodontal pathogen. *Porphyromonas gingivalis*. *Journal of Biological Chemistry* **273**, 6611–6614.

Callard, R.E., O'Neill, L., Fitzgerald, K., and Gearing, A. (2001). *The Cytokine Factsbook, 2nd edn.* London: Academic Press.

Caron, E., Ross, A., Liautard, J.-P., and Dornand, J. (1996). Brucella species release a specific, protease-sensitive, inhibitor of TNF-α expression, active on human macrophage-like cells. *Journal of Immunology* **156**, 2885–2893.

Contractor, N.V., Bassiri, H., Reya, T., Park, A.Y., Baumgart, D.C., Wasik, M.A.,

Emerson, S.G., and Carding, S.R. (1998). Lymphoid hyperplasia, autoimmunity, and compromised intestinal intraepithelial lymphocyte development in colitis-free gnotobiotic IL-2-deficient mice. *Journal of Immunology* **160**, 385–394.

COPE, http://www.copewithcytokines.de/cope.cgi?3267

Denis, M., Campbell, D., and Gregg, E.O. (1991a). Cytokine stimulation of parasitic and microbial growth. *Research in Microbiology* **142**, 979–983.

Denis, M., Campbell, D., and Gregg, E.O. (1991b). Interleukin-2 and granulocyte-macrophage colony-stimulating factor stimulate growth of a virulent strain of *Escherichia coli*. *Infection and Immunity* **59**, 1853–1856.

Dinarello, C.A. (1989). Was the original pyrogen interleukin-1? In *Interleukin-1 Inflammation and Disease*, ed. R. Bomford and B. Henderson, pp. 17–28, North Holland: Elsevier.

Eaves-Pyles, T., Murthy, K., Liaudet, L., Virag, L., Ross, G., Soriano, F.G., Szabo, C., and Salzman, A.L. (2001). Flagellin, a novel mediator of Salmonella-induced epithelial activation and systemic inflammation: I kappa B alpha degradation, induction of nitric oxide synthase, induction of proinflammatory mediators, and cardiovascular dysfunction. *Journal of Immunology* **166**, 1248–1260.

Fletcher, J., Reddi, K., Poole, S., Nair, S., Henderson, B., Tabona, P., and Wilson, M. (1997). Interactions between periodontopathogenic bacteria and cytokines. *Journal of Periodontology* **32**, 200–205.

Fletcher, J., Nair, S., Poole, S., Henderson, B., and Wilson, M. (1998). Cytokine degradation by biofilms of *Porphyromonas gingivalis*. *Current Microbiology* **36**, 216–219.

Gresser, I. (1997). Wherefore interferon? *Journal of Leukocyte Biology* **61**, 567–574.

Henderson, B. (2000). Therapeutic control of cytokines: Lessons from microorganisms. In *Novel Cytokine Inhibitors*, ed G.A. Higgs and B. Henderson, pp. 243–261. Birkhauser Verlag.

Henderson, B., Poole, S., Wilson, M., and Henderson, B. (1996a). Bacterial modulins: A novel class of virulence factor which causes host tissue pathology by inducing cytokine synthesis. *Microbiology Reviews* **60**, 316–341.

Henderson, B., Poole, S., and Wilson, M. (1996b). Bacterial/Host interactions in health and disease: Who controls the cytokine network? *Immunopharmacology* **35**, 1–21.

Henderson, B., Wilson, M., and Wren, B. (1997). Are bacterial exotoxins cytokine network regulators? *Trends in Microbiology* **5**, 454–458.

Henderson, B. and Wilson, M. (1998). Commensal communism in the mouth. *Journal of Dental Research* **77**, 1674–1683.

Henderson, B., Poole, S., and Wilson, M. (1998). *Bacteria-Cytokine Interactions in Health and Disease*. London: Portland Press.

Hooper, L.V., Wong, M.H., Thelin, A., Hansson, L., Falk, P.G., and Gordon, J.I. (2001). Molecular analysis of commensal host-microbial relationships in the intestine. *Science* **291**, 881–884.

Horvat, R.T., Clabaugh, M., Duval-Jobe, C., and Parmely, M.J. (1989). Inactivation of human gamma interferon by *Pseudomonas aeruginosa* proteases: elastase augments the effects of alkaline protease despite the presence of alpha 2-macroglobulin. *Infection and Immunity* **57**, 1668–1674.

Janeway, C.A. (1992). The immune system evolved to discriminate infectious nonself from noninfectious self. *Immunology Today* **13**, 11–16.

Jubier-Maurin, V., Boigegrain, R.A., Cloeckaert, A., Gross, A., Alvarez-Martinez, M.T., Terraza, A., Liautard, J., Koler, S., Ruout, B., Dornand, J., and Liautard, J.P. (2001). Major outer membrane protein Omp25 of *Brucella suis* is involved in inhibition of tumor necrosis factor alpha production during infection of human macrophages. *Infection and Immunity* **69**, 4823–4830.

Kapur, V., Majesky, M.W., Li, L.-L., Black, R.A., and Musser, J.M. (1993). Cleavage of interleukin-1β (IL-1β) precursor to produce active IL-1β by a conserved extracellular cysteine protease from *Streptococcus pyogenes*. *Proceedings of the National Academy of Sciences USA* **90**, 7676–7680.

Klapproth, J.-M., Donnenberg, M.S., Abraham, J.M., Mobley, H.L.T., and James, S.J. (1995). Products of enteropathogenic *Escherichia coli* inhibit lymphocyte activation and lymphokine production. *Infection and Immunity* **63**, 2248–2254.

Kotwal, G.J. (1999). Virokines: mediators of virus-host interactions and future immunomodulators in medicine. *Archivum immunologiae et therapiae experimentalis (Warsaw)* **47**, 135–138.

Kotwal, G.J. (2000). Poxviral mimicry of complement and chemokine system components: what's the end game? *Immunology Today* **21**, 242–248.

Kuhn, R., Lohler, J., Rennick, D., Rajewsky, K., and Muller, W. (1993). Interleukin-10-deficient mice develop chronic enterocolitis. *Cell* **75**, 263–274.

Kurita-Ochiai, T. and Ochiai, K. (1996). Immunosuppressive factor from *Actinobacillus actinomycetemcomitans* down regulates cytokine production. *Infection and Immunity* **64**, 50–54.

Lourbakos, A., Potempa, J., Travis, J., D'Andrea M.R., Andrade-Gordon, P., Santulli, R., Mackie, E.J., and Pike, R.N. (2001). Arginine-specific protease from *Porphyromonas gingivalis* activates protease-activated receptors on human oral epithelial cells and induces interleukin-6 secretion. *Infection and Immunity* **69**, 5121–5130.

Luo, G., Niesel, D.W., Shahan, R.A., Grimm, E.A., and Klimpel, G.R. (1993). Tumor necrosis factor alpha binding to bacteria: evidence for high affinity receptor and alteration of bacterial virulence properties. *Infection and Immunity* **61**, 830–835.

Mantovani, A., Dinarello, C.A., and Ghezzi, P. (2000). *Pharmacology of Cytokines.* Oxford: Oxford University Press.

Matsui, K. (1996). A purified protein from *Salmonella typhimurium* inhibits proliferation of murine splenic anti-CD3 antibody-activated T-lymphocytes. *FEMS Immunology and Medical Microbiology* **14**, 121–127.

Meager, T. (1998). *The Molecular Biology of Cytokines.* Chichester: John Wiley & Sons.

Mikolajczyk-Pawlinska, J., Travis, J., and Potempa, J. (1998). Modulation of interleukin-8 activity by gingipains from *Porphyromonas gingivalis*: implications for pathogenicity of periodontal disease. *FEBS Letters* **440**, 282–286.

Mintz, C.S., Miller, R.D., Gutgsell, N.S., and Malek, T. (1993). *Legionella pneumophila* protease inactivates interleukin-2 and cleaves CD4 on human T cells. *Infection and Immunity* **61**, 3416–3421.

Moore, P.S., Boshoff C., Weiss, R.A., and Chang, Y. (1996). Molecular mimicry of human cytokine and cytokine response pathways by KSHV. *Science* **274**, 1739–1744.

Mullberg, J., Rauch, C.T., Wolfson, M.F., Castner, B., Fitzner, J.N., Otten-Evans, C., Mohler, K.M., Cosman, D., and Black, R.A. (1997). Further evidence for a common mechanism for shedding of cell surface proteins. *FEBS Letters* **401**, 235–238.

Neish, A.S., Gewirtz, A.T., Zeng, H., Young, A.N., Hobert, M.E., Karmali, V., Rao, A.S., and Madara, J.L. (2000). Prokaryotic regulation of epithelial responses by inhibition of I kappa B-alpha ubiquitination. *Science* **289**, 1560–1563.

Nigou, J., Zelle-Reiser, C., Gileron, M., Thurner, M., and Puzo, G. (2001). Mannosylated lipoarabinomannans inhibit IL-12 production by human dendritic cells: evidence for a negative signal delivered through the mannose receptor. *Journal of Immunology* **166**, 7477–7485.

Oppenheim, J.J. and Feldmann, M. (2000). *Cytokine Reference.* London: Academic Press.

Pahlevan, A.A., Wright, D.J.M., Andrews, C., George, K.M., Small, P.L.C., and Foxwell, B.M. (1999). The inhibitory action of *Mycobacterium ulcerans* soluble factor on monocyte/T cell cytokine production and NF-κB function. *Journal of Immunology* **163**, 3928–3935.

Parker, A.E. and Bermudez, L.E. (2000). Sequence and characterization of the glyceraldehyde-3-phosphate dehydrogenase of *Mycobacterium avium*: correlation with an epidermal growth factor binding protein. *Microbial Pathogenesis* **28**, 135–144.

Parmely, M., Gale, A., Clabaugh, M., Horvat, R., and Zhou, W.W. (1990). Proteolytic inactivation of cytokines by *Pseudomonas aeruginosa*. *Infection and Immunity* **58**, 3009–3014.

Porat, R., Clark, B.D., Wolff, S.M., and Dinarello, C.A. (1991). Enhancement of the growth of virulent strains of *Escherichia coli* by interleukin-1. *Science* **254**, 430–432.

Relman, D.A. and Falkow, S. (2001). The meaning and impact of the human genome sequence for microbiology. *Trends in Microbiology* **9**, 206–208.

Sadlack, B., Merz, H., Schorle, H., Schimpl, A., Feller, A.C., and Horak, I. (1993). Ulcerative colitis-like disease in mice with disrupted interleukin-2 gene. *Cell* **75**, 253–261.

Sellon, R.K., Tonkonogy, S., Schultz, M., Dieleman, L.A., Grenther, W., Balish, E., Rennick, D.M., and Sartor, R.B. (1998). Resident enteric bacteria are necessary for development of spontaneous colitis and immune system activation in interleukin-10-deficient mice. *Infection and Immunity* **66**, 5224–5231.

Seymour, R.M. and Henderson, B. (2001). Pro-inflammatory – anti-inflammatory cytokine dynamics mediated by cytokine-receptor dynamics in monocytes. *IMA Journal of Mathematics Applied in Medicine and Biology* **18**, 159–192.

Sharp, L., Poole, S., Reddi, K., Fletcher, J., Nair, S., Wilson, M., Curtis, M., Henderson, B., and Tabona, P. (1998). A lipid A-associated protein of *Porphyromonas gingivalis*, derived from the haemagglutinating domain of the R1 protease gene family, is a potent stimulator of interleukin-6 synthesis. *Microbiology* **144**, 3019–3026.

Tanner, G.W. (1995). *Normal Microflora*. London: Chapman and Hall.

Toki, S., Agatsuma, T., Ochiai, K., Saitoh, Y., Ando, K., Nakanishi, S., Lokker, N.A., Giese, N.A., and Matsuda, Y. (2001). RP-1776, a novel cyclic peptide produced by *streptomyces* spp., inhibits the binding of PDGF to the extracellular domain of its receptor. *Journal of Antibiotic (Tokyo)* **54**, 405–414.

Vollmer, P., Ealev, I., Rose-John, S., and Bhakdi, S. (1996). Novel pathogenic mechanism of microbial metalloproteinases: Liberation of membrane-anchored molecules in biologically active form exemplified by studies with the human interleukin-6 receptor. *Infection and Immunity* **64**, 3646–3651.

Wilson, M., Seymour, R., and Henderson, B. (1998). Bacterial perturbation of cytokine networks. *Infection and Immunity* **66**, 2401–2409.

Yao, Z., Fanslow, W.C., Seldin, M.F., Rosseau, A.M., Painter, S.L., Comeau, M.R. et al. (1995). Herpesvirus Saimiri encodes a new cytokine, IL-17, which binds to a novel cytokine receptor. *Immunity* **3**, 811–821.

Zhang, J., Dong, H., Kashket, S., and Duncan, M.J. (1999). IL-8 degradation by *Porphyromonas gingivalis* proteases. *Microbial Pathogenesis* **26**, 275–280.

CHAPTER 11

Enterotoxins: Adjuvants and immune inhibitors

Jan-Michael A. Klapproth and Michael S. Donnenberg

11.1 INTRODUCTION

Toxins are defined as "soluble substances that alter the normal metabolism of host cells with deleterious effects on the host" (Schlessinger and Schaechter, 1993). Enterotoxins in particular elicit their primary effect in the intestinal tract, initiating a metabolic cascade that results in excessive fluid and electrolyte secretion. The uniform host response is the development of diarrhoea. However, at a cellular and subcellular level, certain enterotoxins induce sophisticated and fascinating metabolic alterations, which can also affect the local immune system in a characteristic fashion. Occasionally, enterotoxins induce disease even outside the gastrointestinal tract, affecting other organ systems. Enterotoxins can be stimulatory and inhibitory at the same time, depending on the encountered cell type. Further, the same toxin can have more than one effect, either inducing or suppressing the immune cascade. Modulation of the immune cascade with either induction or suppression of local and systemic immunocompetent cell populations is initiated at the level of the mucosal immune systems of the lungs, urogenital tract, cornea, and gut. Gut-associated lymphoid tissue (GALT) is a mixture of immunocompetent cells in the intestinal lining, constantly exposed to foreign antigens and tightly regulated to prevent continuous activation. However, even the massive, continuous exposure to foreign antigens in the intestinal lumen does not, under normal circumstances, lead to a measurable immune response. This lack of response to antigens is known as oral tolerance.

How exactly do antigens and enterotoxins gain access to the host immune system and what are the tolerance-defining cell populations in the GALT? There is evidence for the presence of at least two distinct pathways of antigen processing. The more important pathway is through M cells, specialized epithelial cells (Owen, 1999), which continuously sample the intestinal lumen and present antigens to immunocompetent cells in Peyer's patches, a loose aggregation of lymphocytes in various stages of differentiation (McGhee et al., 1999). The second pathway involves epithelial paracellular transport of antigens to the lamina propria, processing by antigen-presenting cells, and presentation to T cells. T lymphocytes and antigen-presenting cells account for more than 60% of all cells in the lamina propria, with the T-cell population being predominantly helper–inducer CD4+/TCR$\alpha\beta$ lymphocytes (James and Kiyono, 1999). Beside macrophages and T cells, the lamina propria also contains a large number of plasma cells producing IgA (McGhee et al., 1999). The various populations of immunocompetent cells in GALT require multiple, simultaneous signals to overcome the tightly negatively regulated state of activation.

Tolerance in the tightly controlled GALT can be broken with the use of adjuvants. Adjuvants are molecules that invoke an immune response to a bystander antigen if applied at the same time, resulting in significantly elevated specific antibody and cytokine response to both adjuvant and antigen. Adjuvants are now utilized in the development of vaccines to induce antigen- or pathogen-specific and protective immune responses.

Enteric pathogens and their products modify the immune response and alter the local environment to gain an advantage that usually allows proliferation and persistence in the host, either in the intestinal lumen or in distal sites. This alteration targets key regulatory elements of defense mechanisms, as enteric pathogens not only fend off the systemic and local host immune response, but also other bacterial species sharing the same environment. Microorganisms have adapted and evolved an array of mechanisms that allow massive down-regulation of the local and systemic immune response, producing factors that have antibiotic activities. Immunosuppression and immunoinduction by enteric pathogens and their products is an emerging field of investigation, allowing further insight into immune pathways. Even though enterotoxins are usually detrimental to the host as a whole, they have aided microbiologists, immunologists, and clinicians to delineate immune pathways for targeted therapeutic modulation of the immune cascade. In addition to playing an important role in the pathogenesis of infections, enterotoxins have proven to be valuable tools for identifying critical regulatory immunological pathways and may be used therapeutically, for example, in the treatment of autoimmune disorders. The biology of enterotoxins that are also superantigens is described in Chapter 8.

The following is a summary of the metabolic alterations that selected enterotoxins have on the immune system, including the regulation of proliferation, cytokine and cell surface marker expression *in vitro*, the effects on individual cell populations, and the complex interactions that occur *in vivo*. A table at the end of the chapter summarizes these effects.

11.2 CHOLERA TOXIN

11.2.1 Introduction

The disease cholera has afflicted the human race in a series of pandemics and in endemic forms. *Vibrio cholerae* produces cholera toxin (CT), which can cause a severe, cramping form of voluminous and life-threatening diarrhoea (Gorbach et al., 1971; Rowe et al., 1970). Interestingly, mucosal inflammation is absent in cholera. CT is a multimeric protein produced by *V. cholerae* serotypes O1 and O139, consisting of a single A subunit (CT-A) and a pentamer of B subunits (CT-B), with masses of 28.8 kDa (for the A subunit) and 55 kDa (for the B pentamer) (Bastiaens et al., 1996). CT is similar in structure and function to heat-labile toxin (LT) from enterotoxigenic *E. coli* (ETEC) with about 80% sequence homology (Dallas and Falkow, 1980; Zhang et al., 1995). The B subunit specifically binds to the monosialoganglioside GM_{1b} present on almost all eukaryotic cell types (Kuziemko et al., 1996), mediating uptake of the holotoxin (Fukuta et al., 1988; Holmgren et al., 1982). Both CT subunits can be detected intracellularly and, after dissociation, are found in different cellular compartments (Bastiaens et al., 1996). The holotoxin is transported from the plasma membrane into the Golgi compartments, where A and B subunits are separated. The B subunits remain in the Golgi compartment, whereas the A subunit is redirected to the plasma membrane by retrograde transport through the endoplasmic reticulum. During this microtubule-dependent transport, CT-A is proteolytically cleaved between residues 192 and 194, followed by reduction of a single disulfide between cysteine 187 and 199 to generate free A_1 and A_2 peptides (Orlandi and Fishman, 1993; Majoul et al., 1996). A_2 is responsible for binding to the B subunit, whereas A_1 carries the enzymatic activity. A_1 is an ADP-ribosyltransferase that uses nicotinamide adenine dinucleotide to catalyze the ADP-ribosylation of the α chain of the heterotrimeric G protein $G_{s\alpha}$ (Gill and King, 1975; Vedia et al., 1988). The subsequent dissociation of the α chain from the $G_{s\beta\gamma}$ dimer leads to a persistent activation of adenylate cyclase, resulting in an increase in cytoplasmic cAMP. The elevated intracellular concentrations of cAMP in turn cause activation of protein kinase A

(PKA), which phosphorylates and opens the cystic fibrosis transmembrane channel, leading to efflux of Cl⁻ ions (Field and Chang, 1989). In addition to the effect on water and ion homoeostasis, CT and its subunits elicit a profound effect on a range of immune and nonimmune cells, inducing a local and systemic immune response (Spangler, 1992).

11.2.2 Effect of CT on antigen-presenting cells

CT has a direct effect on macrophages. Early studies examining the effect of CT on the immune system determined that the toxin increases IL-1α expression and promotes T-cell proliferation (Bromander et al., 1991). It appears that both CT and CT-B have a negative effect on macrophage antigen processing and MHC-II-restricted presentation. Pretreatment of macrophages with CT inhibits the presentation of soluble and bacteria-expressed antigens to T cells (Matousek et al., 1996). Delaying macrophage exposure to CT, but not CT-B, after priming with antigens, results in an enhanced antigen presentation for both types of antigens, suggesting this effect to be dependant on the A subunit of CT. These findings concur with results that indicate LT-B and CT-B alone enhance expression of preexisting peptide-MHC-II complexes on macrophages, whereas the holotoxins inhibit intracellular antigen processing (Matousek et al., 1998). See Chapter 2 for a discussion of antigen presentation by CD1.

CT promotes maturation of monocytes to macrophages, as determined by an increase in HLA-DR expression on the cell surface (Gagliardi et al., 2000). Further, CT has an effect on the expression of CD86 (B7.2), an important co-stimulatory molecule present on the cell surface of activated macrophages. CD86 binds to CD28 on T cells, allowing lymphocyte proliferation. CT, in conjunction with IFN-γ, selectively induces expression of the co-stimulatory molecule CD86, mediated by an increase in intracellular cAMP, but does not affect closely related molecules, such as CD80 or ICAM-1 (Cong et al., 1997). Induction of CD86 leads to increased functional co-stimulatory activity for anti-CD3- and antigen-stimulated T cells, as well as an induction of IgG1, consistent with the development of the Th2-dominant immune response. Reinforcing the development of a Th2 immune response is the suppressive effect of CT on antigen-, antibody-, or cytokine-stimulated macrophages and expression of IL-12 and its receptor, as well as TNFα, and chemokines such as CCR1, CCR5, RANTES, MIP-1α, and MIP-1β. Concomitantly, CT leads to the induction of CXCR4, CCR7, IL-6, and IL-10, the latter being a lymphokine critical for suppression of a Th1 response (Burkart et al., 1999; Braun et al.,

1999). However, addition of exogenous IFNγ, a Th1 defining lymphokine, can reverse the suppressive effect of CT-B, restoring normal reactivity of macrophages to LPS, as determined by an increase in TNFα (Gagliardi et al., 2000). Other monokines such as IL-8 or TGF-β_1 are not affected by either holotoxin or CT-B alone.

In addition to a direct effect on macrophages, CT modifies differentiation and antibody expression by B cells. CT is a strong inhibitor of IgM production and increases the number and frequency of IgG$_1$- and IgA-producing cells by acting predominantly on membrane-IgM+/IgG−/IgA− cells, leading to isotype switching (Lycke and Strober, 1989). CT or CT-B increase the number of IgA− (Kim et al., 1998), and in conjunction with IL-2, IL-4, and TGFβ_1, the formation of IgG$_1$ plasma cells (Lycke et al., 1990; Lycke, 1992). Ironically, IL-2, IL-4, TGFβ_1 by themselves can be responsible for an increase in intracellular concentrations of cAMP, stabilizing germ line γ1-RNA transcripts, and allowing an increase in IgG$_1$.

Beside macrophages and B cells, epithelial cells are able to function as antigen-presenting cells. Interestingly, CT strongly enhances antigen presentation by intestinal epithelial cells (Bromander et al., 1993). Epithelial cells exposed to CT triggered allogen-specific T-cell proliferation through an increase in CD1a antigen expression when compared to cells treated with IFN-γ (see Chapter 2 for a discussion of CD1). Enhancing the process of antigen presentation by epithelial cells, CT promotes expression of IL-1 and IL-6 by epithelial cells.

In summary, CT promotes monocyte maturation to macrophages and on one side inhibits presentation of concomitantly applied antigens, while enhancing presentation of preexisting peptide-MHC-II complexes on the other. Further, CT suppresses expression of both IL-12 subunits, p35 and p40, TNFα, and CCR1, initiating a Th2-dominant immune response, characterized by the inhibition of IgM, and inducing production of IgG$_1$ and IgA.

11.2.3 Effect of CT on T lymphocytes

The effect of CT is not limited to antigen-presenting cells, as CT elicits a direct and profound effect on T cells. IL-2, a central mediator in the activation of T cells, functions as an autocrine growth factor. CT suppresses expression of IL-2 in TCR-activated lymphocytes, without influencing the expression of the high affinity IL-2 receptor (Szamel et al., 1997). Similar to lymphostatin (see below), CT suppresses IL-2 expression by inhibiting the TCR-mediated increase in both peak and sustained cytoplasmic calcium release, blocking the generation of inositol triphosphate (Imboden et al., 1986), presumably through an effect on G-binding proteins (Anderson and Tsoukas, 1989). Suppression of cytoplasmic calcium release is followed by a down-regulation of CD3 cell surface expression, without affecting CD2 (Nunes et al., 1989). The decrease in cytoplasmic calcium is also responsible, at least in part, for the selective inhibition of protein kinase C-α, without affecting protein kinase C-β (Eriksson et al., 2000). Decreased cytoplasmic calcium concentrations and selective inhibition of protein kinase C-α results in a decreased proliferative response in TCR-stimulated CD4+ and CD8+ naïve, as well as memory T cells. The anti-proliferative response, and IL-2 suppression, is specific for Th1 cells, as CT does not interfere with TCR- or PMA/ionomycin-mediated Th2 cell proliferation or expression of IL-4 (Munoz et al., 1990). It appears that the anti-proliferative response is mediated by blocking *c-fos* expression, a transcription factor critical for the formation of AP-1 heterodimer, required for IL-2 expression. In the same study, CT increased cAMP in both Th1 and Th2 cells. However, only Th1 cells failed to proliferate in response to mitogens, indicating that Th2 cells might not use PKC as a major signal transmission pathway after TCR-mediated stimulation, as implicated above. The negative regulatory effect of CT is not absolute and can be overcome. Addition of exogenous IL-2 does not reverse the suppressive effect of CT or CT-B (Woogen et al., 1987), but binding of anti-CD28 restores the proliferative response in naïve, but not activated memory T cells and naïve memory T-cell populations respond with an increased secretion of IL-2. However, only memory CD45RO+ T cells produce significant amounts of IL-4 and IFNγ (Gagliardi et al., 2000). This effect on T cells is dependent on an intact holotoxin, as the B subunit alone has no effect on T-cell proliferation. These studies indicate a separate effect of CT on signaling pathways for lymphokine production and proliferation. Further, these results allow the assumption that CT interacts with CD45-specific, but does not inhibit CD28-specific pathways. In contrast to an earlier study, in this particular study, CT also inhibited expression of the IL-2 receptor (CD25) (Szamel et al., 1997).

CT selectively targets lymphokine expression in stimulated lymphocytes, but also has an effect on T-cell survival. CT induces apoptosis specifically in CD4+ cells after TCR activation (Yamamoto et al., 1999). This process is mediated by an increase in cAMP. However, it is less clear how T-cell subpopulations respond differently to the same mediator.

Thus, CT is a potent immunomodulator, inhibiting intracellular calcium release and *c-fos* expression, leading to suppression of IL-2 release and T-cell proliferation. Parallel to a decrease in IL-2, CT induces IL-4 expression in

CD45RO+ memory cells, enhancing the Th2-immune pathway. Further, CT also prevents expression of the IL-2 receptor and induces apoptosis in CD4+ lymphocytes.

11.2.4 Effect of CT *in vivo*

With the possible exception of LT, CT is the most important and best characterized adjuvant known. Oral feeding of foreign protein antigens alone fails to induce a protective mucosal IgA or serum IgG antibody response, unless accompanied by an adjuvant, a phenomenon known as oral tolerance (Jackson et al., 1993). Conjugation of CT with a protein antigen induces significantly higher level of IgA and IgG1 when compared to a mixture of unconjugated CT plus antigen or either molecule given alone (McKenzie and Halsey, 1984). Nontoxic mutant forms of CT, devoid of ADP-ribosyltransferase activity and the ability to induce fluid accumulation, elicit an immune response with high serum titers of antigen-specific IgG and IgA antibodies, when given orally or parenterally (Yamamoto et al., 1997).

CT abrogates oral tolerance, leading to the induction of secretory IgA, and serum IgG1, IgG2b, and IgE antibody responses after oral or intranasal administration (Tamura et al., 1994), but only when administered at the same time as the antigen (Elson and Ealding, 1984a and b). Helper T-cell activity for IgA and IgG synthesis is present in all lymphoid tissues examined: spleen, mesenteric lymph nodes, and Peyer's patches. Plasma cells secreting anti-CT IgG are markedly elevated in cultures from spleen, whereas anti-CT IgA plasma cells are identified predominantly in cultures from intestinal Peyer's patches.

Several studies have addressed the question of how CT abrogates oral tolerance. Oral immunization with CT does not affect the overall composition of gut-associated lymphoid tissue, but causes a marked depletion of intraepithelial lymphocytes, mainly CD8+ suppressor cells located in the dome epithelium of Peyer's patches (Elson et al., 1995). In the same study, CD8+ cells bound more CT *in vitro* in comparison to CD4+ cells, resulting in a preferential reduction during culture with a polyclonal activator. Unlike other adjuvants, CT induces a significant local IgA antibody response that is independent of IL-12, consistent with a Th2-dominant mucosal immune response (Grdic et al., 1999), but dependant on IL-4, as IL-4-/- mice failed to respond to CT given with a soluble protein antigen, which is different in comparison to LT (see below) (Vajdy et al., 1995). Systemic and local IgA immune responses after oral immunization with CT are heavily dependent on the presence of CD4+ cells in spleen, mesenteric lymph nodes, and Peyer's patches (Hornqvist et al., 1991), as challenge of CD4+-depleted animals with CT blocked the ability to mount a protective gut immune response (Benedetti et al., 1998). In addition to an antibody switch from IgM to IgG1 and IgA, CT significantly increases serum IgE in response to immunization with tetanus toxoid, ovalbumin, or hen egg lysozyme (Marinaro et al., 1995). Further, it appears that additional genetic factors play a significant role in the induction of a protective immune response. The degree of immune response to CT, as determined by IgA expression, is determined by the genes located in the *H-2* haplotype of the MHC complex, especially the *I-A* region (Elson and Ealding, 1985, 1987). Congenic animal strains of the H-2^b and H-2^q haplotype are identified as high responders, whereas H-2^s, H-2^k, and H-2^d expressing strains are low responders.

Analogous to findings *in vitro*, the immune response to oral CT plus antigen is dominated by a Th2 cytokine profile *in vivo* (Marinaro et al., 1995; Xu-Amano et al., 1993). Specifically, the analysis of cytokine production from CD4+ lymphocytes isolated from Peyer's patches and spleen revealed high numbers of IL-4, IL-5, and IL-6 producing CD4+ T cells. IFNγ and IL-2 concentrations did not change in comparison to negative controls. Further, animals exhibited decreased expression of IL-12 p40 after intraperitoneal injection of CT and subsequent LPS challenge.

CT also induces a cytotoxic T cell response (Bowen et al., 1994). It facilitates antigen uptake into target cells and presentation to allow sensitization and priming for antigen-specific class-I restricted CD8+-mediated lysis. This response is antigen specific in the spleen of animals receiving either CT or CT-B.

Oral immunization, priming with CT and a protein antigen, induces long-lasting immunological memory in the intestinal lamina propria (Vajdy and Lycke, 1992), extraintestinal sites, such as Peyer's patches, mesenteric lymph nodes, and spleen (Lycke and Holmgren, 1986), as well as in peripheral blood (Lycke et al., 1987). A single dose of priming with CT and antigen is sufficient to induce long-term memory in the intestinal mucosa. There is evidence that the antigen-specific memory response lasts up to five years after the initial oral vaccination, if not for the lifetime of test animals (Jertborn et al., 1988). The memory response to the immunized antigen is dominated by IgA and IgM isotypes and does not require simultaneous application of CT (Lycke et al., 1987).

Early studies investigating the route of CT and antigen administration revealed interesting results. Application of high-dose CT intravenously results in involution of lymphoid organs, especially the spleen and thymus. Within days of toxin administration, the spleen and thymus decrease in weight, with only the spleen recovering its weight after seven days (Holmgren et al., 1974). The spleen and thymus exhibit a marked reduction in cellularity, affecting stromal elements in the thymus and red pulp of the spleen (Chisari and Northrup, 1978). Interestingly, intravenous immunization with CT induced predominantly IgM and IgG in the spleen, without an effect on antibody production in the gastrointestinal immune system (Xu-Amano et al., 1994). In contrast to intravenous immunization, oral or intramuscular vaccination with CT induced a secretory IgA response in the intestine, as determined by lavage (Svennerholm et al., 1982). Quantification of the immune response was comparable for both intravenous or oral routes of immunization, but the response persisted longer following oral vaccination. CT-specific IgG-secreting peripheral blood lymphocytes appeared within four days after intravenous injection and persisted for about two weeks, after which they became undetectable (Lycke et al., 1985). Further, intranasal application of recombinant CT-B induced a specific IgA and IgG response locally and in vaginal secretions and induced protective systemic immunity (Bergquist et al., 1997). Transcutaneous immunization with unmodified CT leads to induction of a specific and protective serum and mucosal IgG and IgA immune response that is comparable to other routes of immunization (Glenn et al., 1998a). Application of CT to the skin induced a mixed Th1 and Th2 phenotype (Hammond et al., 2001), reflected by induction of a broader IgG response with increased concentrations of IgG1, IgG2a, IgG2b, and IgG3, usually observed with LT (Glenn et al., 1998b).

In summary, the Th2-dominant immune response observed after the simultaneous application of CT plus antigen *in vivo* can be explained with results obtained from *in vitro* experiments. Initiation of the immune response appears to occur at the level of monocytes, specifically stimulating CD4+ memory T lymphocytes in intestinal and extra-intestinal lymphoid tissues. Stimulation of CD4+ cells leads to an increased expression of IL-4. IL-4 is a critical factor in CT-induced switching of antibody production from IgM to IgA and IgG1. Even though CT suppresses expression of IL-12 in macrophages, IL-12 is not critical for the generation of a mucosal IgA response.

11.2.5 Clinical application of CT

CT and its subunits have been utilized in animal models for a variety of infectious and immunological disorders. Immunization with mutant CT (S61F, E112K), lacking ADP-ribosyltransferase activity, and pneumococcal surface protein A, induces significant protection against subsequent challenge with live *Streptococcus pneumoniae* (Yamamoto et al., 1998). In contrast to LT-B, intranasal application of CT-B plus herpes simplex virus type 1 (HSV-1) glycoprotein does not induce protective immunity against subsequent challenge with live HSV-1 (Richards et al., 2001). Interestingly, LT-B and CT-B induced a Th2-biased immune response with increased secretion of IgG1, IL-4, IL-10, and IFNγ, as well as T-cell proliferation. CT has been tested in animals with abnormal immune regulation. The nonobese diabetic mouse model of insulin-dependent diabetes mellitus is considered a T-cell-mediated process, characterized by an imbalance of immunoregulatory and anti-islet effector cells. In this model, the administration of CT-B prevented the onset of diabetes; only 18% of CT-B-treated mice developed diabetes in comparison to 75% of saline-treated animals (Sobel et al., 1998). Further, CT-B as a fusion protein with myelin basic protein has been utilized in the prevention of autoimmune encephalitis in animals (Sun et al., 1996). In this model, expression of IL-2 was reduced, IFNγ increased, and leukocyte infiltration of the spinal cord markedly decreased in comparison to animals treated with repeated feedings of large doses of myelin basic protein alone. In this animal model of autoimmune disease, CT-B effectively suppresses delayed-type hypersensitivity reactions in systemically immune animals. So far, CT or its subunits have only been tested in animals, but not in human disease.

Thus far, concerns about the toxicity of CT have precluded the use of the holotoxin or its derivatives in humans. CT induces diarrhoea and leads to the development of a hypersensitivity reaction with high serum concentrations of histamine (Snider et al., 1994). An additional concern is that CT results in a suppressive effect on FcγR expression on polymorphonuclear leukocytes, a critical component for phagocytosis of bacterial components, resulting in decreased migration and intracellular killing of staphylococci (Niemialtowski et al., 1993). Therefore, attempts at maintaining adjuvant characteristics, but deleting motifs accounting for toxicity, are under way.

11.2.6 Summary

CT and its derivatives break oral tolerance to allow induction of an antigen- and CT-specific immune response. The immune response is Th2-dominant, characterized by an increase in IL-4, IL-5, and IL-10 with simultaneous decrease in cytokines defining the Th1 response, namely IL-2 and IL-12. Increased expression of Th2 cytokines allows for an antibody switch from IgM to protective and specific IgG1 and IgA expression. The protective immune response is long lasting, with the persistence of antigen-specific memory cells years after the initial immunization. If concerns regarding potential toxicity can be overcome, CT may prove promising as an adjuvant that helps to induce protection against pathogens and may be used for the treatment of immunological disorders.

11.3 HEAT-LABILE TOXIN

11.3.1 Introduction

Heat-labile toxins (LT) are a group of toxins produced by various strains of ETEC (Kunkel and Robertson, 1979), an important agent of childhood diarrhoea in developing countries and diarrhoea in travelers who visit these areas (Rowe et al., 1970; Sack et al., 1971). LT from ETEC induces a systemic antibody response after the acute infection (Cushing and Smart, 1985). LT can be divided into two groups: LT-I and LT-II. LT-I proteins are encoded by structural genes located on plasmids and can be neutralized by an antibody against CT. LT-II proteins are encoded by structural genes located on the chromosome and cannot be neutralized by antibodies against CT. CT and LT share about 80% amino acid sequence homology and nearly identical structures and mechanisms of action (Dallas and Falkow, 1980). Analogous to the structure of CT, LT consists of a single A subunit noncovalently bound to a homopentamer of B subunits (Zhang et al., 1995; Holmes et al., 1986; Sixma et al., 1991). Like CT-B, the B subunit of LT is responsible for binding the GM_{1b} ganglioside, a glycosphingolipid present in eukaryotic cell membranes. However, CT only binds to GM_{1b} ganglioside, whereas LT also binds, with lower affinity, to asialo-GM1, GM2, and polyglycosylceramides (Fukuta et al., 1988; Holmgren et al., 1982). Similar to CT, LT A and B subunits both account for significant toxicity, immunogenicity, and oral adjuvanticity (Guidry et al., 1997; Nashar et al., 1996), but it is the A subunit that possesses the ADP-ribosyltransferase enzymatic activity essential for the induction of an LT-A-specific mucosal and serum antibody response (Spangler, 1992; DeHaan et al., 1999). In comparison to CT, little is known about the intracellular fate of LT after binding to, and internalization by, eukaryotic cells (Bastiaens et al., 1996).

Like CT, LT is a potent adjuvant for the induction of antigen-specific IgA responses *in vivo* (Klipstein et al., 1982). Induction of a protective immune response with tetanus toxoid as an antigen and LT as an adjuvant is dependant on increased intracellular concentrations of cAMP (Cheng et al., 1999). In comparison to CT, the increase in cAMP induced by LT is slower in onset, lower in magnitude, and slower decaying, suggesting a lower ADP-ribosyltransferase activity of LT (Matousek et al., 1998). Site-directed mutagenesis of the A subunit, replacing arginine at position 7 with lysine, renders LT deficient in ADP-ribosylating activity, but still allows cell-specific binding and the generation of a specific immune response against ovalbumin or tetanus toxoid (Douce et al., 1995). LT-B can act as a mucosal adjuvant after intranasal administration in comparison to CT-B, which is unable to act as an adjuvant through the same route. Oral immunization with LT-B leads to a rapid up-regulation of IL-13, which is essential for up-regulation of intestinal IgA and serum IgA and IgG production (Bost et al., 1996). Treatment of LT-B immunized animals with antibodies against IL-13 leads to a marked reduction of LT-B-specific antibody production in the intestinal mucosa and serum compartments, probably mediated via a decrease in IL-4 expression.

11.3.2 Effect of LT on antigen-presenting cells

CT and LT enhance CD86 (B7.2) expression on B cells and macrophages, a co-stimulatory molecule for T-cell activation (Yamamoto et al., 2000). CD86 is critical for induction of a proliferative response of CD4+ T cells in response to LT or CT, which is abrogated in the presence of anti-CD86 (Cong et al., 1997). This effect is independent of the interaction of CD40 with its ligand CD40L *in vitro*. However, oral immunization of CD40L-/- mice failed to induce antigen-specific immune response, implicating CD40L as a critical component for the generation of antigen-specific antibodies (Renshaw et al., 1994). LT increases expression of CD86 on macrophages, but LT-B alone induces additional markers of activation on B cells that are essential for differentiation, including increased expression of MHC-II molecules, ICAM-1, CD40, and CD25 on the cell surface, without increasing antigen-presenting cell proliferation (Nashar et al., 1997). However, if examined more carefully, it appears that both LT-B and CT-B enhance expression of preexisting peptide-MHC-II complexes on macrophages, whereas both holotoxins inhibit intracellular antigen processing, although the extent of inhibition is less with LT (Matousek et al., 1998). The inhibitory effect does not affect previously processed antigens or exogenous preprocessed synthetic peptides. This indicates that CT- and LT-mediated inhibition of antigen processing in macrophages is dependent on the ADP-ribosyltransferase activity.

11.3.3 Effect of LT on T cells

In a similar manner to CT, LT induces proliferation of anti-CD3-stimulated CD4+ cells, but unlike CT, LT enhances expression of both Th1- and Th2-characteristic lymphokines and antibody profile (Yamamoto et al., 2000). CT selectively inhibits activation of the Th1 pathway, whereas LT maintains the Th1 response through direct suppression of IL-4 gene expression in CD4+ cells. LT selectively depletes CD8+ lymphocytes from lymph node tissue through induction of apoptosis, tipping the immunological balance toward activation of CD4+ lymphocytes (Nashar et al., 1996; Nahar et al., 1996).

11.3.4 Effect of LT *in vivo*

CT and LT act as mucosal adjuvants. As opposed to CT, which induces a Th2-dominant immune response, LT elicits a mixed Th1 and Th2 response with increased expression of IFNγ, IL-4, and IL-5, as well as IgG1, IgG2a, IgG2b, IgM, and fecal IgA (Yamamoto et al., 2000; Takahashi et al., 1996). An explanation for this effect is that CT directly inhibits Th1 lymphokine expression, whereas LT selectively suppresses IL-4 expression in CD4+ lymphocytes. Remarkably, in IL-4-/- animals immunized with LT or CT, neither IgG1 or IgE are detectable, suggesting IL-4 to be a critical mediator of antigen-specific serum antibody production. However, LT induces significant IgA production in IL-4-/- animals without affecting IL-5, IL-6, or IL-10 expression, implicating these cytokines as being of critical importance for mucosal IgA immunity. Comparing CT and LT side-by-side in inducing a protective systemic immune response, both appeared similar in their ability to generate enhanced IgG-specific serum antibody concentrations against the antigen fimbrillin from *Porphyromas gingivalis* (Connell et al., 1998).

Because of the toxicity of the holotoxins in humans, B subunits of CT and LT were investigated alone for the ability to generate a protective immune response. Even though CT-B and LT-B are close homologues, their ability to function as a mucosal adjuvant to inducing high-titer antibody responses against various antigens is remarkably different. LT-B and CT-B, after intranasal application or directly onto bare skin, differ in eliciting an immune response both systemically and at distant mucosal sites (Millar et al., 2001; Beignon et al., 2001). Significant differences exist for serum IgG1, faecal IgA, and nasal IgG1; after intranasal immunization with CT and LT, the latter being clearly the more potent adjuvant. An explanation for these findings is the fact that LT binds to a wider range of cell surface molecules present on immune and nonimmune cells for induction of a specific immune response. Further, these studies suggest that application of ADP-ribosylating exotoxins to bare skin is safe and well tolerated in mice and should be further examined as a potentially safe and efficient alternative route for immunization, analogous to experiments performed with CT (Glenn et al., 1999).

In addition to the use of only the B subunits of either CT or LT as mucosal adjuvants, these toxins have been genetically altered to render them devoid or markedly reduced in ADP-ribosylating enzymatic activity. LTK63, a LT derivative without any ADP-ribosyltransferase activity and LTR72, retaining 1% enzymatic activity, have been extensively studied *in vivo* and *in vitro*. The absence of enzymatic activity in LTR63 allows for induction of a Th1-Th2 mixed immune response with elevated concentrations of IL-12, IFNγ, and TNFα after intranasal application (Ryan et al., 1999; Ryan et al., 2000). The minimally active mutant form of LT, LTR72, induces a Th2-dominant response, characterized by significantly enhanced production of IL-4 and IL-5, but inhibition of LPS-induced IL-12 expression. It is interesting to note that when LTR72 is used as an adjuvant instead of the native toxin, test animals have delayed clearing of *Bordetella pertussis* from their lungs, indicating the Th1 immune response of critical importance in clearing the infection *in vivo*. Induction of a mixed T-cell response appears to be related to the increase in cAMP and resulting suppression of TNFα and IL-12 expression in macrophages. When administered with ovalbumin, LTK63 leads to induction of a long-lasting, protective local and systemic immune response when given by the intravaginal route (DiTommaso et al., 1996) and enhances proliferative and cytotoxic T-cell response to synthetic peptides (Partidos et al., 1999). LT-G192 is another interesting mutant of LT (Douce et al., 1999). This mutant increases stability of the holotoxin structure, as cleavage at position 192 in the A subunit is essential for LT toxicity and amino acid substitution in this location alters proteolytic susceptibility. However, this mutant retains some of its toxic activity when tested in Y1 adrenal cell assays or rabbit ileal loop experiments. When tested *in vivo*, LT-G192, like wild type LT, induces a significantly higher KLH-specific antibody response when compared to LTK63 (Douce et al., 1999). In addition, T-cell proliferation, IL-5, and IL-12 production were markedly up-regulated, suggesting a mixed Th1/Th2 immune response. LT-G192 in conjunction with killed *Salmonella* or *Shigella* O-antigen conveys protection against subsequent oral challenge with bacteria (Chong et al., 1998; Hartman et al., 1999). Test animals immunized with LT-G192 had higher IFNγ, IL-2, and IgG responses in comparison to animals immunized with *Salmonella* alone.

11.3.5 Clinical application of LT

In comparison to CT-B, LT-B can function as a mucosal adjuvant alone (de Haan et al., 1998). This difference appears to be due to the fact that LT-B exhibited a higher degree of stability and maintains its quarternary structure in a range of conditions, resulting in less degradation than expected for other, similar proteins (Ruddock et al., 1995). In addition to the classical antigens, like β-galactosidase, hen egg lysozyme, or tetanus toxoid, more clinically relevant antigens have been tried *in vivo* in animal and human subjects. Immunization of mice with LT-B and HSV-1 glycoprotein, followed by challenge with live HSV-1 virus, resulted in markedly reduced lid disease with zosteriform spread (Richards et al., 2001; Hazama et al., 1993). Another example is the combination of LT with respiratory syncytial virus M2 matrix protein or oval-bumin peptide and induction of a specific, IFN-γ-independent CD8+ T cell response (Simmons et al., 1999, 2001). Further, significant concentrations of salivary IgA against influenza virus hemagglutinin and neuraminidase were observed using LT as an adjuvant (Gluck et al., 1999; Verweij et al., 1998; Hashigucci et al., 1996; Katz et al., 1997). Similar observations have been made for LT mutants LTR63 and LTR72, both of which induced a systemic CTL response against HIV-1 gag-p55 protein after intranasal immunization (Neidleman et al., 2000). LT has also found application in an animal model for autoimmune disease. Simultaneous immunization of mice with collagen II and LT-B reduces the proportion of test animals that developed arthritis by 80% (Williams et al., 1997). LT-B immunized animals show markedly reduced leukocyte infiltration in the joints, synovial hyperplasia, and cartilage degeneration. Moreover, LTR72 induced a systemic CTL response that was not observed utilizing LTR63 as an adjuvant.

11.3.6 Summary

LT is produced by ETEC, the toxin proteins being similar in structure and function to CT. However, important differences exist. Unlike CT, LT induces an immune response characterized by increase of both Th1 and Th2 lymphokine expression, especially IFNγ and IL-5. This broader profile of lymphokine induction by LT is reflected in the qualitative antibody response with up-regulation of IgG1, IgG2a, IgG2b, IgM, and IgA. In contrast to CT, LT-mediated IgA induction is independent of IL-4. Various mutant forms of LT are under investigation and results consistently ascribe the induction of a Th1 and Th2 response to the presence of ADP-ribosyltransferase activity, whereas mutation of this critical motif allows development of a Th2-dominant immune response. Depending on the applied antigen, LT and its mutant forms induce a specific immune response allowing rapid clearance of the offending agent.

11.4 SHIGA TOXINS

11.4.1 Introduction

Enterohemorrhagic *E. coli* (EHEC) belong to the class of Shiga toxin-producing *E. coli*. Shiga toxin-producing *E. coli*, particularly the most important serotype, O157:H7, cause bloody diarrhea in humans (Karmali, 1989). Furthermore, infections with these bacteria can be complicated by acute renal failure, thrombocytopenia, and microangiopathic hemolytic anemia, a clinical entity known as hemolytic uremic syndrome (HUS; Karmali et al., 1985). HUS is thought to result from endothelial cell death leading to a thrombotic microangiopathy and tissue destruction from ischemia.

Shiga toxins consist of toxin type I (Stx-1) and toxin II (Stx-2), encoded lysogenic bacteriophages related to lambda (Smith et al., 1983). Stx-1 is virtually identical to Shiga toxin produced by *Shigella dysenteriae* type I, whereas Stx-2 is only 56% identical to Shiga toxin and Stx-1 on the deduced amino acid sequence level (Jackson et al., 1987). Analogous to CT and LT, Stx-1 and Stx-2 share an AB_5 holotoxin molecular structure, consisting of a single enzymatically active A subunit and a noncovalently associated pentamer of receptor-binding B subunits. Stx-1 and Stx-2 bind to the glycolipids globotriaosylceramide (Gb_3) or globotetraosylceramide (Gb_4) (O'Brien et al., 1992). Biochemically, Stx-1 and Stx-2 target ongoing protein synthesis and are highly specific N-glycosidases that depurinate a single adenosine residue in the 28S rRNA subunit of eukaryotic ribosomes, rendering them inactive (Ogasawara et al., 1988).

11.4.2 Effect of Stx on antigen-presenting cells

Stx-1 and Stx-2 can induce apoptosis in immunocompetent cells that have high concentrations of Gb_3, including undifferentiated monocytes and B lymphocytes. Undifferentiated monocytes in particular have decreased viability and are unable to produce cytokines upon stimulation with Stx-1 (Ramegowada and Tesh, 1996). Macrophages, however, have decreased concentration of Gb_3 in the cell membrane compared to monocytes, and subsequently decreased sensitivity to the lethal effect of Stx. In comparison to macrophages stimulated with LPS alone, macrophages stimulated with LPS and Stx have delayed production of TNFα, IL-1α, IL-1β, and IL-6 (Tesh et al., 1994). Stx-1 alone increased synthesis of TNFα and IL-1β, which is thought to be a critical component in the development of HUS as both TNFα and IL-1β up-regulate the expression of Gb_3 on endothelial cell, sensitizing target cells to the effect of Stx (van de Kar et al., 1992).

11.4.3 Effect of Stx on mixed mononuclear cell populations

Low concentrations of Stx-1 block the mitogen-induced enhancement of metabolic activity in bovine peripheral blood mononuclear cells (PBMC) *in vitro* (Menge et al., 1999). High concentrations of Stx-1 nearly completely abolish the stimulatory effect of mitogens, including phytohemagglutinin, concavalin A, or pokeweed mitogen after four days of incubation. Even after prolonged incubation with Stx-1, there was no evidence of increased cytotoxicity in bovine PBMCs. In comparison, the same study found that in the presence of LPS or T cell mitogens, Stx-1 leads to marked apoptosis of a B lymphoma cell line (Mangeney et al., 1991). It is thought that Gb$_3$ functions as the receptor for the induction of apoptosis in B cells. Therefore, it appears that Stx-1 and Stx-II induce apoptosis in individual cells and cell lines, but exposure in mixed populations can prevent this effect.

11.4.4 Effect of Stx *in vivo*

Recently, the systemic immune response to Stx-2 was explored (Sugatani et al., 2000). Intravenous injection of Stx-2 into mice lead to a decrease of CD4+ T cells in peripheral blood and spleen within hours with the nadir after 24 hours, while the concentration of CD45+ splenocytes continued to decline until 45 hours. Near the time of death the concentration of CD4+ and CD8+ cells in the thymus decreased even further. In addition, LPS-responsive mice strains, such as C57BL/6 and C3H/HeN, exhibited an increased susceptibility to co-administration with LPS, reducing the lethal dose of Stx-2. Similar observations were reported using gnotobiotic pigs (Christopher-Hennings et al., 1993). Oral inocculation with EHEC continuously decreases absolute lypmphocyte counts in peripheral blood. In addition, mitogen-induced lymphocyte proliferation from infected pigs *in vitro* is significantly impaired in comparison to proliferation observed after infection with Stx-2 negative EHEC strains.

Few studies have investigated immunological changes in humans during the acute infection with EHEC. Pro-inflammatory cytokines are detected in some, but not all blood and urine samples from infected children (Inward et al., 1997). A more consistent result is the detection of IL-1β and IL-8 in the plasma of eight out of nineteen children tested. Even though total neutrophil cell count was elevated, this did not correlate with cytokine concentration determined in serum from infected patients. Unfortunately, the lymphocyte response was not reported in this study.

11.4.5 Summary

Stx-1 and Stx-2 induce apoptosis in immune cells that have high concentrations of Gb$_3$ as well as the expression of TNFα and IL-1β in macrophages.

TNFα and IL-1β further up-regulate Gb$_3$, leading to a local mucosal and systemic state of immunosuppression with depletion of B cells and suppression of T lymphocyte proliferation. The clinical relevance of these changes is unclear.

11.5 *CLOSTRIDIUM DIFFICILE* TOXIN A AND B

11.5.1 Introduction

Clostridium difficile is the most common cause of antibiotic-associated diarrhea and pseudomembranous colitis in humans. It is thought that antibiotics cause disturbance of the normal enteric microflora with subsequent growth of opportunistic *C. difficile,* leading to serious clinical complications such as fluid loss, protein-losing enteropathy, and toxic megacolon (Kelly et al., 1994). The organism releases two exotoxins, toxin A and toxin B, which affect metabolic functions including tight-junction permeability in a wide variety of eukaryotic cells (Donta et al., 1982; Hecht et al., 1988). Both clostridial cytotoxins disaggregate the microfilament cytoskeleton by catalyzing the transfer of the sugar moiety from UDP-glucose to a conserved threonine in small GTP-binding proteins of the Rho family (RhoA, Rac1, Cdc42), which regulate the cytoskeleton, rendering these proteins inactive (Just et al., 1995a, 1995b). Other bacterial proteins that target the Rho family of proteins are discussed in Chapter 7.

It is currently unclear how toxin A and B gain access to the mucosa and submucosa with its different immunocompetent cells populations, as an increase in monolayer permeability caused by the clostridial cytotoxins is limited to molecules with hydrodynamic radii of less than 5.7 Å, smaller than either toxin A or toxin B (Hecht et al., 1988). Multiple carbohydrates with the core structure Galβ1–4GlcNAc have been identified as ligands for toxin A (Tucker and Wilkins, 1988). Carbohydrates with this core structure, designated I, Y, and X, are found on intestinal epithelial cells, among other cell types. In rabbits, the brush border enzyme sucrase-isomaltase bearing this critical carbohydrate structure has been identified as a receptor for the C-terminal portion of toxin A (Pothoulakis et al., 1996; Dove et al., 1990).

11.5.2 Effect of *C. difficile* toxins on antigen-presenting cells

Toxin B causes the dose-dependant expression of IL1, IL-6, and TNFα by human monocytes *in vitro* (Flegel et al., 1991). However, concentrations greater than 5 ng/ml decrease monokine expression. Additionally, toxin A induces IL-8 expression in macrophages and epithelial cells (Branka et al., 1997). The details of this activation have been further investigated. After uptake by monocytes, toxin A activates nuclear translocation of the NF-κB heterodimer p50/p65 and AP-1, as well as intracellular calcium release and calmodulin activation for the transcriptional up-regulation of IL-8 mRNA (Jefferson et al., 1999). In addition, toxin A activates ERK and p38 MAP kinases as critical components in the pathway for IL-8 expression and subsequent monocyte necrosis (Warny et al., 2000). Carefully performed kinetic studies further suggested that the glucosylation of Rho is a late event in the activation of monocytes. Thus, the mechanism by which the toxin induces these effects remains unclear. Expression of IL-8 leads to a dramatic increase in neutrophil migration and expression of CD11b/CD18, and ICAM-1 on endothelial cells (Linevsky et al., 1997). In these studies, toxin A was also able to induce monokine expression; however, in comparison to toxin B, concentrations required were about 1,000-fold higher. Toxin A and toxin B did not exhibit synergism regarding their effect on monokine expression and exposure to high toxin concentrations resulted in cell death. After an initial synthesis and release of cytokines, toxin A induces cell death in human colonic lamina propria cells (Mahida et al., 1998). High concentrations of toxin A lead to the total loss of macrophages from the lamina propria within 72 hours, followed by apoptosis of eosinophilic granulocytes and memory T cells at 120 hours. Earlier observations made in human PBMC indicating an inhibition of mitogen-activated T-cell proliferation in the presence of clostridial toxins are most likely the result of toxin-mediated apoptosis in monocytes and T cells and not to a specific immunosuppressive effect (Daubener et al., 1988).

11.5.3 Effect of *C. difficile* toxins *in vivo*

Injection of toxin A *in vivo* into rat peritoneal cavities induces a dose-dependent neutrophil migration that, at least in part, is dependent on IL-1β, TNFα, and leukotriene expression from migrated macrophages (Rocha et al., 1997). Further, toxin A induces expression of pro-inflammatory cytokines in nonimmune cells. The rat homologue of human IL-8, MIP-2, produced in macrophages and in intestinal epithelial cells, is induced within hours of toxin exposure after administration into rat ileal loops (Castagliuolo et al., 1998a). In addition, toxin A mediates infiltration of PMN into the mucosa through the release of substance P and its interaction with the neurokinin-1 receptor from dorsal root ganglia. NK-1R-/- deficient animals showed dramatically attenuated PMN infiltration and histological scores (Castagliuolo et al., 1998b).

11.5.4 Summary

Clostridial toxins initially target macrophages for the expression and release of chemotactic cytokines, such as IL-8 and MIP-2. The release of cytokines leads to the rapid accumulation of predominantly PMNs, and subsequent apoptosis of macrophages and T cells, having a negative effect on immune function. The clinical significance of these effects are not known.

11.6 LYMPHOSTATIN

11.6.1 Introduction

The number of immunoregulatory genes and factors characterized in enteric bacteria is likely to increase, given the fact that the genomes of an increasing number of pathogens have been sequenced completely. Recently, the sequence for the prototype strain of EHEC became available and this information has led to the identification of more than 1,000 additional genes encoding for metabolic capacities and potential virulence factors (Perna et al., 2001). One of these emerging immunoregulatory factors, a large toxin called lymphostatin that specifically targets subsets of mononuclear cells, has recently been identified and characterized in enteropathogenic *E. coli* (EPEC).

11.6.2 Effect of lymphostatin on T cells

Lymphostatin was initially identified in bacterial lysates and subsequently in bacterial supernatant from an EPEC strain. Lymphostatin induces profound suppression of lymphokine expression in activated lymphocytes (Klapproth et al., 1995). Lymphostatin has a negative regulatory effect on Th1 and Th2 lymphokine expression, in particular IL-2, IL-4, IFNγ, and to a lesser degree IL-5, in lymphocytes isolated from peripheral blood and the gastrointestinal mucosa. Lymphostatin appears to elicit an effect on transcriptional lymphokine regulation, as it affects IL-2, IL-4, IL-5, and IFN-γ mRNA expression. The marked inhibition of lymphokine expression is reflected in the complete absence of proliferation in the presence of lymphostatin and mitogens. The negative regulatory effect is not accompanied by induction of cell death, selective depletion of a particular T-cell population, or increased expression of either IL-10 or TGF-β. Studies examining the inhibitory effect on T cells further revealed that the antigen-driven pathway appears to be down-regulated as ovalbumin- or KLH-specific activation was down-regulated (Malstrom and James, 1998). Further, markers of activation, such as CD25 on CD4+, CD8+, or CD45RO+ cells, were not affected by lymphostatin and cell surface expression, as determined by flow cytometry, did not decrease (Klapproth et al., 1996).

The *lifA* gene encoding the lymphostatin protein has been identified and is located on the EPEC bacterial chromosome (Klapproth et al., 2000). The 9,669 bp gene, encoding for a putative protein size of approximately 366 kDa, is the largest open reading frame identified thus far in *E. coli*. Interestingly, the sequence near the N-terminus shares significant homology with the catalytic domains of the large clostridial cytotoxins, which have glycosyltransferase activity. These enzymes transfer the sugar moiety from a UDP-sugar to a small GTP-binding protein, rendering it inactive. For the large clostridial cytotoxins, these target molecules have been identified as Ras, Rac, Rho, and Cdc42, which are critical regulators of the cytoskeleton. For lymphostatin the target has not been identified yet. It is further of interest to note that a similar gene encoding for a lymphostatin-like activity has been identified in closely related organisms, including *Citrobacter rodentium*, EHEC, and other EPEC strains. In EHEC the orthologous gene is located on the large pathogenicity plasmid, which can confer inhibitory activity if transformed into nonpathogenic laboratory strains (Klapproth et al., 2000). The gene encoding lymphostatin is present in bacteria that have the ability to cause attaching and effacing lesions, including EPEC, *Citrobacter rodentium*, and EHEC strains. However, the association between attaching and effacing and lymphostatin activity is not absolute and is not a requirement for the observed inhibitory activity on T cells.

11.6.3 Effect of lymphostatin on antigen-presenting cells

In contrast to the suppressive effect on lymphokines, lymphostatin does not affect monokine expression. The presence of bacterial products from EPEC and mitogens does not prevent expression of monokines, such as IL-1β, IL-6, IL-8, IL-10, or IL-12. In addition, no inhibitory effect is observed when bacterial products are tested in epithelial cell culture, as determined by proliferation or morphology. Interestingly, lymphostatin was independently identified as an adherence factor for epithelial cells in an EHEC strain of serotype O111:H-(Nicholls et al., 2000). The sequence for the genes from the two strains is identical. Perhaps it is not too surprising that a protein as large as lymphostatin may have multiple parallel functions.

11.6.4 Summary

Lymphostatin is a large toxin encoded by *lifA*, a gene present in pathogenic *E. coli* strains on either chromosome or plasmid. Lymphostatin has a remarkable suppressive effect on T-cell activation, specifically

Table 11.1. *Summary of immunological activities of enterotoxins*

| Toxin | APC | Effect on | | |
		Th1 responses	Th2 responses	Antibody responses
Cholera toxin	↓ IL-12, IL-12R ↑ IL-1α, CD86	↓ IL-2	(+) IL-4	↓ IgM ↑ IgA, IgG1,
Heat-labile toxin	↑ CD86	↑ IFN-γ	↓ IL-4 ↑ IL-5	↑ IgM, IgA, IgG1, IgG2a, IgG2b
Clostridial toxin A/B	↑ IL-8	↑ apoptosis	↑ apoptosis	
Shiga toxin	↑ apoptosis ↑ IL-1β, TNF-α	↓ proliferation and metabolic activity	↓ proliferation and metabolic activity	
Lymphostatin		↓ IL-2, IFN-γ	↓ IL-4, IL-5	

down-regulating expression of both Th1 and Th2 lymphokines. Low concentrations of IL-2 result in absent T-cell proliferation without affecting other markers of activation, such as CD25 or monokine expression. Lymphostatin may also function as an adherence factor for epithelial cells. The role of lymphostatin in disease has yet to be established.

REFERENCES

Anderson, D. and Tsoukas, C. (1989). Cholera toxin inhibits resting human T cell activation via a cAMP-independent pathway. *Journal of Immunology* **143**, 3647–3652.

Bastiaens, P.I., Majoul, I.V., Verveer, P.J., Soling, H.D., and Jovin, T.M. (1996). Imaging the intracellular trafficking and state of the AB5 quaternary structure of cholera toxin. *European Molecular Biology Organisation Journal* **15**, 4246–4253.

Beignon, A.S., Briand, J.P., Muller, S., and Partidos, C.D. (2001). Immunization onto bare skin with heat-labile enterotoxin of *Escherichia coli* enhances immune responses to coadministered protein and peptide antigens and protects mice against lethal toxin challenge. *Immunology* **102**, 344–351.

Benedetti, R., Lev, P., Massouh, E., and Flo, J. (1998). Oral administration of one dose of cholera toxin induces a systemic immune response prior to a mucosal immune response by a direct presentation in the spleen. *Immunology Letters* **60**, 149–156.

Bergquist, C., Johansson, E.L., Lagergard, T., Holmgren, J., and Rudin, A. (1997). Intranasal vaccination of humans with recombinant cholera toxin B subunit induces systemic and local antibody responses in the upper respiratory tract and the vagina. *Infection and Immunity* **65**, 2676–2684.

Bost, K.L., Holton, R.H., Cain, T.K., and Clements, J.D. (1996). *In vivo* treatment with anti-interleukin-13 antibodies significantly reduces the humoral immune response against an oral immunogen in mice. *Immunology* **87**, 633–641.

Bowen, J.C., Nair, S.K., Reddy, R., and Rouse, B.T. (1994). Cholera toxin acts as a potent adjuvant for the induction of cytotoxic T-lymphocyte responses with non-replicating antigens. *Immunology* **81**, 338–342.

Branka, J.E., Vallette, G., Jarry, A., Bou-Hanna, C., Lemarre, P., Van, P.N., and Laboisse, C.L. (1997). Early functional effects of *Clostridium difficile* toxin A on human colonocytes. *Gastroenterology* **112**, 1887–1894.

Braun, M.C., He, J., Wu, C.Y., and Kelsall, B.L. (1999). Cholera toxin suppresses interleukin (IL)-12 production and IL-12 receptor beta1 and beta2 chain expression. *Journal of Experimental Medicine* **189**, 541–552.

Bromander, A., Holmgren, J., and Lycke, N. (1991). Cholera toxin stimulates IL-1 production and enhances antigen presentation by macrophages in vitro. *Journal of Immunology* **146**, 2908–2914.

Bromander, A.K., Kjerrulf, M., Holmgren, J., and Lycke, N. (1993). Cholera toxin enhances alloantigen presentation by cultured intestinal epithelial cells. *Scandinavian Journal of Immunology* **37**, 452–458.

Burkart, V., Kim, Y., Kauer, M., and Kolb, H. (1999). Induction of tolerance in macrophages by cholera toxin B chain. *Pathobiology* **67**, 314–317.

Castagliuolo, I., Keates, A.C., Wang, C.C., Pasha, A., Valenick, L., Kelly, C.P., Nikulasson, S.T., LaMont, J.T., and Pothoulakis, C. (1998a). *Clostridium difficile* toxin A stimulates macrophage-inflammatory protein-2 production in rat intestinal epithelial cells. *Infection and Immunity* **160**, 6039–6045.

Castagliuolo, I., Riegler, M., Pasha, A., Nikulasson, S., Lu, B., Gerard, C., Gerard, N.P., and Pothoulakis, C. (1998b). Neurokinin-1 (NK-1) receptor is required in *Clostridium difficile*-induced enteritis. *Journal of Clinical Investigation* **101**, 1547–1550.

Cheng, E., Cardenas-Freytag, L., and Clements, J. (1999). The role of cAMP in mucosal adjuvanticity of *Escherichia coli* heat-labile enterotoxin (LT). *Vaccine* **18**, 38–49.

Chisari, F. and Northrup, R. (1978). Pathophysiologic effects of lethal and immunoregulatory doses of cholera enterotoxin in the mouse. *Journal of Immunology* **113**, 740–749.

Chong, C., Friberg, M., and Clements, J. (1998). LT(R192G), a non-toxic mutant of the heat-labile enterotoxin of *Escherichia coli*, elicits enhanced humoral and cellular immune responses associated with protection against lethal oral challenge with *Salmonella* spp. *Vaccine* **16**, 732–740.

Christopher-Hennings, J., Willgohs, J.A., Francis, D.H., Raman, U.A., Moxley, R.A., and Hurley, D.J. (1993). Immunocompromise in gnotobiotic pigs induced by verotoxin-producing *Escherichia coli* (O111:NM). *Infection and Immunity* **61**, 2304–2308.

Cong, Y., Weaver, C., and Elson, C. (1997). The mucosal adjuvanticity of cholera toxin involves enhancement of costimulatory activity by selective up-regulation of B7.2 expression. *Journal of Immunology* **159**, 5301–5308.

Connell, T.D., Metzger, D., Sfintescu, C., and Evans, R.T. (1998). Immunostim- ulatory activity of LT-IIa, a type II heat-labile enterotoxin of *Escherichia coli*. *Immunology Letters* **62**, 117–120.

Cushing, A. and Smart, J. (1985). Gastrointestinal carriage of toxigenic bacteria: relation to diarrhea and to serum immune response. *Journal of Infectious Disease* **151**, 114–123.

Dallas, W. and Falkow, S. (1980). Amino acid sequence homology between cholera toxin and *Escherichia coli* heat-labile toxin. *Nature* **288**, 499–501.

Daubener, W., Leiser, E., von Eichel-Streiber, C., and Hadding, U. (1988). *Clostrid- ium difficile* toxins A and B inhibit human immune response *in vitro. Infection and Immunity* **56**, 1107–1112.

de Haan, L., Feil, I.K., Verweij, W.R., Holtrop, M., Hol, W.G., Agsteribbe, E., and Wilschut, J. (1998). Mutational analysis of the role of ADP-ribosylation activity and GM1-binding activity in the adjuvant properties of the *Escherichia coli* heat-labile enterotoxin towards intranasally administered keyhole limpet hemocyanin. *European Journal of Immunology* **28**, 1243–1250.

de Haan, L., Holtrop, M., Verweij, W.R., Agsteribbe, E., and Wilschut, J. (1999). Mucosal immunogenicity and adjuvant activity of the recombinant A subunit of the *Escherichia coli* heat-labile enterotoxin. *Immunology* **97**, 706–713.

DiTommaso, A., Saletti, G., Pizza, M., Rappuoli, R., Dougan, G., Abrignani, S., Douce, G., and De Magistris, M.T. (1996). Induction of antigen-specific an- tibodies in vaginal secretions by using a nontoxic mutant of heat-labile en- terotoxin as a mucosal adjuvant. *Infection and Immunity* **64**, 974–979.

Donta, S., Sullivan, N., and Wilkins, T. (1982). Differential effects of *Clostridium difficile* toxins on tissue-cultured cells. *Journal of Clinical Microbiology* **15**, 1157–1158.

Douce, G., Turcotte, C., Cropley, I., Roberts, M., Pizza, M., Domenghini, M., Rappuoli, R., and Dougan, G. (1995). Mutants of *Escherichia coli* heat-labile toxin lacking ADP-ribosyltransferase activity act as nontoxic, mucosal adju- vants. *Proceedings of the National Academy of Sciences USA* **92**, 1644–1648.

Douce, G., Giannelli, V., Pizza, M., Lewis, D., Everest, P., Rappuoli, R., and Dougan, G. (1999). Genetically detoxified mutants of heat-labile toxin from

Escherichia coli are able to act as oral adjuvants. *Infection and Immunity* **67**, 4400–4406.

Dove, C.H., Wang, S.Z., Price, S.B., Phelps, C.J., Lyerly, D.M., Wilkins, T.D., and Johnson, J.L. (1990). Molecular characterization of the *Clostridium difficile* toxin A gene. *Infection and Immunity* **58**, 480–488.

Elson, C. and Ealding, W. (1984a). Cholera toxin feeding did not induce oral tolerance in mice and abrogated oral tolerance to an unrelated protein antigen. *Journal of Immunology* **133**, 2892–2897.

Elson, C. and Ealding, W. (1984b). Generalized systemic and mucosal immunity in mice after mucosal stimulation with cholera toxin. *Journal of Immunology* **136**, 2736–2741.

Elson, C. and Ealding, W. (1985). Genetic control of the murine immune response to cholera toxin. *Journal of Immunology* **135**, 930–932.

Elson, C. and Ealding, W. (1987). Ir gene control of the murine secretory IgA response to cholera toxin. *European Journal of Immunology* **17**, 425–428.

Elson, C.O., Holland, S.P., Dertzbaugh, M.T., Cuff, C.F., and Anderson, A.O. (1995). Morphologic and functional alterations of mucosal T cells by cholera toxin and its B subunit. *Journal of Immunology* **154**, 1032–1040.

Eriksson, K., Nordstrom, I., Czerkinsky, C., and Holmgren, J. (2000). Differential effect of cholera toxin on CD45RA+ and CD45RO+ T cells: specific inhibition of cytokine production but not proliferation of human naive T cells. *Clinical Experimental Immunology* **121**, 283–288.

Field M., Rao, M.C., and Chang, E.B. (1989). Intestinal electrolyte transport and diarrheal disease (1). *New England Journal of Medicine* **321**, 800–806.

Flegel, W.A., Mullerm, F., Daubenerm, W., Fischerm, H.G., Haddingm, U., and Northoff, H. (1991). Cytokine response by human monocytes to *Clostridium difficile* toxin A and toxin B. *Infection and Immunity* **59**, 3659–3666.

Fukuta, S., Magnani, J.L., Twiddy, E.M., Holmes, R.K., and Ginsburg, V. (1988). Comparison of the carbohydrate-binding specificities of cholera toxin and *Escherichia coli* heat-labile enterotoxins LTh-I, LT-IIa, and LT-IIb. *Infection and Immunity* **56**, 1748–1753.

Gagliardi, M.C., Sallusto, F., Marinaro, M., Langenkamp, A., Lanzavecchia, A., and De Magistris, M.T. (2000). Cholera toxin induces maturation of human dendritic cells and licences them for Th2 priming. *European Journal of Immunology* **30**, 2394–2403.

Gill, D. and King, C. (1975). The mechanism of action of cholera toxin in pigeon erythrocyte lysates. *Journal of Biological Chemistry* **250**, 6424–6432.

Glenn, G.M., Rao, M., Matyas, G.R., and Alving, C.R. (1998a). Skin immunization made possible by cholera toxin. *Nature* **391**, 851.

Glenn, G.M., Scharton-Kersten, T., Vassell, R., Mallett, C.P., Hale, T.L., and Alving, C.R. (1998b). Transcutaneous immunization with cholera toxin protects mice against lethal mucosal toxin challenge. *Journal of Immunology* **161**, 3211–3214.

Glenn, G.M., Scharton-Kersten, T., Vassell, R., Matyas, G.R., and Alving, C.R. (1999). Transcutaneous immunization with bacterial ADP-ribosylating exotoxins as antigens and adjuvants. *Infection and Immunity* **67**, 1100–1106.

Gluck, U., Gebbers, J., and Gluck, R. (1999). Phase 1 evaluation of intranasal virosomal influenza vaccine with and without Escherichia coli heat-labile toxin in adult volunteers. *Journal of Virology* **73**, 7770–7776.

Gorbach, S.L., Banwell, J.G., Chatterjee, B.D., Jacobs, B., and Sack, R.B. (1971). Acute undifferentiated human diarrhea in the tropics. I. Alterations in intestinal microflora. *Journal of Clinical Investigation* **50**, 881–889.

Grdic, D., Smith, R., Donachie, A., Kjerrulf, M., Hornquist, E., Mowat, A., and Lycke, N. (1999). The mucosal adjuvant effects of cholera toxin and immune-stimulating complexes differ in their requirement for IL-12, indicating different pathways of action. *European Journal of Immunology* **29**, 1774–1784.

Guidry, J.J., Cardenas, L., Cheng, E., and Clements, J.D. (1997). Role of receptor binding in toxicity, immunogenicity, and adjuvanticity of *Escherichia coli* heat-labile enterotoxin. *Infection and Immunity* **65**, 4943–4950.

Hammond, SA., Walwender, D., Alving, C.R., and Glenn, G.M. (2001). Transcutaneous immunization: T cell responses and boosting of existing immunity. *Vaccine* **19**, 2701–2707.

Hartman, A., Verg, L.V.D., and Venkatesan, M. (1999). Native and mutant forms of cholera toxin and heat-labile enterotoxin effectively enhance protective efficacy of live attenuated and heat-killed Shigella vaccines. *Infection and Immunity* **67**, 5841–5847.

Hashigucci, K., Ogawa, H., Ishidate, T., Yamashita, R., Kamiya, H., Watanabe, K., Hattori, N., Sato, T., Suzuki, Y., Nagamine, T., Aizawa, C., Tamura, S., Kurata, T., and Oya, A. (1996). Antibody responses in volunteers induced by nasal influenza vaccine combined with *Escherichia coli* heat-labile enterotoxin B subunit containing a trace amount of the holotoxin. *Vaccine* **14**, 113–119.

Hazama, M., Mayumi-Aono, A., Miyazaki, T., Hinuma, S., and Fujisawa, Y. (1993). Intranasal immunization against herpes simplex virus infection by using a recombinant glycoprotein D fused with immunomodulating proteins, the B subunit of *Escherichia coli* heat-labile enterotoxin and interleukin-2. *Immunology* **78**, 643–649.

Hecht, G., Pothoulakis, C., LaMont, J.T., and Madara, J.L. (1988). *Clostridium difficile* toxin A perturbs cytoskeletal structure and tight junction permeability of cultured human intestinal epithelial monolayers. *Journal of Clinical Investigation* **82**, 1516–1524.

Holmes, R., Twiddy, E., and Pickett, C. (1986). Purification and characterization of type II heat-labile enterotoxin of *Escherichia coli*. *Infection and Immunity* **53**, 464–473.

Holmgren, J., Lindholm, L., and Lonnroth, I. (1974). Interaction of cholera toxin and toxin derivatives with lymphocytes. I. Binding properties and interference with lectin-induced cellular stimulation. *Journal of Experimental Medicine* **139**, 801–819.

Holmgren, J., Fredman, P., Lindblad, M., Svennerholm, A.M., and Svennerholm, L. (1982). Rabbit intestinal glycoprotein receptor for *Escherichia coli* heat-labile enterotoxin lacking affinity for cholera toxin. *Infection and Immunity* **38**, 424–433.

Hornqvist, E., Goldschmidt, T.J., Holmdahl, R., and Lycke, N. (1991). Host defense against cholera toxin is strongly CD4+ T cell dependent. *Infection and Immunity* **59**, 3630–3638.

Imboden, J.B., Shoback, D.M., Pattison, G., and Stobo, J.D. (1986). Cholera toxin inhibits the T-cell antigen receptor-mediated increases in inositol trisphosphate and cytoplasmic free calcium. *Proceedings of the National Academy of Sciences, USA* **83**, 5673–5677.

Inward, C.D., Varagunam, M., Adu, D., Milford, D.V., and Taylor, C.M. (1997). Cytokines in haemolytic uraemic syndrome associated with verocytotoxin-producing *Escherichia coli* infection. *Archives of Diseases of Children* **77**, 145–147.

Jackson, M.P., Newland, J.W., Holmes, R.K., and O'Brien, A.D. (1987). Nucleotide sequence analysis and comparison of the structural genes for Shiga-like toxin I and Shiga-like toxin II encoded by bacteriophages from *Escherichia coli* 933. *FEMS Microbiology Letter* **44**, 109–114.

Jackson, R.J., Fujihashi, K., Xu-Amano, J., Kiyono, H., Elson, C.O., and McGhee, J.R. (1993). Optimizing oral vaccines: induction of systemic and mucosal B-cell and antibody responses to tetanus toxoid by use of cholera toxin as an adjuvant. *Infection and Immunity* **61**, 4272–4279.

James, S. and Kiyono, H. (1999). Gastrointestinal lamina propria cells. In *Mucosal Immunology*, ed. P. Ogra, pp. 381–396. San Diego: Academic Press.

Jefferson, K., Smith, M., and Bobak, D. (1999). Roles of intracellular calcium and NF-kappa B in the *Clostridium difficile* toxin A-induced up-regulation and secretion of IL-8 from human monocytes. *Journal of Immunology* **160**, 6039–6045.

Jertborn, M., Svennerholm, A., and Holmgren, J. (1988). Five-year immunologic memory in Swedish volunteers after oral cholera vaccination. *Journal of Infectious Disease* **157**, 374–377.

Just, I., Wilm, M., Selzer, J., Rex, G., von Eichel-Streiber, C., Mann, M., and Aktories, K. (1995a). The enterotoxin from *Clostridium difficile* (ToxA) monoglucosylates the Rho proteins. *Journal of Biological Chemistry* **270**, 13,932–13,936.

Just, I., Selzer, J., Wilm, M., von Eichel-Streiber, C., Mann, M., and Aktories, K. (1995b). Glucosylation of Rho proteins by *Clostridium difficile* toxin B. *Nature* **375**, 500–503.

Karmali, M.A., Petric, M., Lim, C., Fleming, P.C., Arbus, G.S., and Lior, H. (1985). The association between idiopathic hemolytic uremic syndrome and infection by verotoxin-producing *Escherichia coli*. *Journal of Infectious Disease* **151**, 775–782.

Karmali, M. (1989). Infection by verocytotoxin-producing *Escherichia coli*. *Clinical Microbiological Review* **2**, 15–38.

Katz, J.M., Lu, X., Young, S.A., and Galphin J.C. (1997). Adjuvant activity of the heat-labile enterotoxin from enterotoxigenic *Escherichia coli* for oral administration of inactivated influenza virus vaccine. *Journal of Infectious Disease* **175**, 352–363.

Kelly, C., Pothoulakis, C., and LaMont, J. (1994). *Clostridium difficile* colitis. *New England Journal of Medicine* **330**, 257–262.

Kim, P.H., Eckmann, L., Lee, W.J., Han, W., and Kagnoff, M.F. (1998). Cholera toxin and cholera toxin B subunit induce IgA switching through the action of TGF-beta 1. *Journal of Immunology* **160**, 1198–1203.

Klapproth, J.M., Donnenberg, M.S., Abraham, J.M., Mobley, H.L., and James, S.P. (1995). Products of enteropathogenic *Escherichia coli* inhibit lymphocyte activation and lymphokine production. *Infection and Immunity* **63**, 2248–2254.

Klapproth, J.M., Donnenberg, M.S., Abraham, J.M., and James, S.P. (1996). Products of enteropathogenic *E. coli* inhibit lymphokine production by gastrointestinal lymphocytes. *American Journal of Physiology* **271**, G841–G848.

Klapproth, J.M., Scaletsky, I.C., McNamara, B.P., Lai, L.C., Malstrom, C., James, S.P., and Donnenberg, M.S. (2000). A large toxin from pathogenic *Escherichia coli* strains that inhibits lymphocyte activation. *Infection and Immunity* **68**, 2148–2155.

Klipstein, F., Engert, R., and Clements, J. (1982). Arousal of mucosal secretory immunoglobulin A antitoxin in rats immunized with *Escherichia coli* heat-labile enterotoxin. *Infection and Immunity* **37**, 1086–1092.

Kunkel, S. and Robertson, D. (1979). Purification and chemical characterization of the heat-labile enterotoxin produced by enterotoxigenic *Escherichia coli*. *Infection and Immunity* **25**, 586–596.

Kuziemko, G., Stroh, M., and Stevens, R. (1996). Cholera toxin binding affinity and specificity for gangliosides determined by surface plasmon resonance. *Biochemistry* **35**, 6375–6384.

Linevsky, J.K., Pothoulakis, C., Keates, S., Warny, M., Keates, A.C., Lamont, J.T., and Kelly, C.P. (1997). IL-8 release and neutrophil activation by *Clostridium difficile* toxin-exposed human monocytes. *American Journal of Physiology* **273**, G1333–G1340.

Lycke, N. (1992). Cholera toxin promotes B cell isotype switching by two different mechanisms. cAMP induction augments germ-line Ig H-chain RNA transcripts whereas membrane ganglioside GM1-receptor binding enhances later events in differentiation. *Journal of Immunology* **150**, 4810–4821.

Lycke, N., Lindholm, L., and Holmgren, J. (1985). Cholera antibody production *in vitro* by peripheral blood lymphocytes following oral immunization of humans and mice. *Clinical Experimental Immunology* **62**, 39–47.

Lycke, N. and Strober, W. (1989). Cholera toxin promotes B cell isotype differentiation. *Journal of Immunology* **142**, 3781–3787.

Lycke, N., Severinson, E., and Strober, W. (1990). Cholera toxin acts synergistically with IL-4 to promote IgG1 switch differentiation. *Journal of Immunology* **145**, 3316–3324.

Lycke, N. and Holmgren, J. (1986). Intestinal mucosal memory and presence of memory cells in lamina propria and Peyer's patches of mice 2 years after oral immunization with cholera toxin. *Scandinavian Journal of Immunology* **23**, 611–615.

Lycke, N., Hellstrom, U., and Holmgren, J. (1987). Circulating cholera antitoxin memory cells in the blood one year after oral cholera vaccination in humans. *Scandinavian Journal of Immunology* **26**, 207–211.

Mahida, Y.R., Galvin, A., Makh, S., Hyde, S., Sanfilippo, L., Borriello, S.P., and Sewell, H.F. (1998). Effect of *Clostridium difficile* toxin A on human colonic lamina propria cells: early loss of macrophages followed by T-cell apoptosis. *Infection and Immunity* **66**, 5462–5469.

Majoul, I., Bastiaens, P., and Soling, H. (1996). Transport of an external Lys-Asp-Glu-Leu (KDEL) protein from the plasma membrane to the endoplasmic reticulum: studies with cholera toxin in Vero cells. *Journal of Cellular Biology* **133**, 777–789.

Malstrom, C. and James, S. (1998). Inhibition of murine splenic and mucosal lymphocyte function by enteric bacterial products. *Infection and Immunity* **66**, 3120–3127.

Mangeney, M., Richard, Y., Coulaud, D., Tursz, T., and Wiels, J. (1991). CD77: an antigen of germinal center B cells entering apoptosis. *European Journal of Immunology* **21**, 1131–1140.

Marinaro, M., Staats, H.F., Hiroi, T., Jackson, R.J., Coste, M., Boyaka, P.N., Okahashi, N., Yamamoto, M., Kiyono, H., and Bluethmann, H. (1995). Mucosal adjuvant effect of cholera toxin in mice results from induction of T helper 2 (Th2) cells and IL-4. *Journal of Immunology* **155**, 4621–4629.

Matousek, M., Nedrud, J., and Harding, C. (1996). Distinct effects of recombinant cholera toxin B subunit and holotoxin on different stages of class II MHC antigen processing and presentation by macrophages. *Journal of Immunology* **156**, 4137–4145.

Matousek, M.P., Nedrud, J.G., Cieplak, W., and Harding, C.V. (1998). Inhibition of class II major histocompatibility complex antigen processing by *Escherichia coli* heat-labile enterotoxin requires an enzymatically active A subunit. *Infection and Immunity* **66**, 3480–3484.

McGhee, J., Lamm, M., and Strober, W. (1999). Mucosal immune response: an overview. In *Mucosal Immunology*, ed. P. Ogra, pp. 485–506. San Diego: Academic Press.

McKenzie, S. and Halsey, J. (1984). Cholera toxin B subunit as a carrier protein to stimulate a mucosal immune response. *Journal of Immunology* **133**, 1818–1824.

Menge, C., Wieler, L.H., Schlapp, T., and Baljer, G. (1999). Shiga toxin 1 from *Escherichia coli* blocks activation and proliferation of bovine lymphocyte subpopulations *in vitro*. *Infection and Immunity* **67**, 2209–2217.

Millar, D., Hirst, T., and Snider, D. (2001). *Escherichia coli* heat-labile enterotoxin B subunit is a more potent mucosal adjuvant than its closely related homologue, the B subunit of cholera toxin. *Infection and Immunity* **69**, 3476–3482.

Munoz, E., Zubiaga, A.M., Merrow, M., Sauter, N.P., and Huber, B.T. (1990). Cholera toxin discriminates between T helper 1 and 2 cells in T cell receptor-mediated activation: role of cAMP in T cell proliferation. *Journal of Experimental Medicine* **172**, 95–103.

Nahar, T., Williams, N., and Hirst, T.R. (1996). Cross-linking of cell surface ganglioside GM1 induces the selective apoptosis of mature CD8+ T lymphocytes. *International Immunology* **8**, 731–736.

Nashar, T.O., Webb, H.M., Eaglestone, S., Williams, N.A., and Hirst, T.R. (1996). Potent immunogenicity of the B subunits of *Escherichia coli* heat-labile enterotoxin: receptor binding is essential and induces differential modulation of lymphocyte subsets. *Proceedings of the National Academy of Sciences USA* **93**, 226–230.

Nashar, T., Hirst, T., and Williams, N. (1997). Modulation of B-cell activation by the B subunit of *Escherichia coli* enterotoxin: receptor interaction up-regulates MHC class II, B7, CD40, CD25 and ICAM-1. *Immunology* **91**, 572–578.

Neidleman, J.A., Vajdy, M., Ugozzoli, M., Ott, G., and O'Hagan, D. (2000).

Genetically detoxified mutants of heat-labile enterotoxin from *Escherichia coli* are effective adjuvants for induction of cytotoxic T-cell responses against HIV-1 gag-p55. *Immunology* **101**, 154–160.

Nicholls, L., Grant, T., and Robins-Browne, R. (2000). Identification of a novel genetic locus that is required for in vitro adhesion of a clinical isolate of enterohaemorrhagic *Escherichia coli* to epithelial cells. *Molecular Microbiology* **35**, 275–288.

Niemialtowski, M., Klucinski, W., Malicki K., and de Faundez, I.S. (1993). Cholera toxin (choleragen)-polymorphonuclear leukocyte interactions: effect on migration *in vitro* and Fc gamma R-dependent phagocytic and bactericidal activity. *Microbiology and Immunology* **37**, 55–62.

Nunes, J., Bagnasco, M., Lopez, M., Olive, D., and Mawas, C. (1989). Cholera toxin inhibits the increase in cytoplasmic free calcium induced via the CD2 pathway of human T-lymphocyte activation. *Journal of Cellular Biology* **39**, 391–400.

O'Brien, A.D., Tesh, V.L., Donohue-Rolfe, A., Jackson, M.P., Olsnes, S., Sandvig, K., Lindberg, A.A., and Keusch, G.T. (1992). Shiga toxin: biochemistry, genetics, mode of action, and role in pathogenesis. *Current Topics in Microbiology and Immunology* **180**, 65–94.

Ogasawara, T., Ito, K., Igarashi, K., Yutsudo, T., Nakabayashi, N., and Takeda, Y. (1988). Inhibition of protein synthesis by a Vero toxin (VT2 or Shiga-like toxin II) produced by *Escherichia coli* O157:H7 at the level of elongation factor 1-dependent aminoacyl-tRNA binding to ribosomes. *Microbial Pathogenesis* **4**, 127–135.

Orlandi, P. and Fishman, P. (1993). Orientation of cholera toxin bound to target cells. *Journal of Biological Chemistry* **268**, 17,038–17,044.

Owen, R. (1999). Uptake and transport of intestinal macromolecules and microorganisms by M cells in Peyer's patches – a personal and historical perspective. *Seminars in Immunology* **11**, 157–163.

Partidos, C.D., Salani, B.F., Pizza. M., and Rappuoli, R. (1999). Heat-labile enterotoxin of *Escherichia coli* and its site-directed mutant LTK63 enhance the proliferative and cytotoxic T-cell responses to intranasally co-immunized synthetic peptides. *Immunology Letters* **67**, 209–216.

Perna, N.T., Plunkett, G., Burland, V., Mau, B., Glasner, J.D., Rose, D.J., Mayhew, G.F., Evans, P.S., Gregor, J., Kirkpatrick, H.A., Posfai, G., Hackett, J., Klink, S., Boutin, A., Shao, Y., Miller, L., Grotbeck, E.J., Davis, N.W., Lim, A., Dimalanta, E.T., Potamousis, K.D., Apodaca, J., Anantharaman, T.S., Lin, J., Yen, G., Schwartz, D.C., Welch, R.A., and Blattner, F.R. (2001). Genome sequence of enterohaemorrhagic *Escherichia coli* O157:H7. *Nature* **409**, 529–533.

Pothoulakis, C., Gilbert, R.J., Cladaras, C., Castagliuolo, I., Semenza, G., Hitti, Y., Montcrief, J.S., Linevsky, J., Kelly, C.P., Nikulasson, S., Desai, H.P., Wilkins, T.D., and LaMont, J.T. (1996). Rabbit sucrase-isomaltase contains a functional intestinal receptor for *Clostridium difficile* toxin A. *Journal of Clinical Investigation* **98**, 641–649.

Ramegowada, B. and Tesh, V. (1996). Differentiation-associated toxin receptor modulation, cytokine production, and sensitivity to Shiga-like toxins in human monocytes and monocytic cell lines. *Infection and Immunity* **64**, 1173–1180.

Renshaw, B.R., Fanslow, W.C., Armitage, R.J., Campbell, K.A., Liggitt, D., Wright, B., Davison, B.L., and Maliszewski, C.R. (1994). Humoral immune responses in CD40 ligand-deficient mice. *Journal of Experimental Medicine* **180**, 1889–1900.

Richards, C.M., Aman, A.T., Hirst, T.R., Hill, T.J., and Williams, NA. (2001). Protective mucosal immunity to ocular herpes simplex virus type 1 infection in mice by using *Escherichia coli* heat-labile enterotoxin B subunit as an adjuvant. *Journal of Virology* **75**, 1664–1671.

Rocha, M.F., Maia, M.E., Bezerra, L.R., Lyerly, D.M., Guerrant, R.L., Ribeiro, R.A., and Lima, A.A. (1997). *Clostridium difficile* toxin A induces the release of neutrophil chemotactic factors from rat peritoneal macrophages: role of interleukin-1beta, tumor necrosis factor alpha, and leukotrienes. *Infection and Immunity* **65**, 2740–2746.

Rowe, B., Taylor, J., and Bettelheim, K. (1970). An investigation of traveller's diarrhoea. *Lancet* **1**, 1–5.

Ruddock, L.W., Ruston, S.P., Kelly, S.M., Price, N.C., Freedman, R.B., and Hirst, T.R. (1995). Kinetics of acid-mediated disassembly of the B subunit pentamer of *Escherichia coli* heat-labile enterotoxin. Molecular basis of pH stability. *Journal of Biological Chemistry* **270**, 29,953–29,958.

Ryan, E.J., McNeela, E., Murphy, G.A., Stewart, H., O'Hagan, D., Pizza, M., Rappuoli, R., and Mills, K.H. (1999). Mutants of *Escherichia coli* heat-labile toxin act as effective mucosal adjuvants for nasal delivery of an acellular pertussis vaccine: differential effects of the nontoxic AB complex and enzyme activity on Th1 and Th2 cells. *Infection and Immunity* **67**, 6270–6278.

Ryan, E.J., McNeela, E., Pizza, M., Rappuoli, R., O'Neill, L., and Mills, K.H. (2000). Modulation of innate and acquired immune responses by *Escherichia coli* heat-labile toxin: distinct pro- and anti-inflammatory effects of the nontoxic AB complex and the enzyme activity. *Journal of Immunology* **165**, 5750–5759.

Sack, R.B., Gorbach, S.L., Banwell, J.G., Jacobs, B., Chatterjee, B.D., and Mitra, R.C. (1971). Enterotoxigenic *Escherichia coli* isolated from patients with severe cholera-like *disease. Journal of Infectious Disease* **123**, 378–385.

Schlessinger, D. and Schaechter, M. (1993). Bacterial toxins. In *Mechanisms of Microbial Disease*, ed. M. Schachter, G. Medoff, and B. Eisenstein, pp. 162–175. Baltimore: William and Wilkins.

Simmons, C.P., Mastroeni, P., Fowler, R., Ghaem-maghami, M., Lycke, N., Pizza, M., Rappuoli, R., and Dougan, G. (1999). MHC class I-restricted cytotoxic lymphocyte responses induced by enterotoxin-based mucosal adjuvants. *Journal of Immunology* **163**, 6502–6510.

Simmons, C.P., Hussell, T., Sparer, T., Walzl, G., Openshaw, P., and Dougan, G. (2001). Mucosal delivery of a respiratory syncytial virus CTL peptide with enterotoxin-based adjuvants elicits protective, immunopathogenic, and immunoregulatory antiviral CD8+ T cell responses. *Journal of Immunology* **166**, 1106–1113.

Sixma, T.K., Pronk, S.E., Kalk, K.H., Wartna, E.S., van Zanten, B.A., Witholt, B., and Hol, W.G. (1991). Crystal structure of a cholera toxin-related heat-labile enterotoxin from *E. coli*. *Nature* **351**, 371–377.

Smith, H., Green, P., and Parsell, Z. (1983). Vero cell toxins in *Escherichia coli* and related bacteria: transfer by phage and conjugation and toxic action in laboratory animals, chickens and pigs. *Journal of General Microbiology* **129**, 3121–3137.

Snider, D.P., Marshall, J.S., Perdue, M.H., and Liang, H. (1994). Production of IgE antibody and allergic sensitization of intestinal and peripheral tissues after oral immunization with protein Ag and cholera toxin. *Journal of Immunology* **153**, 647–657.

Sobel, D.O., Yankelevich, B., Goyal, D., Nelson, D., and Mazumder, A. (1998). The B-subunit of cholera toxin induces immunoregulatory cells and prevents diabetes in the NOD mouse. *Diabetes* **47**, 186–191.

Spangler, B. (1992). Structure and function of cholera toxin and the related *Escherichia coli* heat-labile enterotoxin. *Microbiological Reviews* **56**, 622–647.

Sugatani, J., Igarashi, T., Shimura, M., Yamanaka, T., Takeda, T., and Miwa, M. (2000). Disorders in the immune responses of T- and B-cells in mice administered intravenous verotoxin 2. *Life Sciences 2000* **67**, 1059–1072.

Sun, J.B., Rask, C., Olsson, T., Holmgren, J., and Czerkinsky, C. (1996). Treatment of experimental autoimmune encephalomyelitis by feeding myelin basic protein conjugated to cholera toxin B subunit. *Proceedings of the National Academy of Sciences USA* **93**, 7196–7201.

Svennerholm, A.M., Sack, D.A., Holmgren, J., and Bardhan, P.K. (1982). Intestinal antibody responses after immunization with cholera B subunit. *Lancet* **1**, 305–308.

Szamel, M., Ebel, U., Uciechowski, P., Kaever, V., and Resch, K. (1997). T cell antigen receptor dependent signalling in human lymphocytes: cholera toxin inhibits interleukin-2 receptor expression but not interleukin-2 synthesis by preventing activation of a protein kinase C isotype, PKC-alpha. *Biochimica et Biophysica Acta* **1356**, 237–248.

Takahashi, I., Marinaro, M., Kiyono, H., Jackson, R.J., Nakagawa, I., Fujihashi, K., Hamada, S., Clements. J,D., Bost, K.L., and McGhee, J.R. (1996). Mechanisms for mucosal immunogenicity and adjuvancy of *Escherichia coli* labile enterotoxin. *Journal of Infectious Disease* **173**, 627–635.

Tamura, S., Shoji Y., Hasiguchi, K., Aizawa, C., and Kurata, T. (1994). Effects of cholera toxin adjuvant on IgE antibody response to orally or nasally administered ovalbumin. *Vaccine* **12**, 1238–1240.

Tesh, V., Ramegowda, B., and Samuel, J. (1994). Purified Shiga-like toxins induce expression of proinflammatory cytokines from murine peritoneal macrophages. *Infection and Immunity* **62**, 5085–5094.

Tucker, K. and Wilkins, T. (1988). Toxin A of *Clostridium difficile* binds to the human carbohydrate antigens I, X, and Y. *Infection and Immunity* **56**, 1107–1112.

Vajdy, M. and Lycke, N. (1992). Cholera toxin adjuvant promotes long-term immunological memory in the gut mucosa to unrelated immunogens after oral immunization. *Immunology* **75**, 488–492.

Vajdy, M., Kosco-Vilbois, M.H., Kopf, M., Kohler, G., and Lycke, N. (1995). Impaired mucosal immune responses in interleukin 4-targeted mice. *Journal of Experimental Medicine* **181**, 41–53.

van de Kar, N.C., Monnens, L.A., Karmali, M.A., and van Hinsbergh, V.W. (1992). Tumor necrosis factor and interleukin-1 induce expression of the verocytotoxin receptor globotriaosylceramide on human endothelial cells: implications for the pathogenesis of the hemolytic uremic syndrome. *Blood* **80**, 2755–2764.

Vedia, L.M., Reep, B., and Lapetina, E. (1988). Platelet cytosolic 44-kDa protein is a substrate of cholera toxin-induced ADP-ribosylation and is not recognized by antisera against the alpha subunit of the stimulatory guanine nucleotide-binding regulatory protein. *Proceedings of the National Academy of Sciences USA* **85**, 5899–5902.

Verweij, W.R., de Haan, L., Holtrop, M., Agsteribbe, E., Brands, R., van Scharrenburg, G.J., and Wilschut, J. (1998). Mucosal immunoadjuvant activity of recombinant *Escherichia coli* heat-labile enterotoxin and its B subunit: induction of systemic IgG and secretory IgA responses in mice by intranasal immunization with influenza virus surface antigen. *Vaccine* **16**, 2069–2076.

Warny, M., Keates, A.C., Keates, S., Castagliuolo, I., Zacks, J.K., Aboudola, S., Qamar, A., Pothoulakis, C., LaMont, J.T., and Kelly, C.P. (2000). p38 MAP kinase activation by *Clostridium difficile* toxin A mediates monocyte necrosis, IL-8 production, and enteritis. *Journal of Clinical Investigation* **105**, 1147–1156.

Williams, N.A., Stasiuk, L.M., Nashar, T.O., Richards, C.M., Lang, A.K., Day, M.J., and Hirst, T.R. (1997). Prevention of autoimmune disease due to lymphocyte modulation by the B-subunit of *Escherichia coli* heat-labile enterotoxin. *Proceedings of the National Academy of Sciences USA* **94**, 5290–5295.

Woogen, S., Ealding, W., and Elson, C. (1987). Inhibition of murine lymphocyte proliferation by the B subunit of cholera toxin. *Journal of Immunology* **139**, 3764–3770.

Xu-Amano, J., Kiyono, H., Jackson, R.J., Staats, H.F., Fujihashi, K., Burrows, P.D., Elson, C.O., Pillai, S., and McGhee, J.R. (1993). Helper T cell subsets for immunoglobulin A responses: oral immunization with tetanus toxoid and cholera toxin as adjuvant selectively induces Th2 cells in mucosa associated tissues. *Journal of Experimental Medicine* **178**, 1309–1320.

Xu-Amano, J., Jackson, R.J., Fujihashi, K., Kiyono, H., Staats, H.F., and McGhee, J.R. (1994). Helper Th1 and Th2 cell responses following mucosal or systemic immunization with cholera toxin. *Vaccine* **12**, 903–911.

Yamamoto, S., Takeda, Y., Yamamoto, M., Kurazono, H., Imaoka, K., Yamamoto, M., Fujihashi, K., Noda, M., Kiyono, H., and McGhee, J.R. (1997). Mutants in the ADP-ribosyltransferase cleft of cholera toxin lack diarrheagenicity but retain adjuvanticity. *Journal of Experimental Medicine* **185**, 1203–1210.

Yamamoto, M., Briles, D.E., Yamamoto, S., Ohmura, M., Kiyono, H., and McGhee J.R. (1998). A nontoxic adjuvant for mucosal immunity to pneumococcal surface protein A. *Journal of Immunology* **161**, 4115–4121.

Yamamoto, M., Kiyono, H., Yamamoto, S., Batanero, E., Kweon, M.N., Otake, S., Azuma, M., Takeda, Y., and McGhee, J.R. (1999). Direct effects on antigen-presenting cells and T lymphocytes explain the adjuvanticity of a nontoxic cholera toxin mutant. *Journal of Immunology* **162**, 7015–7021.

Yamamoto, M., Kiyono, H., Kweon, M.N., Yamamoto, S., Fujihashi, K., Kurazono, H., Imaoka, K., Bluethmann, H., Takahashi, I., Takeda, Y., Azuma, M., and McGhee, J.R. (2000). Enterotoxin adjuvants have direct effects on T cells and antigen-presenting cells that result in either interleukin-4-dependent or-independent immune responses. *Journal of Infectious Disease* **182**, 180–190.

Zhang, R.G., Scott, D.L., Westbrook, M.L., Nance, S., Spangler, B.D., Shipley, G.G., and Westbrook, E.M. (1995). The three-dimensional crystal structure of cholera toxin. *Journal of Molecular Biology* **251**, 563–573.

CHAPTER 12

Type III protein secretion and inhibition of NF-κB

Klaus Ruckdeschel, Bruno Rouot, and Jürgen Heesemann

12.1 INTRODUCTION

When faced with a bacterial pathogen, the multicellular organism raises a series of defense responses involving both the innate and adaptive immune systems. The innate immune system refers to the first-line host defense, which confers immediate antimicrobial activities. Professional phagocytes, such as neutrophils and resident macrophages, largely constitute the cellular components of innate immunity. These cells directly attack the invading pathogen, mediate secretion of proinflammatory cytokines (see Chapter 10), and mount a protective inflammatory response. In addition to these early defense responses, the innate immune system facilitates the maturation of subsequent adaptive immunity. Both the innate and adaptive immune responses serve to eliminate the infectious challenge. However, to prevail within the host, pathogenic bacteria have evolved sophisticated strategies for evasion or neutralization of host defense mechanisms. A broad range of pathogenic Gram-negative bacteria, including phytopathogens, utilize the type III protein secretion system as a powerful tool to modulate immune responses of the host, which first enables disease.

The type III protein secretion systems are complex weapons that specifically mediate polarized delivery of bacterial virulence proteins directly inside eukaryotic cells (Galan and Collmer, 1999; Donnenberg, 2000). These secretion machineries are composed of approximately twenty proteins, which form a needle complex that protrudes from the inner bacterial membrane to the attached eukaryotic cell. The type III protein secretion systems are first activated when the bacteria come into contact with the host cell. They act as syringes, translocating bacterial virulence factors with high efficiency from inside the bacterium into the host cell cytoplasm. Once inside the host cell, the injected virulence proteins interfere with key processes involved in cellular activation. Although all known type III protein secretion systems share a set of core structural components, the injected virulence proteins largely differ between diverse pathogens, thus displaying a multitude of effector functions. For instance, numerous pathogenic bacteria use their type III secretion-associated functions to modulate the host cell actin cytoskeleton, but the effects on the host cell function are quite different. *Yersinia* spp. and *Pseudomonas aeruginosa* translocate effector proteins to avoid being ingested, whereas *Salmonella* and *Shigella* spp. intend to gain access to eukaryotic cells by delivering their virulence proteins. In this chapter, we will discuss the impact of type III protein secretion systems on one of the most ancient defense systems in eukaryotic cells, the NF-κB signaling networks. For a more detailed discussion of type III secretion systems the reader is referred to Chapter 7.

12.2 TRANSCRIPTION FACTOR NF-κB: THE MASTER REGULATOR OF THE HOST IMMUNE RESPONSE

NF-κB is a heterodimeric transcription factor, that is central to innate immunity. It acts as a key regulator of the inflammatory response (Hatada et al., 2000; Perkins, 2000; Baldwin, 2001). NF-κB is composed of the transcriptionally inactive p50 subunit and the p65 subunit, which contains the transactivation domains necessary for gene induction. The NF-κB signaling pathway is highly conserved in the fruit fly *Drosophila* and in mammals. NF-κB and its *Drosophila* counterparts, Dorsal, Dif, and Relish, induce activation of genes that encode for potent antimicrobial factors. In mammals, NF-κB rapidly upregulates the synthesis of cytokines (such as TNFα, IL-1, IL-6, IL-8), acute-phase proteins (e.g., serum amyloid A), adhesion molecules (e.g., VCAM-1, ICAM-1), and inducible enzymes (e.g., iNOS, Cox-2). In addition, NF-κB is implicated in the prevention of apoptosis in mammalian cells. Activation of NF-κB occurs in response to a wide range of agents and cellular stress conditions, including treatment with cytokines, growth factors, and bacterial and viral infections. The activation of NF-κB involves liberation of NF-κB from its inhibitory proteins IκBα, IκBβ, and IκBε, which sequester preformed NF-κB in the cytoplasm. Upon stimulation, the IκBs are phosphorylated and subsequently degraded through the ubiquitin-proteasome pathway, thereby releasing NF-κB. Free NF-κB translocates to the nucleus, where it binds to its target sequences and activates transcription. The critical step in NF-κB activation is phosphorylation of the IκBs, which is conferred by

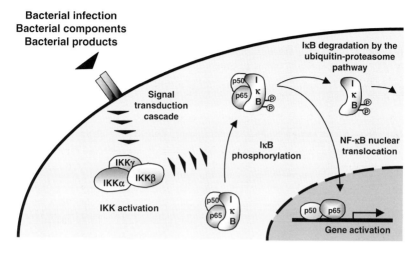

Figure 12.1. The mechanism of NF-κB activation.

a multi-subunit protein kinase, called the IκB kinase (IKK) complex. This complex is composed of at least three proteins, IKK, IKKβ, and IKKγ. IKKα and IKKβ encode catalytic kinase subunits, which mediate IκB phosphorylation. In contrast, IKKγ (also named NF-κB essential modulator NEMO) has a regulatory role and is required for activation of the complex by upstream signaling pathways. In both insects and mammals, IKKβ appears to be the master regulator of NF-κB and its homologues (Hatada et al., 2000; Perkins, 2000; Baldwin, 2001). Figure 12.1 schematically summarizes the mechanisms that lead to activation and nuclear translocation of NF-κB.

Particular attention has been paid in the last few years to the exploration of signaling pathways, that lead to activation of NF-κB. Numerous independent lines of studies imply an intriguing role of Toll receptors and associated signals in activation of the innate immune response (Aderem and Ulevitch, 2000; Anderson, 2000; Brightbill and Modlin, 2000; Medzhitov and Janeway, 2000; Zhang and Ghosh, 2001). For more details of Toll receptors, see Chapter 1. Evolutionary conserved Toll receptors are expressed on the surface of cells of the innate immune system in insects and mammals. They serve to identify specific microbial cell wall components, such as LPS from Gram-negative bacteria or peptidoglycan and lipoteichoic acid from Gram-positive bacteria. Once a microbial pathogen is recognized, the Toll membrane receptors initiate an intracellular kinase cascade that ultimately leads to nuclear translocation of NF-κB and other pro-inflammatory transcription factors such as AP-1. The signal relay downstream from the mammalian receptor and upstream from the IKK complex involves molecules of the IL-1 signaling cascade, namely MyD88 (myeloid differentiation factor 88), IRAK (IL-1-receptor-associated kinase), and TRAF6 (TNF-α-receptor-associated factor 6). According to its role as a positive regulator of the inflammatory response, exposure of eukaryotic cells to numerous bacterial pathogens, such as *Listeria monocytogenes,* group B *Streptococci, Neisseria gonorrhoeae, Pseudomonas aeruginosa, Borrelia burgdorferi, Helicobacter pylori,* or *Chlamydia pneumoniae,* (Ebnet et al., 1997; Hauf et al., 1997; Munzenmaier et al., 1997; Naumann et al., 1997; DiMango et al., 1998; Molestina, et al., 2000; Vallejo et al., 2000) activates the NF-κB signaling pathway. However, the activation of NF-κB is not only implicated in mediating a protective inflammatory response against an invading pathogen, but also is part of the pathogenic strategies of bacteria.

K. RUCKDESCHEL, B. ROUOT, AND J. HEESEMANN

12.3 TARGETING OF THE NF-κB PATHWAY BY PATHOGENIC AND NONPATHOGENIC BACTERIA IN THE GUT

Various reports, especially of enteropathogenic bacteria, show that the inflammatory response greatly contributes to the pathology of infection. In general, intestinal epithelial cells are largely refractory to bacteria and bacterial products, which is essential to maintain the physiological ecosystem of the gut. Colonic epithelial cells live in close contact with the resident bacterial microflora and have necessarily evolved mechanisms to prevent or limit the activation of inflammatory responses to nonpathogenic bacteria. Otherwise, the recognition of harmless bacterial commensals as pathogens by the intestinal mucosa would result in constitutive expression of proinflammatory cytokines and uncontrolled intestinal inflammation. In contrast to nonpathogenic bacteria, enteropathogenic bacteria are capable of activating an inflammatory response by intestinal epithelia. For instance, it is a characteristic feature of the pathogenesis of most salmonellae to provoke a profuse intestinal secretory and inflammatory response. This response is important for the establishment of the inflammatory diarrhoea that follows *Salmonella* infection (Galan, 1999). Similarly, two other enteric pathogens, *Shigella* and enteropathogenic *Escherichia coli* (EPEC), cause an intense and acute inflammatory diarrhoea. All three enteropathogenic bacteria share the common capacity to activate NF-κB in intestinal epithelial cells, which stimulates the production of proinflammatory cytokines, such as IL-8 and TNFα (Hobbie et al., 1997; Savkovic et al., 1997; Philpott et al., 2000). The released cytokines in turn contribute to the typical clinical picture of the respective disease. In both *Salmonella* and *Shigella*, NF-κB activation largely depends on an invasive bacterial phenotype, which is conferred by type III protein secretion system-delivered virulence factors (Dyer et al., 1993; Hobbie et al., 1997; Philpott et al., 2000). Accordingly, the *Salmonella* effector protein SopE simultaneously stimulates nuclear as well as cytoskeletal responses by directly activating the small GTPases Cdc42 and Rac (Hardt et al., 1998). EspB, an EPEC type III secretion-dependent effector protein, appears to be crucially involved in NF-κB activation by EPEC (Savkovic et al., 1997).

In contrast to these intestinal bacterial pathogens, which activate NF-κB to mediate disease, some nonpathogenic bacteria apparently actively inhibit NF-κB activation in gut epithelial cells. A report from Neish et al. (2000) revealed that interaction of nonpathogenic *Salmonella* strains with epithelia attenuates synthesis of inflammatory effector molecules elicited by diverse proinflammatory stimuli (Neish et al., 2000). This is accomplished by blockage of ubiquitination and degradation of IκBα by an unknown mechanism, preventing nuclear translocation of NF-κB in response to the nonpathogenic bacterium. The authors speculate that this may be a mechanism of intestinal immune tolerance to a commensal prokaryote, explaining the absence of inflammation in the gut mucosa despite its constant exposure to a variety of indigenous micoorganisms. However, the strategy to inhibit activation of NF-κB and to avoid inflammation is not exclusively followed by nonpathogenic bacteria in order to prevent illness: it is also an ingenious tactic of pathogenic microorganisms to counteract or to escape the host immune response and to mediate disease. A number of bacterial and viral pathogens provide examples by which subversion of the NF-κB pathway supports development of infection (Ruckdeschel et al., 1998; Yuk et al., 2000; Hiscott et al., 2001). The contribution of NF-κB inhibition to bacterial pathogenesis is probably best characterized for the Gram-negative genus *Yersinia*, that actively paralyzes the host immune response in order to survive and multiply within the host tissue. In contrast to *Salmonella, Shigella,* and EPEC, the pathogenic *Yersinia,* target professional phagocytes rather than intestinal epithelial cells for NF-κB inhibition.

12.4 THE *YERSINIA* TYPE III PROTEIN SECRETION MACHINERY: A PROTOTYPICAL TOOL FOR SUBVERTING THE PHAGOCYTE

The major virulence determinant of pathogenic *Yersinia* is a type III protein secretion system, which represents a paradigm for these type III secretion machineries (Cornelis et al., 1998). The *Yersinia* type III secretion apparatus is encoded on a 70 kb virulence plasmid that is common to the three human pathogenic *Yersinia* species: *Y. pestis, Y. enterocolitica, and Y. pseudotuberculosis. Y. pestis* is the causative agent of plague, whereas *Y. enterocolitica* and

Y. pseudotuberculosis are enteric pathogens. They can cause gastrointestinal syndromes, lymphadenitis, and septicaemia. The three pathogenic *Yersinia* species exhibit a common tropism for lymphatic tissue. *Y. enterocolitica* and *Y. pseudotuberculosis* invade the intestinal mucosa in the terminal ileum through lymphoid follicle-associated M cells and subsequently penetrate the lymphoid tissue of the Peyer's patches. This represents a first critical step in the lifestyle of enteropathogenic *Yersinia*. The bacteria come in contact with phagocytes, thus challenging the first-line host defense. Accordingly, *Yersinia* developed a complex strategy to disarm phagocytic cells and to disrupt their response to infection (Cornelis et al., 1998). Type III-dependent mechanisms are central to this antihost response and crucially determine the outcome of *Yersinia* infection. Virulence plasmid-harbouring yersiniae resist the phagocytic attack, whereas virulence plasmid-cured yersiniae are rapidly killed and eliminated in the Peyer's patches by phagocytes.

Yersinia uses its type III secretion machinery to deliver of a set of at least six effector "*Yersinia* outer proteins" (Yops; YopE, YopH, YopM, YopT, YopO/YpkA, YopP/J) inside the eukaryotic cell. The diverse Yops act at distinct cellular levels to neutralize a sequence of programmed phagocyte effector functions (Cornelis et al., 1998). By rapid injection of YopH, YopE, YopT, and presumably YopO/YpkA, *Yersinia* paralyzes the phagocytic machinery, thereby preventing phagocytosis by macrophages and polymorphonuclear leukocytes. The activities of these Yops, which are described in detail by Forberg and coworkers in Chapter 7, interfere with the actin cytoskeleton dynamics of the host cell. In addition to their antiphagocytic effect, YopH and YopE prevent killing of the bacteria by the phagocytic oxidative burst. They act in concert to suppress the generation of bactericidal reactive oxygen intermediates by macrophages and polymorphonuclear leukocytes. The molecular mechanisms of this inhibition are not yet completely understood. By the action of these Yops, *Yersinia* resists the initial attack of the phagocytes. This in turn prevents clearance of the pathogen from the lymphoid tissue and enables its extracellular survival and proliferation. From the local lymphoid tissue, the yersiniae may subsequently disseminate to the adjacent lymph nodes, liver, and spleen, and to establish a systemic infection.

However, the impact of *Yersinia* on phagocytes is more subtle than only mediating its primary survival during the first interaction. *Yersinia* additionally modulates long-term phagocyte antibacterial activities, which accounts for the later stages of infection, especially for the spread of the bacteria into deeper lymphatic tissues (Monack et al., 1998). One of the key mechanisms to maintain the infectious process is the down-regulation of NF-κB activity in eukaryotic cells, in particular in macrophages. The disruption of the

NF-κB pathway is mediated by the effector protein YopP (*Y. enterocolitica*) or its homologue YopJ (*Y. pseudotuberculosis, Y. pestis*) in a type III secretion-dependent manner (Cornelis et al., 1998; Schesser et al., 1998). According to the global regulatory function of NF-κB, the suppression of NF-κB by a bacterial pathogen exerts multiple effects on the host immune response.

12.5 SUPPRESSION OF THE HOST CELL CYTOKINE PRODUCTION

In vivo studies using infected mice first demonstrated that *Yersinia* inhibit the release of the proinflammatory cytokines TNFα and IFNγ (Nakajima and Brubaker, 1993). These cytokines are crucial in limiting the severity of *Yersinia* infection, because additional administration of cytokine-neutralizing antibodies to *Yersinia*-infected mice seriously exacerbates disease (Autenrieth and Heesemann, 1992). Macrophages are the major source for TNFα in the compromised host. The suppression of the TNFα production by *Yersinia* can be demonstrated *in vitro* on cultivated macrophages. *Yersinia* inhibit TNFα mRNA expression and TNFα synthesis within as short a time as 120 minutes after onset of infection (Ruckdeschel et al., 1997a). The TNFα-suppressive effect has been shown to be mediated by YopP/J (Boland and Cornelis, 1998; Palmer et al., 1998). In correlation, *Yersinia* and YopP/J down-regulate the activation of NF-κB: in the initial phase, the macrophages respond with NF-κB induction to *Yersinia* infection, but after 30–60 minutes, a lag time necessary for Yop translocation and direction to their intracellular targets, the nuclear translocation of NF-κB is blocked (Ruckdeschel et al., 1998). Orth et al. (1999) discovered a potential mechanism by which YopP/J exerts its effect on the NF-κ B pathway. They revealed an interaction of YopP/J with over-expressed IKKβ in epithelial cells. Indeed, YopP/J selectively interacts with IKKβ but not IKKα in infected macrophages and simultaneously suppresses IKKβ activities (Ruckdeschel et al., 2001). IKKβ is the major LPS-responsive NF-κB-activating kinase in monocytes and macrophages, as compared to IKKα (O'Connell et al., 1998). Thus, *Yersinia* specifically target the immediate NF-κB activator, which is implicated in the response to bacterial infection and consequently disrupt the NF-κB pathway. In addition, the cytokine-suppressive effect of YopP/J is not restricted to its potential primary target cell, the macrophage. The subversion of NF-κB by YopP/J also contributes to abrogation of IL-8 synthesis in epithelial HeLa cells (Schesser et al., 1998).

Exploring different signal relays potentially modulated by YopP/J, a series of studies demonstrated that YopP/J not only targets the NF-κ B pathway, but also the signaling cascades of the mitogen-activated protein kinases (MAPK; Ruckdeschel et al., 1997a; Boland and Cornelis, 1998; Palmer et al., 1998). The

MAPK cascades, which include the ERK, p38, and JNK pathways, are down-regulated by *Yersinia* in parallel to NF-κB. This results in suppression of activation of additional transcription factors, such as CREB and AP-1 (Meijer et al., 2000). The activities of the NF-κB and MAPK pathways in common determine the production of TNFα. They synergistically control the synthesis of TNFα at the transcriptional (NF-κB, ERK, JNK) and translational level (p38, JNK). Accordingly, the promotor of the TNFα gene contains several enhancer sequences and maximal activation requires at least two cis-acting regulatory elements, such as NF-κB and c-Jun (Yao et al., 1997). Experiments with specific synthetic inhibitors of the NF-κB and MAPK signaling pathways suggest that *Yersinia* suppresses the TNFα production in a multimodal manner, by simultaneous inhibition of both the NF-κB and MAPK signaling pathways (Ruckdeschel et al., 1998). In agreement with these observations, Orth et al. (1999) found targeting and inhibition of members of the MAPKK family by YopP/J in addition to IKKβ. The MAPKKs are the direct upstream activators of the MAPKs. They possess homology to the IKKs in the kinase activation loop, providing the idea that YopP/J interferes with kinase activation by upstream signaling pathways. However, a recent study from Orth et al. (2000) suggests a cysteine protease activity of YopP/J on proteins modified with the ubiquitin-like molecule SUMO-1. A link between reduced SUMO-1 conjugation of proteins and inhibition of IKKβ or MAPKKs is hitherto unclear. Orth et al. speculate that SUMO-1 conjugation may be an important posttranslational modification associated with MAPK and related signaling complexes, directing their intracellular processing. However, the precise mechanism by which YopP/J modifies IKKβ and MAPKK activities still awaits clarification.

Together, the series of studies dealing with the impact of YopP/J on eukaryotics cells elucidates the disruption of NF-κB communication as an efficient mechanism to interfere with the host cell cytokine production. Interestingly, the abrogation of NF-κB by a microbial pathogen exerts another profound effect on the host immune response besides cytokine suppression. This phenomenon can be attributed to the dual function of NF-κB, on the one hand acting as a mediator of an inflammatory response, and on the other hand, enabling cellular survival by the prevention of apoptosis.

12.6 TRIGGERING MACROPHAGE CELL DEATH BY APOPTOSIS

Several independent reports demonstrate that macrophages affected by *Yersinia* enter an irreversible and final step in their lifecycle: after incubation for between 5 and 20 hours, infected macrophages exhibit characteristic membrane blebbing, shrinkage of the cytoplasm, chromatin condensation and DNA fragmentation, and ultimately the cells die (Mills et al., 1997; Monack et al., 1997; Ruckdeschel et al., 1997b). These features are classical morphological signs of programmed cell death and provide evidence that macrophages exposed to Yops die by apoptosis. Apoptosis is an active physiological process of cell suicide in the multicellular organism. It mediates the ablation of unwanted, damaged, or potentially neoplastic cells without inducing inflammation or damage of contiguous cells. There is increasing evidence that this form of cell death also plays a role in the pathogenesis of infectious diseases. The occurrence of apoptosis after *Yersinia* infection depends on delivery of the YopP/J effector protein into macrophages. As NF-κB plays a role in the regulation of apoptotic processes, the impairment of the NF-κB pathway by YopP/J appears to be implicated in macrophage cell death by *Yersinia*. In general, the onset of apoptosis is tightly controlled by a variety of distinct signals. The cell integrates signals delivered by death-inducing and survival-promoting receptors. Secondary intracellular responses regulate the entry into apoptosis. The precise regulation of cell proliferation and cell death by these complex signaling networks maintains the homeostasis of the multicellular organism. Activation of NF-κB has been found to provide protection against apoptosis under multiple stress-induced conditions (Baichwal and Baeuerle, 1997; Perkins, 2000). Certain extracellular stimuli, such as TNFα, genotoxic agents, or ionizing radiation, simultaneously activate death-inducing and death-preventing signals in eukaryotic cells. NF-κB is implicated in the death-preventing pathways by transcriptionally upregulating the synthesis of proteins that counteract the pathways leading to apoptosis. Such proteins are members of the IAP (inhibitor of apoptosis protein), TRAF or Bcl-2 family (Perkins, 2000). Accordingly, the activation of NF-κB provides protection against apoptotic killing otherwise induced by these stimuli.

In *Yersinia*-infected macrophages, the overexpression of the transcriptionally active p65 subunit of NF-κB exerts a protective effect against YopP/J-induced apoptosis (Ruckdeschel et al., 2001). This provides evidence that subversion of the NF-κB survival pathways by YopP/J is a crucial part of *Yersinia*'s strategy to modulate the machinery of apoptosis. However, the subversion of NF-κB alone is, generally, not sufficient to trigger apoptosis in eukaryotic cells. Normally, apoptosis requires a secondary stimulus, such as TNFα, which activates the intrinsic cytotoxic pathway that leads to apoptosis when NF-κB is inhibited. In fact, *Yersinia* selectively trigger apoptosis in macrophages, but not in epithelial HeLa cells. In contrast, NF-κB inhibition occurs in both cell types. Interestingly, *Yersinia* infection mediates HeLa cell apoptosis when the cells are subsequently treated with TNFα (Ruckdeschel et al., 1998). As the macrophage is directly compelled to undergo apoptosis upon infection with *Yersinia*, this suggests activation of an indispensable pro-apoptotic signal by *Yersinia* specifically in macrophages. This signal apparently mediates macrophage apoptosis together with the YopP/J NF-κB suppressive effect. Recently, LPS from Gram-negative bacteria has been shown to provide the necessary apoptotic signal: the initiation of LPS signaling cooperates with the action of YopP/J to induce macrophage cell death (Ruckdeschel et al., 2001). Thus, *Yersinia* target the NF-κB pathway by injection of YopP/J and simultaneously take advantage of pro-apoptotic LPS signaling to kill the macrophage. It is not yet clear whether other *Yersinia* molecules have the same synergistic action with YopP on apoptosis as does LPS.

In agreement with these findings, there is increasing evidence that bacteria and bacterial components indeed activate pro-apoptotic pathways in macrophages (Kitamura, 1999). NF-κB activation counteracts these cytotoxic signals and mediates survival of macrophages in response to these components. Two recent reports from Aliprantis et al. indicate a crucial role for macrophage Toll-like receptors (TLRs) not only in NF-κB activation, but also in signaling a pro-apoptotic pathway (Aliprantis et al., 1999; 2000). They demonstrate engagement of TLR2 by bacterial lipoproteins, which triggers an apoptotic signal in monocytic cells. The activation of TLR2 is coupled to the death-promoting pathway through activation of caspase-8. The caspases are a family of cysteine aspartate proteases that are the central effectors of the apoptotic machinery. They function to activate and execute the core apoptotic program in a proteolytic cascade. The bifurcation of the pro- and antiapoptotic signals downstream from TLR2 occurs at the level of the adaptor protein MyD88 (Aliprantis et al., 2000). Although the single members of the TLR family exhibit specificity in the recognition of diverse bacterial components, they apparently use the same transmitter molecules to relay the intracellular signal. Thus, the signaling by LPS, which preferentially involves TLR4, may engage the same apoptotic pathways as bacterial lipoproteins.

The current findings suggest that the TLRs are key inducers of innate immunity. They sense invading bacteria and initiate a protective inflammatory and self defense-related response by NF-κB activation. Simultaneously, they signal macrophage apoptosis. By injection of YopP/J, *Yersinia* directly affect the heart of these innate signaling networks. The disruption of NF-κB activation makes the macrophage unable to respond to the bacterial infection. Finally, the continuous transmission of upstream pro-apoptotic signals compels the macrophage to undergo apoptosis. This reflects the endpoint of a multistep strategy of a bacterial pathogen to undermine the host immune response. Figure 12.2 summarizes the impact of YopP/J on signaling pathways and effector functions of macrophages challenged with LPS.

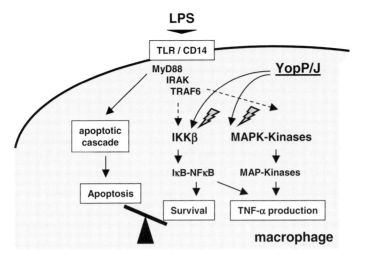

Figure 12.2. Model on the impact of YopP/J on macrophage signal transduction pathways and effector functions.

12.7 IS YopP/J A UNIQUE BACTERIAL EFFECTOR PROTEIN?

By affecting the NF-κB pathway, YopP/J reverses the LPS-induced defense response of macrophages, which ultimately leads to macrophage cell death. Interestingly, YopP/J exhibits high levels of sequence similarities to two bacterial proteins, which are also secreted by a type III dependent mechanism. One of these is AvrRxv, a so-called avirulence protein from the plant pathogenic bacterium *Xanthomonas campestris*. The translocated avirulence proteins trigger a phenomenon termed hypersensitive response in susceptible host plants (Baker et al., 1997). The hypersensitive response is a local defense mechanism that results in cell and tissue death and constrains further spread of the infection. This form of cell death is likely comparable to apoptosis in vertebrates. The precise mechanism of AvrRxv action on the plant cell is not known, but there is increasing evidence that plant cellular defense responses are analogous to the innate immune responses of vertebrates and insects. Indeed, the structural components of the plant defense and resistance responses share a striking similarity with the components of the Toll-NF-κB pathway (Baker et al., 1997). This indicates possible homologies between the strategies of plant and animal pathogenic bacteria to modulate the immune responses of their respective hosts.

Another YopP/J-related protein is AvrA from *Salmonella*. But in contrast to AvrRxv from *X. campestris*, AvrA does not share a functional similarity with

YopP/J, despite sharing 56% sequence identity and 87% similarity (Schesser et al., 2000). AvrA affects neither cytokine production nor survival of infected cells. This suggests different roles of the respective proteins in the pathogenesis of yersiniosis and salmonellosis, although they may have derived evolutionarily from a common genetic ancestor.

In contrast, the pathogen *Bordetella bronchiseptica* has recently been shown to use a strategy that resembles that of YopP/J to modulate the host immune response (Yuk et al., 2000). *B. bronchiseptica* causes persistent infection of the respiratory tract, which usually takes a noninflammatory and asymptomatic course. Accordingly, *B. bronchiseptica* affects NF-κB in lung epithelial cells and triggers apoptosis in epithelial cells and macrophages in a type III secretion-dependent manner. In particular, Yuk et al. (2000) observed aggregation of NF-κB into large cytoplasmic complexes after *B. bronchiseptica* infection. These complexes did not respond to subsequent TNF-α stimulation, suggesting impairment of NF-κB activation by *B. bronchiseptica*.

12.8 CONCLUSION

The current findings indicate that a number of Gram-negative bacteria use a type III secretion apparatus to subvert a central defense pathway in distantly related hosts. The signaling cascade of NF-κB is structurally and functionally conserved in vertebrates, insects, and even plants. In view of this striking parallel among evolutionarily divergent organisms, it is not surprising that the NF-κB pathway is a target of immune modulation by distinct bacterial pathogens. Future studies will shed light on whether this may reflect a common strategy of pathogenic bacteria.

REFERENCES

Aderem, A. and Ulevitch, R.J. (2000). Toll-like receptors in the induction of the innate immune response. *Nature* **406**, 782–787.

Aliprantis, A.O., Yang, R.B., Mark, M.R., Suggett, S., Devaux, B., Radolf, J.D., Klimpel, G.R., Godowski, P., and Zychlinsky, A. (1999). Cell activation and apoptosis by bacterial lipoproteins through toll-like receptor-2. *Science* **285**, 736–739.

Aliprantis, A.O., Yang, R.B., Weiss, D.S., Godowski, P., and Zychlinsky, A. (2000). The apoptotic signaling pathway activated by Toll-like receptor-2. *European Molecular Biology Organisation Journal* **19**, 3325–3336.

Anderson, K.V. (2000). Toll signaling pathways in the innate immune response. *Current Opinion in Immunology* **12**, 13–19.

Autenrieth, I.B. and Heesemann, J. (1992). *In vivo* neutralization of tumor necrosis factor-alpha and interferon-gamma abrogates resistance to *Yersinia enterocolitica* infection in mice. *Medical Microbiology and Immunology* **181**, 333–338.

Baichwal, V.R. and Baeuerle, P.A. (1997). Apoptosis: activate NF-κB or die? *Current Biology* **7**, R94–R96.

Baker, B., Zambryski, P., Staskawicz, B., and Dinesh-Kumar, S.P. (1997). Signaling in plant–microbe interactions. *Science* **276**, 726–733.

Baldwin, Jr., A.S. (2001). The transcription factor NF-κB and human disease. *Journal of Clinical Investigation* **107**, 3–6.

Boland, A. and Cornelis, G.R. (1998). Role of YopP in suppression of tumor necrosis factor alpha release by macrophages during *Yersinia* infection. *Infection and Immunity* **66**, 1878–1884.

Brightbill, H.D. and Modlin, R.L. (2000). Toll-like receptors: molecular mechanisms of the mammalian immune response. *Immunology* **101**, 1–10.

Cornelis, G.R., Boland, A., Boyd, A.P., Geuijen, C., Iriarte, M., Neyt, C., Sory, M.P., and Stainier, I. (1998). The virulence plasmid of *Yersinia*, an antihost genome. *Microbiology and Molecular Biology Reviews* **62**, 1315–1352.

DiMango, E., Ratner, A.J., Bryan, R., Tabibi, S., and Prince, A. (1998). Activation of NF-κB by adherent *Pseudomonas aeruginosa* in normal and cystic fibrosis respiratory epithelial cells. *Journal of Clinical Investigation* **101**, 2598–2605.

Donnenberg, M.S. (2000). Pathogenic strategies of enteric bacteria. *Nature* **406**, 768–774.

Dyer, R.B., Collaco, C.R., Niesel, D.W., and Herzog, N.K. (1993). *Shigella flexneri* invasion of HeLa cells induces NF-κB DNA-binding activity. *Infection and Immunity* **61**, 4427–4443.

Ebnet, K., Brown, K.D., Siebenlist, U.K., Simon, M.M., and Shaw, S. (1997). *Borrelia burgdorferi* activates nuclear factor-kappa B and is a potent inducer of chemokine and adhesion molecule gene expression in endothelial cells and fibroblasts. *Journal of Immunology* **158**, 3285–3292.

Galan, J.E. (1999). Interaction of *Salmonella* with host cells through the centisome 63 type III secretion system. *Current Opinion in Microbiology* **2**, 46–50.

Galan, J.E. and Collmer, A. (1999). Type III secretion machines: bacterial devices for protein delivery into host cells. *Science* **284**, 1322–1328.

Hardt, W.D., Chen, L.M., Schuebel, K.E., Bustelo, X.R., and Galan, J.E. (1998). *S. typhimurium* encodes an activator of Rho GTPases that induces membrane ruffling and nuclear responses in host cells. *Cell* **93**, 815–826.

Hatada, E.N., Krappmann, D., and Scheidereit, C. (2000). NF-κB and the innate immune response. *Current Opinion in Immunology* **12**, 52–58.

Hauf, N., Goebel, W., Fiedler, F., Sokolovic, Z., and Kuhn, M. (1997). *Listeria monocytogenes* infection of P388D1 macrophages results in a biphasic

NF-κB (RelA/p50) activation induced by lipoteichoic acid and bacterial phospholipases and mediated by IκBα and IκBβ degradation. *Proceedings of the National Academy of Sciences USA* **94**, 9394–9399.

Hiscott, J., Kwon, H., and Genin, P. (2001). Hostile takeovers: viral appropriation of the NF-κB pathway. *Journal of Clinical Investigation* **107**, 143–151.

Hobbie, S., Chen, L.M., Davis, R.J., and Galan, J.E. (1997). Involvement of mitogen-activated protein kinase pathways in the nuclear responses and cytokine production induced by *Salmonella typhimurium* in cultured intestinal epithelial cells. *Journal of Immunology* **159**, 5550–5559.

Kitamura, M. (1999). NF-κB-mediated self defense of macrophages faced with bacteria. *European Journal of Immunology* **29**, 1647–1655.

Medzhitov, R. and Janeway, C. (2000). The toll receptor family and microbial recognition. *Trends in Microbiology* **8**, 452–456.

Meijer, L.K., Schesser, K., Wolf-Watz, H., Sassone-Corsi, P., and Pettersson, S. (2000). The bacterial protein YopJ abrogates multiple signal transduction pathways that converge on the transcription factor CREB. *Cellular Microbiology* **2**, 231–238.

Mills, S.D., Boland, A., Sory, M.P., van-der-Smissen, P., Kerbourch, C., Finlay, B.B., and Cornelis, G.R. (1997). *Yersinia enterocolitica* induces apoptosis in macrophages by a process requiring functional type III secretion and translocation mechanisms and involving YopP, presumably acting as an effector protein. *Proceedings of the National Academy of Sciences USA* **94**, 12,638–12,643.

Molestina, R.E., Miller, R.D., Lentsch, A.B., Ramirez, J.A., and Summersgill, J.T. (2000). Requirement for NF-κB in transcriptional activation of monocyte chemotactic protein 1 by *Chlamydia pneumoniae* in human endothelial cells. *Infection and Immunity* **68**, 4282–4288.

Monack, D.M., Mecsas, J., Ghori, N., and Falkow, S. (1997). *Yersinia* signals macrophages to undergo apoptosis and YopJ is necessary for this cell death. *Proceedings of the National Academy of Sciences USA* **94**, 10,385–10,390.

Monack, D.M., Mecsas, J., Bouley, D., and Falkow, S. (1998). *Yersinia*-induced apoptosis in vivo aids in the establishment of a systemic infection of mice. *Journal of Experimental Medicine* **188**, 2127–2137.

Munzenmaier, A., Lange, C., Glocker, E., Covacci, A., Moran, A., Bereswill, S., Baeuerle, P.A., Kist, M., and Pahl, H.L. (1997). A secreted/shed product of *Helicobacter pylori* activates transcription factor nuclear factor-kappa B. *Journal of Immunology* **159**, 6140–6147.

Nakajima, R. and Brubaker, R.R. (1993). Association between virulence of *Yersinia pestis* and suppression of gamma interferon and tumor necrosis factor alpha. *Infection and Immunity* **61**, 23–31.

Naumann, M., Wessler, S., Bartsch, C., Wieland, B., and Meyer, T.F. (1997). *Neisseria gonorrhoeae* epithelial cell interaction leads to the activation of the transcription factors nuclear factor kappaB and activator protein 1 and the induction of inflammatory cytokines. *Journal of Experimental Medicine* **186**, 247–258.

Neish, A.S., Gewirtz, A.T., Zeng, H., Young, A.N., Hobert, M.E., Karmali, V., Rao, A.S., and Madara, J.L. (2000). Prokaryotic regulation of epithelial responses by inhibition of IκBα ubiquitination. *Science* **289**, 1560–1563.

O'Connell, M.A., Bennett, B.L., Mercurio, F., Manning, A.M., and Mackman, N. (1998). Role of IKK1 and IKK2 in lipopolysaccharide signaling in human monocytic cells. *Journal of Biological Chemistry* **273**, 30,410–30,414.

Orth, K., Palmer, L.E., Bao, Z.Q., Stewart, S., Rudolph, A.E., Bliska, J.B., and Dixon, J.E. (1999). Inhibition of the mitogen-activated protein kinase kinase superfamily by a *Yersinia* effector. *Science* **285**, 1920–1923.

Orth, K., Xu, Z., Mudgett, M.B., Bao, Z.Q., Palmer, L.E., Bliska, J.B., Mangel, W.F., Staskawicz, B., and Dixon, J.E. (2000). Disruption of signaling by *Yersinia* effector YopJ, a ubiquitin-like protein protease. *Science* **290**, 1594–1597.

Palmer, L.E., Hobbie, S., Galan, J.E., and Bliska, J.B. (1998). YopJ of *Yersinia pseudotuberculosis* is required for the inhibition of macrophage TNF-alpha production and downregulation of the MAP kinases p38 and JNK. *Molecular Microbiology* **27**, 953–965.

Perkins, N.D. (2000). The Rel/NF-κB family: friends and foe. *Trends in Biochemical Sciences* **25**, 434–440.

Philpott, D.J., Yamaoka, S., Israel, A., and Sansonetti, P.J. (2000). Invasive *Shigella flexneri* activates NF-κB through a lipopolysaccharide-dependent innate intracellular response and leads to IL-8 expression in epithelial cells. *Journal of Immunology* **165**, 903–914.

Ruckdeschel, K., Machold, J., Roggenkamp, A., Schubert, S., Pierre, J., Zumbihl, R., Liautard, J.P., Heesemann, J., and Rouot, B. (1997a). *Yersinia enterocolitica* promotes deactivation of macrophage mitogen-activated protein kinases extracellular signal-regulated kinase-1/2, p38, and c-Jun NH2-terminal kinase: Correlation with its inhibitory effect on tumor necrosis factor-alpha production. *Journal of Biological Chemistry* **272**, 15,920–15,927.

Ruckdeschel, K., Roggenkamp, A., Lafont, V., Mangeat, P., Heesemann, J., and Rouot, B. (1997b). Interaction of *Yersinia enterocolitica* with macrophages leads to macrophage cell death through apoptosis. *Infection and Immunity* **65**, 4813–4821.

Ruckdeschel, K., Harb, S., Roggenkamp, A., Hornef, M., Zumbihl, R., Kohler, S., Heesemann, J., and Rouot, B. (1998). *Yersinia enterocolitica* impairs activation of transcription factor NF-κB: involvement in the induction of programmed cell death and in the suppression of the macrophage tumor necrosis factor alpha production. *Journal of Experimental Medicine* **187**, 1069–1079.

Ruckdeschel, K., Mannel, O., Richter, K., Jacobi, C.A., Trulzsch, K., Rouot, B., and Heesemann, J. (2001). *Yersinia* outer protein P of *Yersinia enterocolitica* simultaneously blocks the nuclear factor-kappaB pathway and exploits lipopolysaccharide signaling to trigger apoptosis in macrophages. *Journal of Immunology* **166**, 1823–1831.

Savkovic, S.D., Koutsouris, A., and Hecht, G. (1997). Activation of NF-κB in intestinal epithelial cells by enteropathogenic *Escherichia coli*. *American Journal of Physiology* **273**, 1160–1167.

Schesser, K., Spiik, A.K., Dukuzumuremyi, J.M., Neurath, M.F., Pettersson, S., and Wolf-Watz, H. (1998). The yopJ locus is required for *Yersinia*-mediated inhibition of NF-κB activation and cytokine expression: YopJ contains a eukaryotic SH2-like domain that is essential for its repressive activity. *Molecular Microbiology* **28**, 1067–1079.

Schesser, K., Dukuzumuremyi, J.M., Cilio, C., Borg, S., Wallis, T.S., Pettersson, S., and Galyov, E.E. (2000). The *Salmonella* YopJ-homologue AvrA does not possess YopJ-like activity. *Microbial Pathogenesis* **28**, 59–70.

Vallejo, J.G., Knuefermann, P., Mann, D.L., and Sivasubramanian, N. (2000). Group B *Streptococcus* induces TNF-α gene expression and activation of the transcription factors NF-κB and activator protein-1 in human cord blood monocytes. *Journal of Immunology* **165**, 419–425.

Yao, J., Mackman, N., Edgington, T.S., and Fan, S.T. (1997). Lipopolysaccharide induction of the tumor necrosis factor-alpha promoter in human monocytic cells. Regulation by Egr-1, c-Jun, and NF-κB transcription factors. *Journal of Biological Chemi*stry **272**, 17,795–17,801.

Yuk, M.H., Harvill, E.T., Cotter, P.A., and Miller, J.F. (2000). Modulation of host immune responses, induction of apoptosis and inhibition of NF-κB activation by the *Bordetella* type III secretion system. *Molecular Microbiology* **35**, 991–1004.

Zhang, G. and Ghosh, S. (2001). Toll-like receptor-mediated NF-kappaB activation: A phylogenetically conserved paradigm in innate immunity. *Journal of Clinical Investigation* **107**, 13–19.

Index